PROCEEDINGS OF THE ICOM WATERLOGGED WOOD WORKING GROUP CONFERENCE

15-18th September, 1981.

Canadian Conservation Institute

1030 Innes Rd., Ottawa K1A 0M8

edited by David W. Grattan

special discussions edited by J. Cliff McCawley

Published by, The International Council of Museums (ICOM),

Committee for Conservation, Waterlogged Wood Working Group

Ottawa, 1982

ISBN 0-9691073-0-7

ACKNOWLEDGEMENTS

We would like to give special thanks to several people for help in the editing of these proceedings:

- to the several colleagues in the CCI, especially Gregory Young, who gave assistance by reading and commenting upon manuscripts.
- to Linda Tremblay for her speed, accuracy and efficiency in typing the manuscript.

For aid in publication of these proceedings, we would like to acknowledge the considerable assistance of Parks Canada for use of word processing equipment. With special thanks to Dr. Bruce Fry and Barbara MacIntyre for their invaluable assistance in producing the final copy.

Publication would not have been possible without financial support from:

The Canadian Conservation Institute,
National Museums of Canada

The Samuel and Saidye Bronfman Family Foundation

INTRODUCTION

The ICOM Committee for Conservation, Waterlogged Wood Working Group Conference, held in Ottawa in September, 1981, coincided with Colin Pearsons resignation as Group Coordinator. It is, therefore, doubly fitting that this publication and its contents should be of such high standard. The Working Group is larger and more active now than it has ever been, and that it is in such a healthy position, is due in no small part to Colin's efforts. We would like, therefore, to dedicate these Proceedings to Colin, as recognition for his work on behalf of the group.

Anyone who has been involved in waterlogged wood conservation over the past ten or so years, cannot fail to have been heartened by the conference we attended in Ottawa. For we can only have been impressed by the quantity, diversity and quality of the work that was reported. The renewed enthusiasm shown by the "old hands," and the new interest of archaeologists, historians, curators and scientists of many disciplines, promise an exciting future for the Waterlogged Wood Working Group.

Cliff McCawley
Conference Chairman

TABLE OF CONTENTS

ICOM COMMITTEE FOR CONSERVATION WORKING GROUP ON THE CONSERVATION OF WATERLOGGED WOOD

Colin Pearson
Materials Conservation Section,
Canberra College of Advanced Education,
P.O. Box 1,
Belconnen, A.C.T. 2616, Australia

Abstract

The activities of the ICOM Committee for Conservation Working Group on the Conservation of Waterlogged Wood are described. These include the Waterlogged Wood Newsletter, the involvement of the Group in international conferences in this field, and the programme of special study into the properties and treatments of waterlogged wood.

History

The ICOM Committee for Conservation originated as a Commission for the Care of Paintings in 1948. It has subsequently grown and developed and was formalised as the Committee for Conservation in 1967. It has held meetings in Brussels, Amsterdam, Madrid, Venice and Zagreb, and these are now scheduled on a triennial basis. At present there are Twenty-three Working Groups in the Committee covering a wide range of conservation activities.

The Committee's stated aims are:

a) the achievement and maintenance of the highest standards of conservation and examination of historical works by bringing together from all countries those who are responsible for cultural property: restorers, research workers and curators;
b) to promote research of a scientific or technological nature pertaining thereto:
c) to collect data and information about materials and workshop methods;
d) to make generally available by publications or otherwise the results of such enquiries.

The structure and aims of the Committee are currently under review and this was discussed at the 6th Triennial Meeting in Ottawa.

Working Group on the Conservation of Waterlogged Wood

The Conservation Committee is based on a series of Working Groups which includes that dealing with the Conservation of Waterlogged Wood. This is a specialized Group looking only at waterlogged wood. However, it does overlap with wooden objects covered in other Working Groups such as 'Icons' and 'Ethnographic Materials'.

Each Working Group carries out its own specific programme of research under the direction of a Coordinator. The Coordinator also invites members of ICOM to join the Working Group, however, there is no commitment by that member to contribute to the Working Group's activities.

At each Triennial Meeting the work of the previous three years is presented and discussed and a programme of collaboration and study for the next three years is formulated. Again, it must be stated that it is up to the Coordinator and the members of the Group to make the programme of activities work.

The Working Group on Waterlogged Wood has been in existence for a number of years. It was developed under the coordination of Ruben Munnikendam and was taken over by the author in 1978. It has always been a relatively small specialised Group with no more than fifteen members. The problems of the conservation of waterlogged wood are well known and increasing activities in underwater and wet site archaeology over the past decade has seen a big increase in this field. A Newsletter was started in 1978 with an initial circulation of fourteen. It now has a circulation of over one hundred active or interested conservators, scientists and curators from twenty-two different countries.

At each Triennial Meeting the basic aim of the Working Party has been to identify one or more particular problems or common areas of interest which would then be studied by members of the Group and reported upon at the next meeting. It is always difficult to get such communication and as a Working Group we have not been very successful in this regard. This was an important matter for discussion at the meeting of the Working Group at the ICOM Committee for Conservation Conference.

The programme of activities agreed to at

the meeting of the Working Group in Zagreb in 1978 included the following topics:

1. Use of detergents in the conservation of waterlogged wood.
2. Use of tetraethyl ortho silicate.
3. Problems connected with the salvage of waterlogged wood.
4. Freeze-drying.
5. Methods of analysis of PEG in waterlogged wood.
6. Use of sucrose.
7. Use of Organic polymers.
8. Irradiation techniques.

Further activities have been added via the Newsletter, including:

9. Analysis and research.
10. Treatment of large ships' timbers.
11. Acetone/rosin process.
12. Controlled drying.
13. PEG impregnation.
14. General interest.

This is a wide range of topics indicating the current problems in this field. The Working Group should perhaps restrict its activities to just a few of these topics.

The following list of papers that were presented to the ICOM Meeting in Ottawa demonstrate the wide range of activities in the conservation of waterlogged wood, but there is little evidence of collaboration amongst workers. Perhaps this will always be the case in conservation and research, however, I do believe that a lot of time, effort and money is wasted on duplicating what has either already been done or what can be better done elsewhere. The papers to be presented are:

1. J. de Jong, W. Eenkhoorn & A.J.M. Wevers. Controlled drying as an approach to the conservation of shipwrecks.
2. M-L.E. Florian. A review: Analyses of different states of deterioration of terrestrial waterlogged wood - conservation implications of the analyses.
3. N.G. Gerassimova, E.A. Mikolajchuk & M.I. Kolosova. On the conservation of wet archaeological wood by introduction of wax-like substances into it.
4. D.W. Grattan, J.C. McCawley & C. Cook. The conservation of a waterlogged dug-out canoe using natural freeze drying.
5. G.H. Grosso. Experiments with sugar in conserving waterlogged wood.
6. F.T.T. Pang. The treatment of waterlogged oak timbers from a 17th Century East Indiaman, Batavia, using poly--ethylene glycol.
7. C. Pearson. Recent advances in the conservation of waterlogged wood.
8. M. Sawada. A modified technique for treatment of waterlogged wood employing the freeze drying technique.
9. K.R. Singley. Design of a large-scale PEG treatment facility for the Brown's ferry vessel.

The Working Group has been successful with the Newsletter and also its involvement in two International Waterlogged Wood Conferences. As mentioned earlier, the Newsletter, published twice a year, now has a wide circulation. It is financed by the Unesco Division of Cultural Activities. In addition to the 100 or more active members it is circulated to institutions and organisations with interest in this field, the total circulation being approximately 400 covering 66 different countries.

The aims of the Newsletter are, on an informal basis, to further communiction and to exchange ideas and information between those interested and/or involved with the conservation of waterlogged wood. A list of members' names, addresses and major areas of activity in the conservation of waterlogged wood are published in the Newsletter. It is hoped that members will then communicate direct with others with similar interests.

The success of the Newsletter depends on contributions from its members, and all are urged to send information, however small, so that we can further the study of the conservation of waterlogged wood and hopefully resolve some of the many problems that still exist in this field. Treating waterlogged wood is similar to treating underwater iron -there is a range of recommended techniques, each having its own champion. Why is it that they are not always successful? We still do not know enough about the properties and degradation of waterlogged wood or the processes by which it can be stabilised. More reports of failures and the reasons for failure of different treatments are required, as is more comparison work of different treatments on the same piece of timber. Such information can be made available through the Newsletter instead of waiting the many months for a formal publication.

The Working Group on Waterlogged Wood has been involved with two international conferences in this field. It was represented by the Coordinator, Dr Colin Pearson, at the International Symposium on the Conservation of Large Objects of Waterlogged Wood, organised by the Netherlands National Commission for Unesco and the 'Save the Amsterdam Foundation', in Amsterdam, September 1979. This Symposium looked specifically at the problems of large pieces of timber, including entire ships' hulls, and the proceedings of the Symposium were made available early this year.[1]

Due to the increased activity in the conservation of waterlogged wood, informal discussions were held at the last ICOM Committee for Conservation meeting in Zagreb concerning the possiblity of holding a major conference immediately prior to the next ICOM meeting in Ottawa - this conference is the result. I would like to take the opportunity to thank the organisers, in particular Brian Arthur, David Grattan and Cliff McCawley for arranging a most stimulating, informative and enjoyable conference. However, if as a group of specialists we are to continue the recent progress made in the conservation of waterlogged wood, members must be prepared to actively participate in the work of the Group both now and in the future.

References

1. de Vries-Zuiderbaan, L.H. editor, Conservation of Waterlogged Wood. International Symposium on the Conservation of Large Objects of Waterlogged Wood, Amsterdam, September 1979. (Amsterdam: Netherlands National Commission for Unesco, 1980).

SESSION I

THE CONSERVATION OF SHIPWRECKS

Eric Lawson
Session Chairman

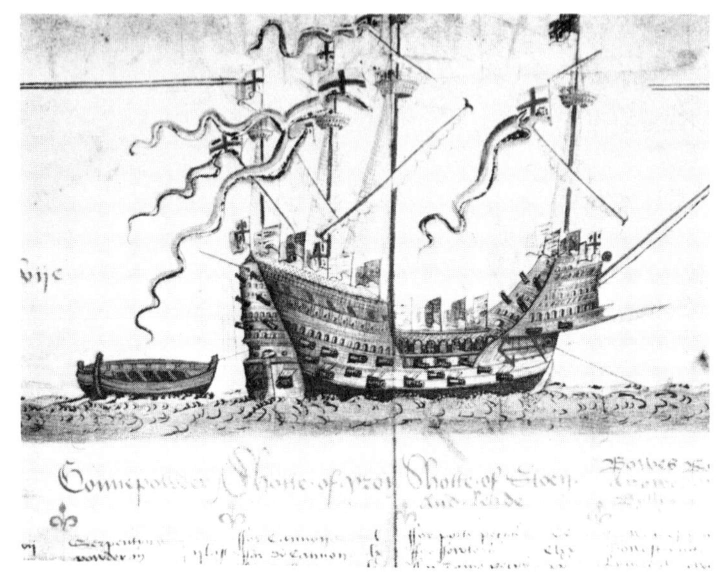

Figure 1: The only known representation of the Mary Rose, from the Anthony Roll compiled in 1546.

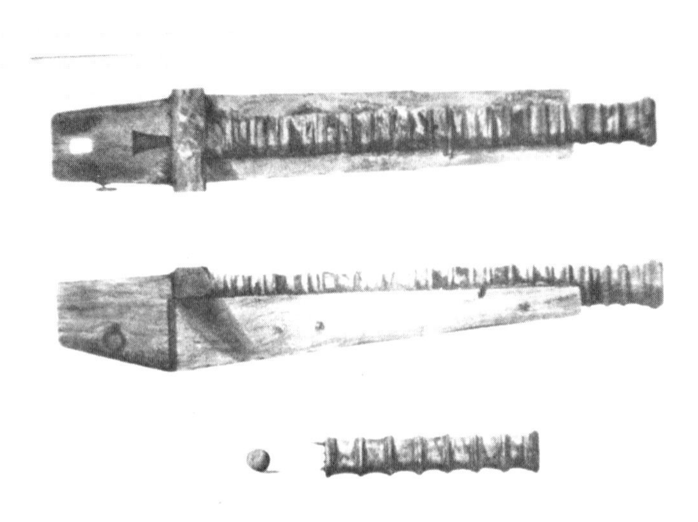

Figure 2: Watercolour of one of the bronze guns found by the Dean brothers in the 19th century.

THE CONSERVATION OF ARTIFACTS FROM THE MARY ROSE

Howard Murray
Mary Rose Trust,
Old Bond Store,
48 Warblington Street,
Old Portsmouth,
Hants, England

Introduction: The Problem

History of the Mary Rose

The Mary Rose was built as part of Henry VIII's naval expansion programme; she was begun at the start of his reign in 1509. There are several references to the Mary Rose, specifically in the records of Henry's early reign: an inventory of guns on board in 1514 exists and in a list of the King's ships of 1525 appears the entry, "The Marye Rosse, of the tonnage of 600 tons, and the age of 14 years." The only known pictorial representation of the Mary Rose, **Figure 1**; comes from the Anthony Roll which is another list of the Kings' ships compiled by the Board of Ordnance in 1546. This shows her as she would have been after her refit of 1536 during which she was uprated from 600 to 700 tons. She campaigned successfully for many years until 1545 when she was involved in an action with the French off Portsmouth, which resulted in her sinking. The circumstances surrounding this event remain a mystery, and it is one of the questions the archaeologists hope to discover during their excavations. Attempts were made immediately to salvage her. These failed, her masts were cut away, some items including guns removed, and she was then abandoned. The site lay undisturbed until well into the 19th century when it was rediscovered accidentally, and more items removed. Among these, were four bronze guns, **Figure 2**; and at least eleven iron guns. Following this brief rediscovery it was simply forgotten about again.

Excavations 1965-1982

The second rediscovery of the Mary Rose was part of an overall survey of the Solent "Project Solent Ships" begun in 1965 led by local historian/diver Alexander McKee. The site of the Mary Rose was eventually identified and in order to protect it from unauthorized salvage attempts the "Mary Rose 1967 Committee" was formed. From 1967 until 1979 the Comittee ran a limited archaeological excavation. A quantity of finds were recovered and passed to the Portsmouth City Museums conservation laboratory. The vast majority of organic material were simply held in what was at the time considered a passive environment, i.e. in water. By the end of 1978 a decision was made to completely excavate, raise and preserve the Mary Rose. To that end a charitable Trust was formed.

Scale of Operation:

When I joined the Trust, the Finds Department, which had been set up the year before, was attempting to deal with the second major excavation season and were in the process of moving to an ex. bonded warehouse, which had been given on extended loan to the Trust. This building consisted of two open plan floors of about 600 sq. meters. The intention was to house finds, organic conservation, drawing, photography and all associated services. Conservation was allocated an area on the first floor for a laboratory and was to share equally the ground floor with the Finds Department. The priorities were:

(i) Building of laboratory and associated works.
(ii) Conservation of material which had been stored from 1967 onwards.
(iii) Supply of conserved artifacts of certain types for archaeological research.
(iv) Supply of prestige items in displayable state for public relations/fund raising purposes.
(v) Consultative role to Finds Department /archaeologists in day to day problems of lifting and storing fragile artifacts.

When I began conservation work I was confronted with a backlog of material (in all some three thousand accessioned artifacts; most of which were organic). In addition to this backlog the archaeological team were engaged in the second major season and finds were being sent ashore at regular intervals.

Passive Holding:

A policy of passive holding had already been instituted by the Finds Department. This consisted of:

(a) Spraying organic material with a 5% solution of sodium borate/boric acid and double wrapping in polyethylene sheet; sealed by a heat sealer.

(b) Immersion of objects in fresh water.

Both these methods soon proved to have serious drawbacks:

In method (a) above, water vapour passed out of the objects through the polyethylene sheeting and thus after a period of six months the objects exhibited signs of drying out (cracking and salt efflorescence on the surface).

Whereas in method (b), the objects absorbed more water than the fibre saturation point and expanded and softened due to the solvent action of the water.

Since both these holding methods caused problems for the material, it was agreed that all organic material needed some conservation treatment within one year of excavation and retrieval from the seabed.

To this end in conjunction with the development of conservation facilities, the conservation team instigated an immediate conservation programme.

Conservation Procedures

Active Conservation

The organic material already in store was assessed on two criteria:

(i) Degree of deterioration caused by storage.

(ii) Importance of material in archaeological and public relations terms.

Batches of material were assembled (e.g. Pikes) and these were treated as one article. In this way it was hoped that as much material as possible could receive at least some of the stages of active conservation (and remove the factors that were causing post excavation deterioration), and also that the small conservation staff would be used more efficiently.

Recording:

Since it was necessary to deal with large quantities of material in batches, a simplified recording system was introduced, allowing essential information to be recorded without taking undue time from the conservation staff, **Figure 3.** This took the form of an envelope in which a stamped card and any other relevant information could be placed. These envelopes are initially filled out by the Finds Department and accompany every object into conservation. The insert cards are completed by conservation staff and the whole returned to the Finds Department with the object on completion of conservation. While in conservation the envelopes are stored in a filing cabinet according to the location of the objects (e.g. PEG Tank, ground floor, Bond Store).

Pretreatment of Wood:

All material on receipt into conservation is washed free of soluble marine salts in a fresh water cascade system, **Figure 4.** Any fragile objects are supported either in plastic netting or foam rubber. After a two week washing period the wood is then placed in a 5% solution of E.D.T.A. di-sodium salt, **Figure 5**; this bath also contains an ultrasonic cleaning device and over a 24-36 hour period as much of the insoluble salt(s) as possible is(are) removed. These salts are mostly iron corrosion products from the metal fittings of the ship. It was found that immersion of the wood for more than 48 hours in the E.D.T.A. solution caused an appreciable softening of it. On removal the objects are again placed in the cascading washing system to remove the E.D.T.A.

PEG Impregnation:

After this pretreatment process the wood's surface detail and colour is much improved and it is ready for wax impregnation. The choice of PEG 3400 is based on my pevious experience in other laboratories. The use of PEG of molecular weight below 3400 i.e. 400 and 1540 necessitates extensive colsolidation of the wood on drying and as the laboratory must deal with very large quantities of material, it is considered advantageous to combine as many steps in the conservation process as possible. The decision to only impregnate up to 50% solution was an attempt to minimize any surface cleaning on drying and remove the difficulties experienced in repairing wax wood with adhesives.

The wax impregnations is carried out in two ways:

(i) A series of baths, **Figure 6**, were built; six contain PEG's solutions at ambient temperature and at the following concentrations: 2 x 5%, 10%, 15%, 20% and 25%. The remaining six are heated to 35-40°C by a domestic hot water system specialy adapted and are at concentrations of 25%, 30%, 35%, 40%,

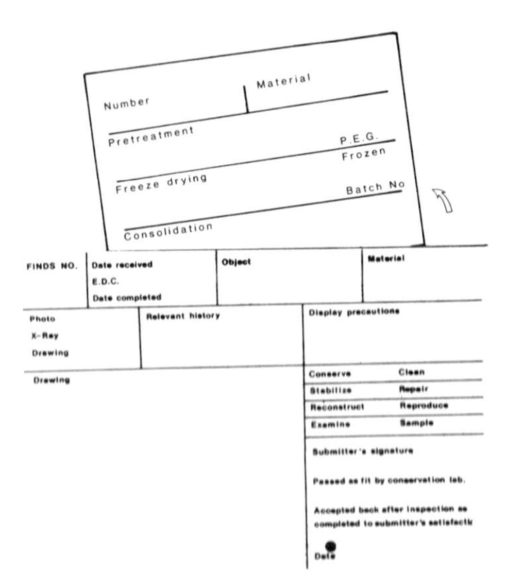

Number | Material

Pretreatment

P.E.G.

Frozen

Freeze drying

Batch No

Consolidation

FINDS NO. | Date received | Object | Material

E.D.C.

Date completed

Photo | Relevant history | Display precautions

X-Ray

Drawing

Drawing

Conserve | Clean

Stabilize | Repair

Reconstruct | Reproduce

Examine | Sample

Submitter's signature

Passed as fit by conservation lab.

Accepted back after inspection as completed to submitter's satisfactio

Date

Figure 3: Recording systems for objects requiring conservation.

Figure 4: Cascade Washing System.

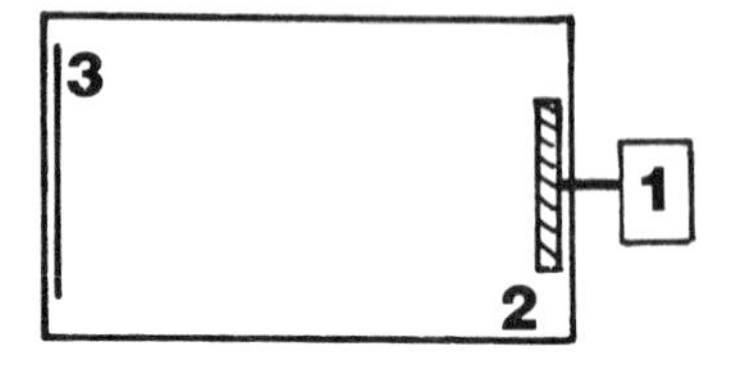

Figure 5: 5% E.D.T.A. bath with ultrasonic transducer. 1) Transducer. 2) Metal Plate. 3) Tank construction. Dexion angle iron with swimming pool liner insert, 2 m x 1.5 m x 1.5 m.

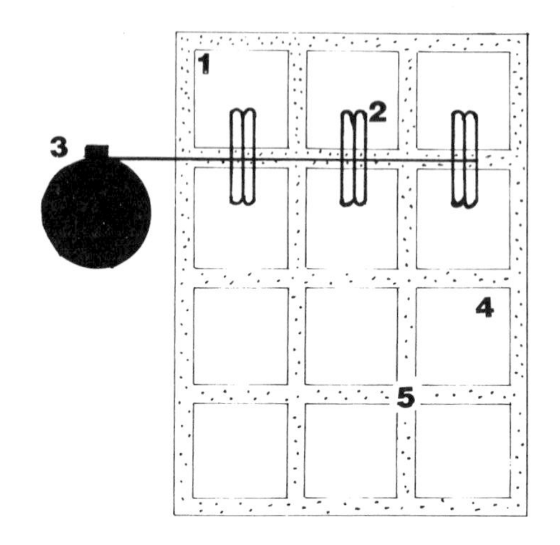

Figure 6: PEG impregnation tanks: 1) Fibreglass water storage tanks 0.75 m x 0.75 m x 0.75 m. 2) Copper heating unit painted with epoxy. 3) Domestic water boiler and pump, thermostatically controlled. 4) Unheated tanks. 5) Insulated with mica fill.

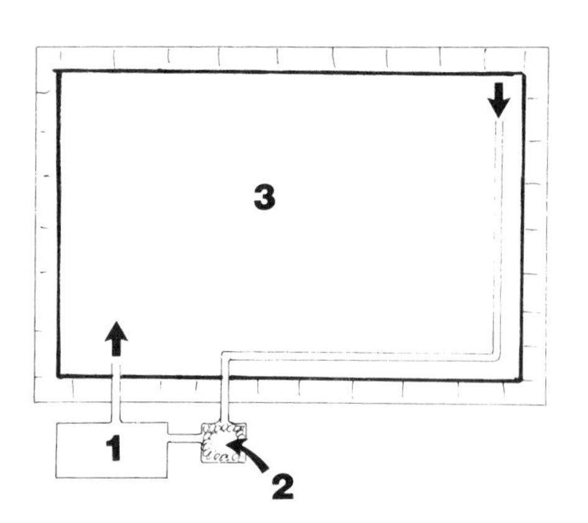

Figure 7: Large (4.5 m x 2.5 m x 1 m) PEG impregnation tanks. 1) Churchill heater - circulator. 2) Removable Filter. 3) Tank constructed of breeze-blocks with concrete skin on inner surface, coated with Quentile Epoxy coat.

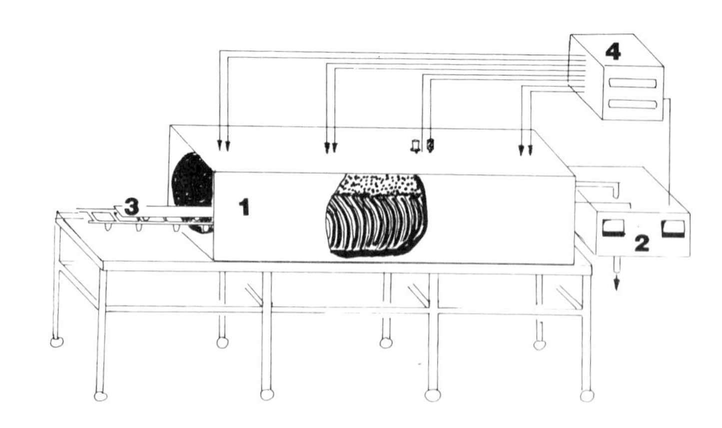

Figure 8: Freeze-Drying Chamber: 1) Stainless styeel chamber, length 2.5 m, diameter 0.5 m). Door with viewing port. Five ports house probes and vacuum equipment. Chamber is wrapped in 80 m of 15 mm copper tubing which is connected to a heater/chiller/circulator (10/CTCV/HP). The whole is wrapped in insulation. 2) Edwards EF4 Modulyo Freeze-Dryer with pirani guages and two stage vacuum pump. 3) Aluminium track and trolley (made for the Trust by HM Dockyard Apprentices). 4) 12 Channel Chart Recorder, monitoring temperature, weight loss and chamber pressure.

45% and 50%. Each week material is moved one bath at a time through a rising sequence of concentrations until after a twelve week period the objects reach 50%.

(ii) A large tank was built, **Figure 7,** to which is attached a heater-circulator which maintains a constant temperature at 35°C. The tank was loaded with wood, and PEG added to 5% concentration. Thereafter, every working day 25 kg of PEG is added until the solution reaches 50%. This process takes about four months.

On reaching a solution strength of 50% the wood is removed, surface PEG washed off and frozen in a deep freeze at -20°C. The wood can remain in the freezer for long periods without further deterioration until it can be freeze-dried.

Pretreatment of Leather:

On receipt all leather is washed free of soluble marine salts in the cascade system for one week; the majority is supported in netting. After the cascade wash the leather is treated for one hour in a 5% Oxalic acid solution again to remove mainly iron corrosion products. It is subjected to a further cascade wash and then placed in a 10% Asak 520S Bavon solution for two weeks. It is then frozen and stored in the same way as the wood. The acid wash and Bavon treatment are new procedures which replace a 5% E.D.T.A. di-sodium salt wash followed by two weeks immersion in 10% PEG 400. The Oxalic acid was found to remove iron salts more efficiently and the Bavon not only acts as a lubricant (the role of the PEG 400) but also combines with the collagen in the leather, thus providing a more effective treatment.

Freeze-Drying:

It was agreed at the outset of the project that freeze-drying offered the best hope to conserve the large quantities of organic material from the Mary Rose. While the pretreatment of material was progressing, I assembled a freeze-drying unit for the immediate needs of the Trust. This was intended to be a stopgap until a much larger facility, (ordered in April of 1980) came into operation in December 1981, **Figure 8.**

Before a run, the drying chamber is cooled to -20°C, it is then loaded with material already prefrozen in a domestic deep freeze. Evacuation is started and the chamber temperature brought up to -10°C. As ice is removed the temperature of the chamber is gradually increased. Approximately one kilogram of ice is removed every 24 hours.

We are experimenting with partial freeze-drying; i.e. drying only the surface of an object then removing it from the chamber and allowing it to air dry. Some good results have been obtained in this way.

Post Drying:

On removal from the drying chamber the objects are allowed to achieve equilibrium with the laboratory relative humidity (50--60%). Any excess PEG is removed using hot air blowers, infra-red lamps and swabs soaked in ethanol (industrial methylated spirits). Reconstruction work is carried out using wooden dowels, nitrocellulose adhesive and melted PEG 6000, and finally the objects are painted with a 50% solution of PEG 6000 as a surface coating.

Suppliers

Freeze-Drying Equipment:
Edwards High Vacuum,
Manor Royal, Crawley,
West Sussex,
RH10 2LW, U.K.

Heater/chiller/circulator:
Conair Churchill
Riverside Way,
Uxbridge, Middlesex,
UB8 2YF, U.K.

Plastic Tanks:
Quentsplass Limited,
Thorp Arch Trading Estate,
Wetherby, West Yorkshire
LS23 7BL, U.K.

Ultrasonic Transducer:
Telesonic Limited,
Cleveland House,
344 Holdenhurst Road,
Bournemouth,
BH8 8BE, U.K.

Plastic Mesh Bags:

Netlon Limited
Mill Hills, Blackburn,
8B2 4PJ, U.K.

Esselte Dymo Limited,
Spur Road,
North Trading Estate,
Feltham, Middlesex,
TW14 0SL, U.K.

Questions

Anon: What types of wood have you encountered?

Howard Murray: Poplar, oak, ash and softwoods. Most of which has been successfully freeze-dried from PEG 3400.

Cliff McCawley: Long storage periods for the waterlogged wood can be devastating to it. How has the wood fared in your "passive" conditions since 1967?

Howard Murray: Most of it has not fared very well. Under either of our storage conditions waterlogged wood survives 6 to 12 months. The storage system is not really passive. If wood is impregnated with PEG and then frozen, it can be held indefinitely without decay.

Colin Pearson: Why is the Mary Rose being dismantled during excavation?

Howard Murray: For safety reasons. Most of the decks are to come out separately, all that will come in one piece is the outer part of the hull.

A THEORETICAL AND COMPARATIVE STUDY OF CONSERVATION METHODS FOR LARGE WATERLOGGED WOODEN OBJECTS

Richard W. Clarke and Jane P. Squirrel
National Maritime Museum,
Greenwich,
London, SE10 9NF

Abstract

Methods for the conservation of waterlogged wood are well represented in the literature but much of this information relates to laboratory scale experiments or the treatment of relatively small objects. Details of the treatment of large composite waterlogged wooden items such as ships hulls are limited to a number of locations around the world. The treatment of such objects presents many problems not least of which are the technical and operational difficulties and the efficiency of certain conservation methods on a large scale. This paper attempts to look objectively at some of the conservation methods for waterlogged wood and their possible application to the treatment of large structures. Each treatment method is assessed in terms of certain criteria related to the research and display requirements for the object, previous successful use of particular treatments and the time involved, safety factors and operational requirements, and the need for specialist operators etc. Amongst the methods discussed will be Dehydration, Freeze-drying, Resin impregnation and polymerisation, PEG treatments and Air-drying. Certain other methods will be considered.

It is not suggested that the criteria outlined in this paper should provide the means for deciding on a particular conservation treatment; experimental investigation is still of prime importance. Such considerations are, however, important in long term proposals and planning when the availability of finance and facilities need to be assessed.

Introduction

The advancement in underwater archaeology in recent years has meant that many wreck sites around the world have been located and identified (Bass, 1972; Muckelroy, 1977; Sledge, 1977). In some cases it has been thought desirable to raise what remains of the ship's hull and to put this on display along with many of the recovered artifacts (Barkman, 1967; Olsen and Crumlin-Pedersen, 1967, Frost, 1972, 1973; Green, 1975; Katsev, 1970). The most recent addition to this list is the excavation of the Mary Rose at present being undertaken in the Solent off Portsmouth, England. Hull structures revealed so far are shown in the **Figure.** (A representation of the Mary Rose is also shown in Howard Murray's paper. ed.) In order to secure the necessary conservation facilities for these large objects much good will or a sound financial agreement has been required. The establishment of a suitable conservation programme with adequate sampling procedures and the provision of associated facilities must not be regarded as the concluding step in the recovery project but as a major consideration in the initial decision as to whether the lifting of the hull structure should be attempted. If this consideration is not made or if it is hoped that by raising the structure sufficient pressure can be brought to bear on those parties responsible for providing finance for storage and conservation facilities, then this is an unsatisfactory approach and is destined to almost total failure (McGrath, 1981).

A recognized and structured research programme is necessary, requiring the co-operation of all concerned with the project - scientists, divers, archaeologists, conservators and administrators. Such a research programme obviously costs money and where the budget is limited, economies are essential. The choice of the conservation method should, ideally, be determined ultimately by the condition of the timber to be treated but with the conservation of large waterlogged objects the optimum method in theory may not always be the most feasible in practice. Other criteria therefore need to be considered in conjunction with the research programme and by recognising these limitations the most suitable techniques for any one project can be decided upon. Time and resources can thus be centred on a small number of praticable methods rather than on a large investigative programme covering a range of possibly unsuitable techniques.

This brief introduction has limited itself to considering underwater sites but the concepts apply equally to those large structures recovered from waterlogged land based sites (Brøgger and Shetelig, 1971; Fenwick, 1978; McGrail, 1981, Milne and Hobley, 1981; Fliedner and Pohl-Weber (undated).

The selection of criteria chosen for this comparative study is by no means comprehensive and obviously other considerations could be included. Also, they are presented in no particular order of priority and the criteria may be weighted differently with respect to a particular project. It is intended only to give an impression of some of the points to be considered when making recommendations concerning the conservation of large waterlogged wooden objects such as ship's hulls where dismantling is undesirable or impracticable. These criteria formed the basis of a preliminary document presented to the Executive Committee of the Mary Rose Trust in connection with the conservation of the Mary Rose hull as part of the research programme being carried out at the National Maritime Museum, Greenwich.

Selected Criteria

(a) Accessibility to staff during treatment:
The need to take samples, examine the structure, post excavation recording, reconstruction work etc. will be an important part of any investigative work that will continue after a large structure has been raised. Direct and continuous accessibility is therefore desirable, if not essential.

(b) Accessibility and viewing by the public during treatment:
Public participation and involvement at an early stage is essential if financial support for a continuing conservation programme is to be obtained. Accessibility with regard to the public is defined as the provision for them to walk around most of the structure but not having direct contact with it. Viewing is where this accessibility is reduced to such an extent that it is limited to viewing through small openings in an encapsulating structure.

(c) Previous successful use of treatment on a large scale:
The replication of experiments on such a large scale is limited as funds and facilities are prohibitive. Any results from previous large scale treatments are therefore of much value and must figure highly in the consideration of a particular technique. It must be remembered however, that not all situations are directly comparable as the condition and state of degredation of the timbers for example, may be extremely variable. Investigative research is therefore necessary prior to making any recommendations.

(d) Safety measures required:
The provision of plant and equipment as well as the use of large quantities of chemicals makes large scale conservation facilities comparable with that of a medium sized industrial establishment. As such, certain safety aspects must be recognized above those normally encountered in the laboratory or with small scale conservation work.

(e) Requirement for specialist staff:
In certain methods particular expertise above that normally expected from the conservator will be essential. Expert assistance may not only be required in the planning and building stages of the facilities but possibly also in the the day-to-day operation of equipment. In others this assistance may only be in an advisory capacity.

(f) Estimated time taken to complete treatments:
Such an estimate is difficult to make but where possible this is based on evidence from similar treatments. Where such comparisons are not possible the estimate is based on a consideration of the operational and material requirements and the physical characterisation of the method involved. Estimates of time are important as one needs to consider the amount of capital investment involved with the project and how this capital is best distributed throughout the conservation programme.

(g) Other considerations:
Other factors such as special requirements during or after treatment, the supply, disposal and the cost of materials required for conservation need to be considered.

Appraisal of Methods

The conservation of waterlogged wooden objects has been well-documented (Christensen, 1970; I.I.C., 1971; Oddy, 1975; Grosso, 1976; de Vries-Zuiderbaan, 1981) but many of the objects treated have been relatively small with a high surface to volume ratio. It is not intended in this paper to detail all these methods but to discuss some of them under a number of broad headings and assess their application to the conservation of large waterlogged wooden structures based on the criteria described earlier and designated as such (a)-(g). The methods covered comprise:

(I) Dehydration and Consolidation
(II) Freeze-drying
(III) Resin impregnation and polymerisation
(IV) Polyethylene glycol (PEG) treatments.
(V) Air drying and controlled air drying.

(I) Dehydration and Consolidation

Acetone-Rosin
Alcohol-Ether-Rosin, Dammar Resin, Beeswax etc.
TEOS (Tetraethoxysilane)
Dehydration alone (Acetone; Alcohol; Ether; Industrial Methylated Spirits i.e. ethanol with 10% methanol
Dewatering Agents

The theory behind such a method is that the replacement of the water within the wood with liquids of low surface tension and a higher volatility than water will, on subsequent evaporation, exert a reduced drying tension on the wood fibres; thus shrinkage and surface deterioration will be minimised. A final treatment with some consolidating material may be required.

(a) Such a treatment would almost certainly require the hull to be immersed in a purpose built tank, thus it would be virtually inaccessible to research and conservation staff until the treatment has been completed.
(b) Similarly, public accessibility and viewing would be impossible because of the strict safety regulations that would need to be enforced.
(c) No such method has been attempted on large scale although there are reports of the use of the TEOS method for the conservation of a 6 metre waterlogged canoe (Bright, 1979).
(d) Because of the large volumes of volatile solvents that would be required for the dehydration and the risk of fire and explosion, extremely stringent safety measures would need to be enforced. Fire certificates and planning permission for museum buildings may be difficult to obtain (McCawley, 1977).
(e) Specialist staff, such as chemical engineers, would almost certainly be required not only for the design and building of the equipment but also for its day to day operation.
(f) Without suitable comparisons it would be difficult to estimate a total treatment time but this might be in the region of 5 to 10 years plus, especially if pretreatment and/or consolidation is required.
(g) The very nature of this group of methods and the fact that they are at present limited to objects of relatively small size makes the possibility of carrying out such techniques on a large scale questionable. The specialist equipment and buildings required, the supply of large volumes of solvent, its disposal and requirements for pollution control make such methods extremely expensive. Dismantling the hull and treatment in small units may be an alternative but in general dehydration is considered unsuitable for large waterlogged wooden objects.

(II) Freeze-drying

Freeze-drying under vacuum
Freeze-drying under reduced temperature air flow
PEG/water/freeze-drying
PEG/t-butyl alcohol (TBA)/freeze-drying
Natural freeze-drying

This technique depends on the sublimation of ice within the wood such that the drying water phase is avoided. Depending on the condition of the timber, pretreatment with a wax or similar bulking material may be required to minimise any shrinkage or distortion which may occur during the drying process.

(a) For freeze-drying the hull would have to be enclosed within a purpose built enclosure unless use could be made of natural climatic conditions (Grattan and McCawley, 1978; Grattan, McCawley and Cook, 1980). If enclosed, staff access to the hull would be very restricted during the freezing and drying phases.

(b) Likewise, public accessibility would be very restricted, although viewing may be possible as in Marseille (see below) where a ship's hull is being conserved inside a specially constructed 'glass tent.'

(c) There is a paucity of documentary evidence dealing with the conservation of waterlogged wooden ships hulls by freeze-drying. Work is at present in progress at the Musee d'Histoire de Marseille (Morel, 1981) with the conservation by freeze drying of a merchant vessel of the 2nd century A.D. measuring 19.2 m x 7.5 m. But results have not yet been published. Natural freeze-drying has been used for the conservation of a waterlogged canoe (McCawley and Grattan, 1980).

(d) The use of liquid or compressed gases and reducing atmospheres plus the need for refrigeration plant would require strict safety measures to be observed. If the timber is pretreated with t-butyl alcohol (TBA) or a similar volatile solvent prior to freezing then the necessary precautions will be required similar to that for dehydration techniques (Jespersen, 1981).

(e) It is envisaged that specialist assistance will be required for the design and building of plant and a suitable freezing chamber and possibly during the initial stages of the operation. Day to day running of the equipment and recording may be taken over by suitably trained conservation staff.

(f) Again, without sufficient comparison, it is difficult to provide an accurate estimate of treatment time. Workers in Marseille give an estimate of 1 year to complete the freeze-drying of that particular vessel (Morel, 1981). A period of 1 to 5 years might be a more reasonable estimate and this might be even longer if the wood requires pre-treatment.

(g) As with dehydration, cost may be the prohibitive factor with this technique; the provision of the freeze-drying chamber being the major component. However, the possibility of having the treated structure on display within a few years and the benefits that this can bring to a museum may favour freeze-drying. If the timber is pretreated with PEG of low molecular weight environmental conditions for display may have to be strictly controlled. Dismantling the hull and treatment in small units may be a possibility and there are conservation centres around the world which have suitable freeze-drying equipment to deal with such an eventuality.

(III) Resin impregnation and polymerisation

All resin impregnation methods.

Polymerisation with gamma-radiation, heat, u.v. light, catalysts

In this group of methods the wood is treated with a monomer or low polymer resin either by immersion or surface appliction. Once satisfactory impregnation has been achieved the resin is polymerised in situ.

(a) Accessibility to the hull by staff would be possible during the impregnation stage if this were by surface application but impossible if the structure were immersed.

(b) Public accessibility would be almost impossible at all stages of the treatment although viewing might be possible at the impregnation stage if this was by surface application.

(c) No treatment of this type has been attempted on a large scale although individual waterlogged ships timbers have been treated using a Cobalt-60 radiation source for polymerisation (de Tassigny, 1981).

(d) If polymerisation of the impregnated wood was by means of a strong radiation source the need for strict safety measures is obvious.

(e) The provision of such radiation equipment would require the assistance of specialist technicians both in the planning and design stages as well as for the day to day operations. The use of other polymerising agents would require specialist advice in the early stages of the process but day to day operation may be taken on by suitably trained conservation staff.

(f) Again, no suitable comparisons exist to make an estimate of treatment time. It would be expected that polymerisation would be relatively rapid but the impregnation of the wood with resin would be the time limiting step. An estimate of 5 to 10 years for the whole treatment would not seem unreasonable.

(g) The provision of safety equipment and the plant required for the polymerisation stage would be extremely costly as would be the supply of the large quantities of resin needed. As with the conservation of small wooden items by resin impregnation the difficulty in reversing the process, if the need should arise, would be almost impossible on such a large scale. Dismantling and treatment as individual units may be an alternative.

(IV) Polyethylene glycol (PEG) treatments
Tank immersion (hot/cold treatments)
Surface application (brushing/hand spraying)
Mechanical spraying
TBA/PEG methods

PEG impregnation is currently the most widely used conservation method for waterlogged wood of all dimensions. The range of molecular weights of PEG available and their methods of application allow for great variability both between laboratories and the type of material treated. In considering these methods four possibilities are envisaged:

(i) Tanking method (whole ship)
(ii) Tanking method (individual units)
(iii) Spraying (whole ship)
(iv) Surface application

(a) Tank immersion of the hull either as a whole ship or as individual units would almost certainly restrict the access of staff for the duration of the treatment programme. With spraying and surface applications however, the hull could be made accessible at any time.

(b) The same distinction can be made concerning accessibility and viewing by the public. If tank immersion was used accessibility would be impossible and viewing could only be maintained as long as an efficient filtration system is used (Hoffman, 1981). With spraying, providing sufficient protection is available accessibility and viewing could be made possible at all times.

(c) Tank immersion of the whole ship in a warmed PEG solution is the conservation method at present in progress for the treatment of the Bremen Cog (Hoffman, 1981). This operation is in its early stages and results will not be available for many years. Dismantling the hull and the treatment of individual units has been used for ship remains of moderate proportions (Katzev, 1979, 1980; Alagna, 1977) and results in the most part have been satisfactory. The impregnation of timbers with PEG by the heated tank immersion method is that used by the National Maritime Museum, Greenwich for the treatment of waterlogged boat remains (Clarke, 1981).
The preservation of the Wasa is the prime example of a large scale conservation treatment that has been shown to work over a period of years and with the back up of much investigative research (Barkman, 1975).
The TBA/PEG technique for the conservation of relatively large waterlogged wooden remains has been investigated in the Netherlands (de Jong, 1977).

(d) No special safety measures would be required other than that expected of standard laboratory practice. However, if biocides are included in the spray solutions the necessary steps would need to be taken to ensure there were no toxic hazards to staff and public. An efficient air extraction system would be needed if the hot PEG immersion methods are used.

(e) Specialist expertise would be required for the design and installation of tanks, pumping equipment and spraying gear but once completed the day to day operation could be taken over by the conservation staff.

(f) PEG treatments of this type take a long time to complete. The Wasa has been undergoing treatment for almost 20 years and 12-15 years has been estimated for the completion of the Bremen Cog. With the Wasa however, the fact that the hull is freely visible and accessible means that the length of treatment time is not so critical as if the hull was completely enclosed for a similar period.

(g) As outlined here the treatment of boat remains with PEG has been widespread, the supply of materials is readily available and reasonably cheap. The building of the treatment tanks as opposed to spraying increases the cost of the operation markedly. Surface application, although not discussed as fully as the other methods is fairly self explanatory, the major requirement being the provision of sufficient personnel to carry out the 'painting.'

(V) Air drying and controlled air drying

Air drying, as the name implies, means that the wood is allowed to dry naturally and is thus akin to the seasoning of green timber. The success of this treatment is very much dependent on the state of degredation of the surviving wood and fastenings; a good state of preservation is required if the wood is to withstand the drying pressures satisfactorily. With controlled air drying the timber is conditioned at progressively lower relative humidities such that the water in the wood is in equilibrium with that of the surrounding atmosphere. In this way it is envisaged that the shrinkage distortion is reduced. Again the condition of the timber is an important consideration, because even with this gradual decline in moisture content the wood must be capable of withstanding the drying pressures.

(a) With air-drying techniques the structure is freely accessible to research staff at all times without the entanglement of conservation equipment. Only if other types of controlled drying are used, such as packing with moist sand or covering with damp sacking as described for the attempted preservation of a number of logboats (McGrail, 1978) does accessibility become difficult.

(b) Similar comments can be made regarding the accessibility and viewing by the public during treatment.

(c) Various, little-degraded hulls recovered from the Polders in the Netherlands have been preserved by air drying (de Jong, 1979). Such a method has been possible because of the good state of preservation of these hulls, the large number of similar types of wreck found and the lack of suitable storage space if the wrecks were to be recovered completely; a problem common to all museums. This has meant that many wrecks have been dried more or less where they stand and apart from some surface cracking and points of obvious shrinkage of the more extensively decayed wood, results are good. The Dutch again have been experimenting with controlled air drying on the hull of the E81, a 17th century merchantman at Ketelhaven, and the results are awaited from these experiments.

An extension of the air drying techniques is where the wood is given the minimum of surface treatment prior to the drying phase as has been described for some boat remains (Rosenqvist, 1959).

(d) Certain safety measures may be required if a biocide or toxic substance are to be used for controlling microbial growths during the drying phase.

(e) No specialist operators would be required other than in an advisory capacity for the design of the humidity control equipment for the controlled air drying techniques.

(f) Treatment time will be determined by the reaction of the wood to the environmental conditions which prevail during the drying programme or to the successive reductions in relative humidity. The fact that the structure is freely accessible means that the length of treatment time is not critical.

(g) An advantage of air drying is that it is cheap although with controlled air drying, the provision of humidity control equipment for use in a large capacity building may add markedly to the cost. A cosmetic treatment may be required if the surface of the wood flakes or becomes friable.

Concluding Remarks

It is appreciated that certain methods have been omitted such as the Alum (Christensen, 1970) and Thessaloniki (Borgin, 1978) processes and sucrose and salt treatments: it will however be apparent to those who are familiar with these techniques how they fit into the general scheme outlined above. The major problem is not that we do not have adequate methods for the conservation of waterlogged wood, it is in recognising the limitations of these methods with respect to the size of object that can be treated within the restrictions of the amount of time, money and facilities available. Such considerations in conjunction with the results from any experimental work, should be taken into account when deciding upon the most suitable conservation method for a large waterlogged wooden object. No mention has been made here of specific costs for these treatments. However, in general, methods that require the structure to be immersed in a purpose built tank or the use of other types of encapsulating structure as well as the provision of specialist personnel and equipment will add markedly to the total cost of project.

References

Algana, P., "The construction of the treatment tanks used in the conservation of the wood of the Marsala Punic ship," Studies in Conservation, vol. 22, 1977, pp. 158-160.

Barkman, L., "On Resurrecting a Wreck," Wasastudier 6, Statens Sudhistoriska Museum, Stockholm. 1967.

Barkman, L., "Preservation of the warship Wasa," in: Problems of the Conservation of Waterlogged Wood, proc. of the symp, 5-6 October 1973, ed. W.A. Oddy (National Maritime Museum: Maritime Monographs and Reports No. 16 1975), pp. 65-106.

Bass, G.F., (ed), A History of Seafaring based on Underwater Archaeology. (London: Thames and Hudson, 1972).

Borgin, K., "Mombasa wreck excavation: Second Preliminary Report, 1978. Appendix 2: Progress report on evaluating the Thessaloniki Process." International Journal of Nautical Archaeology, vol. 7, 1978, pp. 301-319.

Bright, L.S., "Recovery and preservation of a freshwater canoe," International Journal of Nautical Archaeology, vol. 8, 1979, pp. 47-57.

Brøgger, A.W. and Shetelig, H., The Viking Ships. Their Ancestry and Evolution. (London: C. Hurst and Co., 1971).

Christensen, B.B. Conservation of Waterlogged Wood in the National Museum of Denmark. Museum Document No. 1, National Museum of Denmark, Copenhagen, 1970.

Clarke, R.W., "The state of the timber. in: The Brigg 'Raft' and her Prehistoric Environment," ed. S. McGrail, National Maritime Museum, Greenwich Archaeological Series No. 6, British Archaeological Reports British Series 89, 1981, pp 123-129.

Fenwick, V., (ed), The Graveney Boat. National Maritime Museum, Greenwich, Archaeological Reports Series No. 3, British Archaeological Reports, British series 53, 1978.

Fliedner, S. and Pohl-Weber, R., "The Cog of Bremen," Pamphlets of the Focke Museum, Bremen No. 35, 3rd edtn, undated.

Frost, H., "The discovery of a Punic ship," International Journal of Nautical Archaeology, vol. 1, 1972, pp. 113-117.

Frost, H., "First season of excavations on the Punic wreck in Sicily," International Journal of Nautical Archaeology, vol. 2, 1973, pp. 33-49.

Grattan, D.W. and McCawley, J.C., "The potential of the Canadian winter climate for the freeze-drying of degraded waterlogged wood," Studies in Conservation, vol. 23, 1978, pp. 157-167.

Grattan, D.W., McCawley, J.C. and Cook, C. "The potential of the Canadian winter climate for the freeze-drying of degraded waterlogged wood: Part II," Studies in Conservation, vol. 25, 1980, pp. 118-136.

Green, J.N., "The VOC ship Batavia wrecked in 1629 on the Houtman Abrolhos, Western Australia," International Journal of Nautical

Archaoelogy, vol. 4, 1975, pp. 43-63.

Grosso, G., (ed), Pacific Northwest Wet Site Wood Conservation Conference, proc. of the symp, 19-22 September 1976, Neah Bay, Washington, vol. 1 and 2.

Hoffman, J., "Short note on the conservation programme for the 'Bremen Cog', in: Conservation of Waterlogged Wood. International Symposium on the Conservation of Large Objects of Waterlogged Wood, proc. of the symp. 24-28 September 1979, Reporter Lous H. de Vries-Zuiderbaan, (Amsterdam: UNESCO, 1981), pp. 41-44.

IIC., The Conservation of Stone and Wooden Objects, Vol. 2 Conservation of Wooden Objects, proc. of the symp. 7-13 June 1970, 2nd edn, 1971 (London: International Institute for Conservation of Historic and Artisitic Works).

de Jong, J., "Conservation techniques for old waterlogged wood from ship wrecks found in the Netherlands," in: Biodeterioration Investigation Techniques, ed. A.H. Walters (London: Applied Science Publishers, 1977), Chap. 18, pp. 295-338.

de Jong, J., "Protection and conservation of shipwrecks," in: The Archaeology of Medieval Ships and Harbours in Northern Europe, proc. of the symp. ed. S. McGrail, National Maritime Museum, Greenwhich Archaeological Series No. 5, British Archaeological Reports International Series 66, 1979, pp. 247-260.

Jespersen, K., "Conservation of waterlogged wood by use of tertiary butanol, PEG and freeze-drying," in: Conservation of Waterlogged Wood. International Symposium on the Conservation of Large Objects of Waterlogged Wood, proc. of the symp. 24-28 September 1979, Reporter Lous H. de Vries-Zuiderbaan, (Amsterdam: UNESCO, 1981), pp. 69-76.

Katzev, M.L., "Resurrecting the oldest known Greek ship," National Geographic, vol. 137, No. 6, 1970, pp. 841-857.

Katzev, M.L., "Conservation of the Kyrenia ship," 1970-71. National Geographic Society Research Reports, 1970 Projects, 1979, pp. 331-340.

Katzev, M.L., "Conservation of the Kyrenia ship," 1971-71. National Geographic Society Research Reports, vol. 12, 1980, pp. 417-426.

McCawley, J.C., "Waterlogged artifacts: the challenge to Conservation," Journal of the Canadian Conservation Institute, vol. 2, 1977, pp. 17-26.

McCawley, J.C. and Grattan, D.W., "Natural freeze-drying: Saving time, money and a waterlogged canoe," Journal of the Canadian Conservation Institute, vol. 4, 1980, pp. 36-39.

McGrail, S., "Logboats of England and Wales: Part 1, "National Maritime Museum, Greenwich Archaeological Series No. 2, British Archaeological Reports British Series 51(i), 1978.

McGrail, S., (ed), "The Brigg 'Raft' and her Prehistoric Environment," National Maritime Museum, Greenwick Archaeological Series No. 6 British Archaeological Reports British Series 89, 1981.

McGrath, Jr, H.T., "The eventual preservation and stabilization of the U.S.S. Cairo." International Journal of Nautical Archaeology, vol. 10, 1981, pp. 79-94.

Milne, G. and Hobley, B., (ed), Waterfront Archaeology in Britain and Northern Europe, proc. of the symp. 20-22 April 1979, council of British Archaeology Research Report No. 41, 1981.

Morel, M., Personal communication, 1981.

Muckelroy, K., "Historic wreck sites in Britain and their environments," International Journal of Nautical Archaeology, vol. 6, 1977, pp. 47-57.

Oddy, W.A., (ed), "Problems of Conservation of Waterlogged Wood," proc. of the symp. 5-6 October 1973, Maritime Monographs and Reports No. 16, 1975.

Olsen, O. and Crumlin-Pedersen, O., "The Skuldelev Ships," Acta Archaeologica, vol. 38, 1967, pp. 73-174.

Rosenqvist, A.M., "The stabilization of wood found in the Viking ship of Oseberg: Part 1," Studies in Conservation, vol. 4, 1959, pp. 13-22.

Sledge, S., "The wreck inspection programme at the Western Australian Museum: Responsibilities, aims and methods," Papers from the first Southern Hemisphere Conference on Maritime Archaeology, Perth, Western Australia, 1977, pp. 80-90.

de Tassigny C., "The suitability of gamma radiation polymerisation for conservation treatment of large size waterlogged wood," in: Conservation of Waterlogged Wood. International Symposium on the Conservation of Large Objects of Waterlogged Wood, proc. of the symp. 24-28 September 1979, reporter Lous H. de Vries-Zuiderbaan, (Amsterdam: UNESCO, 1981), pp. 77-84.

de Vries-Zuiderbaan, L.H., (reporter), Conservation of Waterlogged Wood. International Symposium on the Conservation of Large Objects of Waterlogged Wood, proc. of the symp. 24-28 September 1979, (The Hague: Government Printing and Publishing Office, 1981).

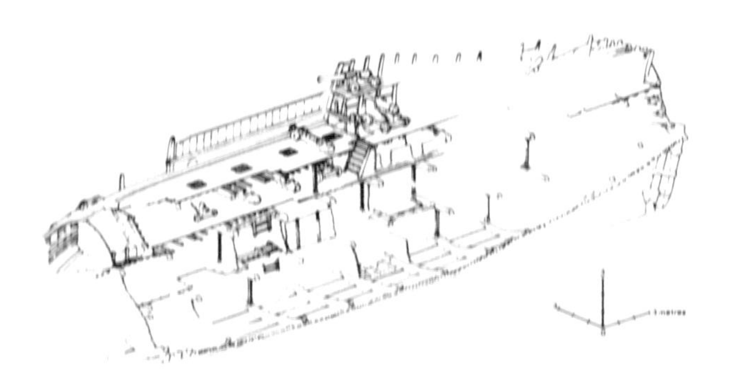

Figure: Hull Structure of Mary Rose, revealed as of 1980.

AN APPROACH TO THE CONSERVATION OF INTACT SHIPS FOUND IN DEEP WATER

Daniel A. Nelson
The Hamilton-Scourge Foundation,
City Hall,
71 Main St. W.,
Hamilton, Ontario,
Canada, L8N 3T4

Abstract

At about 2 a.m. on the morning of August 8, 1913 a violent squall swept across Lake Ontario upsetting two American armed schooners, the Hamilton and the Scourge, sending them to the bottom with heavy loss of life.

The ships remained forgotten and undisturbed until 1973 when a sophisticated search programme organized through the Office of the Chief Archaeologist of the Royal Ontario Museum assisted by the Canada Centre for Inland Waters discovered both ships lying 550 metres apart in 88 metres of water, 10 kilometers off St. Catharines, Ontario. The operation, using magnetometers, side-scan sonar, microwave navigation devices, and automatic data recording equipment, searched about 90 square kilometers of lake bed before the ships were located.

Acoustic, photographic and television probes have revealed both ships resting upright with hulls and spars remarkably intact and guns at the ready on the decks. An exquisite figurehead graces the bow of the Hamilton. The adjacent lake bed is littered with ship-borne objects and the bones of the crews.

Title to the vessels, the last of their kind in existence, was transferred from the U.S. Navy to the Royal Ontario Museum in 1979 and subsequently assigned to the City of Hamilton, Ontario. The Hamilton-Scourge Foundation has been established to administer, on behalf of the City, the on-going programmes aimed at the recovery, conservation, and display of these unique historic opportunities. Innovative approaches are planned that could provide a model for the conservation and display of intact ships found in deep water in the future.

The basic conservation concept is to reproduce as closely as possible the lake bottom environment that has served to preserve and stabilize the ships in excellent condition for 168 years. It is proposed to keep and display the vessels in tanks of cold, conditioned, fresh water.

Background and Current Programmes

Once the "War of 1812" armed schooners Hamilton and Scourge were located and identified at the bottom of Lake Ontario in 1973, we were faced with a number of questions and challenges:

- Although the historic significance of the ships was undeniable, they are the last remaining examples of this type of ship, the fundamental question remained, was their condition good enough, considering the great expense, difficulty and hazard of working in relatively deep water, to warrant further study and perhaps recovery?
- Could the legal morass surrounding title be satisfactorily resolved?
- And most important, was it technically and financially possible to design programmes that would ensure the safety of the vessels through a comprehensive in-situ survey, recovery, transport, and installation into a suitably designed conservation and display facility?

At the present time, we feel all these questions have been, or can be, answered in the affirmative.

As more visual data is accumulated; starting with the initial side-scan sonar images; progressing through a programme of detailed acoustic imagery, to visual glimpses of the Hamilton from remotely controlled television and camera platforms; and culminating in the first direct observations during the Cousteau expedition to the Great Lakes in the autumn of 1980; the ships look better and better. Not only are they remarkably intact, resting upright with only a slight list, and with minimal physical damage (sustained no doubt, during the sinking), but the Hamilton, at least, is graced with an unexpected amount of exquisite carving and ornamentation.

Luis Marden, former chief of the foreign editorial staff of the National Geographic Magazine and for more than 30 years specializing in diving and underwater archaeology reportage for the Geographic, writes "... nor have I observed anything to equal the state of preservation of your ships ...... the condition of the ship's timbers look as though they had been cut and fitted months rather than more than a century and a half ago."

An application to transfer title of both ships to the Royal Ontario Museum from the U.S. Navy was accepted by the Office of the Secretary of the Navy in 1976. For the next three years the document followed a tortuous route though two federal administrations several historic and legal committees and finally, after having been placed before the Congress of the United States and surviving 60 sitting days unchallenged, title was offered to the Royal Ontario Museum. The first time the U.S.A. had relinquished ownership of major historical material. The Royal Ontario Museum however, under an administration different from that requesting the title transfer, and with new priorities, declined to accept title and the matter was held in abeyance. In 1979, demonstrating great courage and foresight, the City of Hamilton sought and was granted title to both ships.

One major question remained, could the technical and financial programmes necessary for the eventual recovery and display of the vessels be successfully planned and executed?

The first operational priority was to conduct an on-site survey in order that everything possible be learned about the ships and their environment. The survey, two years in the planning, would accumulate geotechnical data relating to the sediments in which the ships lie, limited and selected material sampling for conservation studies, and a complete, referenced photographic and television recording of all parts of the ships and the material associated with them on the lake bed. The survey scheduled for 1981 but postponed until 1982 will use a one-man tethered submersible A.D.S. (atmospheric diving system) as the principal research vehicle. This extremely manoeuvrable and versatile mini-sub can accomplish all the photographic and other tasks required. With hard-line communication and surface supplied power, on-site T.V. is continually transmitted to the surface thus allowing monitoring and direction of all operations by surface personnel. Endurance is limited only by the endurance of the pilot. Microwave and acoustic navigational systems position and reference the various programmes both on the surface and on the bottom. All data will be recorded automatically on a real time basis.

The ships, about 22 meters in length (34 meters including the bowsprit), are a handy size to recover. With hulls intact and no structural damage to frames or keels apparent and with strength estimated to be at least 60% of original, the 60 to 70 deadweight tonnes per ship is well within the capacity of current technology to raise gently.

However, raising the Hamilton and Scourge is the last task. First, a complete conservation programme must be developed to preserve them safely and a building must be designed, financed, and built to receive and display the vessels and the thousands of artifacts associated with them. The Hamilton Regional Conservation Authority has allocated a 10 acre lakefront site in Confederation Park for the conservation and display building.

The Hamilton-Scourge Foundation, a non-profit organization, received its charter in 1981 and is expected to provide the Project with the good management so essential to the success of such a diverse and complex operation. The Foundation administrates all Project programmes on behalf of the City of Hamilton.

After funding, which is a constant problem that the Foundation is expected to solve, conservation has been the greatest concern. We feel that the development of a successful conservation programme is paramount to ultimate success.

The ships are currently well stabilized in the deep lake and the suggestion has been put forth that they be left there until a suitable method is developed to preserve the ships in-toto and dry. Unfortunately their specific position is known and, even with fairly elaborate security measures, it is only a matter of time before irreparable damage is done by illegal salvagers using either mini subs, hyperbaric diving equipment or grapnels. The risk is too great to leave the ships on the bottom indefinitely.

After studying all available literature on the subject, we concluded that, although progress is being made, there is still no entirely satisfactory way to preserve complete waterlogged ships by drying them. The main problem results from the difficulty in processing simultaneously widely diverse sizes of timbers of various kinds of wood with substances that are difficult to control and which are often corrosive to metallic hull fastenings and other hardware as well. The enormous costs with no real guarantee of success, and the large facilities required to treat the huge bulk of material are enough to discourage even the most worthy projects.

We took a new approach. It is apparent that the ships and all their contents have survived in exceptional condition after 168 years of immersion in the depths of Lake Ontario. This fresh water site, where no light intrudes, remains constantly at 5°C and is swept by gentle currents of only a few centimeters per second.

Perhaps it would be unwise to try to improve on nature. We propose to reproduce these conditions artifically.

The suggested conservation programme can be divided into three sections:

1. Pre design studies.
2. Design and construction of the conservation and display facility.
3. Recovery and transport of the vessels.

Pre Design Studies

In November 1980, a seminar was held at the Canadian Conservation Institute (CCI) where the proposition was advanced and exhaustively discussed that the Hamilton and Scourge be kept in an environment duplicating, as closely as possible, that at the bottom of the lake. The conclusion was reached that this approach could have validity and the Institute generously offered to assist in several ways. A conservator would be nominated by the Institute to assist in developing, in concert with the Project director and archaeologist, the conservation aspect of the site survey. This would involve what, where, and how samples for conservation study would be obtained. The samples would be scientifically analysed at the CCI and an experimental test programme would follow to ascertain the potential of the proposed method. The ensuing report would make specific recommendations with respect to the requirements, design and operation of the conservation facility.

Conservation and Display Facility

It is proposed that the ships be preserved and displayed in a large fresh water tank. Temperature and light will be controlled to minimize deterioration of organic material from bacterial, fungal or other influences. Since it would be impractical to construct tanks deep enough to cover the masts, dummy masts would be installed either full height through the air-water interface which would occur a few feet above the deck or, the masts could terminate just above the surface if the surface was made opaque by a bubble layer or a covering of some kind. The ships would be viewed through many windows in the sides of the concrete tank and could be described using methods such as sequenced spot lights and acoustic recordings.

Although the concept has been criticized on the basis that any conservation procedure is incomplete unless the artifact is stabilized in the dry state, we feel this approach is valid for the Hamilton and Scourge and has some very distinct advantages.

1. Tremendous financial savings can be realized if the extensive and expensive facilities, staff, equipment and supplies required to dry-preserve complete hulls become redundant.
2. The suggested method is time-proven, relatively inexpensive and kindly to both organic and in organic materials.
3. The display would be unique and popular with the public. The ships could be viewed shortly after recovery thus immediately producing revenue to offset expenses.
4. All environmental conditions (water quality, temperature, light etc.) could be minutely and continually monitored and adjusted to provide optimal protection.
5. Although only fresh water will be used initially, additives (fungicides, antibacterial agents etc.) could be introduced later if it was proven that this would be beneficial.
6. If a satisfactory and economical method to dry preserve complete hulls is devised in the future, the tank could serve as the treatment vat.
7. The hulls would be supported in cradles but a great deal of support would come from the water itself. This would prevent, to a large extent, hogging and other hull distortions.
8. The on-going condition of the hulls could be easily monitored and appropriate

steps quickly taken to counter any problems.

9. Measuring and cleaning the hulls, recovery of detached artifacts, and material monitoring procedures would be conducted in the tank by specialists wearing hot-water diving suits with surface supplied air. Visitors could view all this activity through the view ports. The proposed 6 meter depth of the tank would present no decompression difficulties.
10. With proper insulation and a very large volume of water the energy requirements to maintain the tank at the proper temperature should not be excessive.
11. With the thousands of detached artifacts remaining within the hulls until needed, work on the cataloguing and preservation of this material, using traditional methods, could proceed at leisure over a long period of time by a small conservation staff in an on-site laboratory. The slow, scheduled build up of items in the ancillary display halls of the complex would provide on-going interest for repeat visitors. This approach would also avoid the large "crash" conservation programmes which are invariably excessively expensive and poorly executed.
12. The ships would be safe not only from the treasure hunters of the deep but also from the vandals of the museum hall. Fire would obviously not be a serious hazard.

Recovery and Transport

If the hulls are as strong as expected, the ships will be slung and raised using a floating marine salvage derrick. It is known that the hulls have not settled deeply into the bottom sediments and analysis of core samples and acoustic geotechnical data indicates that the sediments are not consolidated at levels beneath the keels. Messenger lines will be jetted beneath the hulls and a number of wide slings sufficient to adequately support the hulls will be drawn beneath the ships. A spreader-bar frame will be lowered from the surface, slings attached, tensions equalized and masts jury-rigged to the spreader frame. Sediments about the hulls will be water-jetted away and the hulls will be levelled and slowly raised to the surface. Before the hulls are recovered, all deck guns; spars and loose extraneous objects will be recovered. Material within the hulls will not be disturbed. The recovery operation will be accomplished using both A.D.S. and divers in a saturated, hyperbaric mode.

Once the hulls are near the surface the masts will probably be unstepped; however, there is no intention of ever lifting the ships clear of the water. One can imagine the damage sustained by 168 year old paper and other delicate material in sea chests or other locations if the supportive water is suddenly drained away.

The hulls will be supported between two barges during the 35 kilometer trip to the conservation facility at Confederation Park.

Transfer of the ships from the lake to the tank will be either by a locking system or more likely, by a mobile water filled transfer tank.

THE EXCAVATION OF A MID-SIXTEENTH CENTURY BASQUE WHALER IN RED BAY, LABRADOR

E. Willis Stevens
Marine Excavation Unit
Parks Canada, Ottawa

Introduction

The Red Bay Project, under the direction of Robert Grenier, is a co-operative undertaking between the province of Newfoundland and Labrador and Parks Canada. It involves the archaeological investigation of the submerged remains of a wooden vessel presumed to be the Spanish Basque whaler, the San Juan, which was wrecked in Red Bay harbour, Labrador in 1565 (Barkham and Grenier, 1979: 61) (**Figure 1**). The wreck was discovered in the fall of 1978 by Parks Canada's marine excavation unit with the help of archival information provided by Selma Barkham, working in Spain for the National Archives of Canada. The project also includes extensive land site excavations of a Basque whaling station. This work is being conducted by a research team from Memorial University of Newfoundland, under the direction of James A. Tuck. To date several oven structures for rendering blubber into oil, associated structures including a cooperage and several thousand ceramic tile and artifacts have been uncovered. These excavations are being carried out on Saddle Island, located at the entrance to Red Bay harbour. The underwater site was located approximately 15 meters out from the north shore of Saddle Island (**Figure 2**).

The ship was loaded with barrels of whale oil when it was wrecked. The recovery of these containers has led to a major study of sixteenth century coopering technology. A study of the ship as a unique document of the sixteenth century has also been started with the precise and detailed mapping of the hull structure and with the three dimensionsl recording on the surface of loose structural timbers found on the wreck site. A large number of well preserved whale bones were also found around the wreck and indicated the potential of the entire harbour as a major and unique repository of faunal remains which would allow for the study of butchering techniques and carcass disposal patterns utilized by the Basque whalers. Although the ship had been salvaged, as reported in the archival documents, it now appears that enough artifacts will be recovered in and around the ship to possibly reconstruct what life would have been like on board a whaler in the sixteenth century in Terra Nova.

This paper gives a brief summary of Basque whaling in Labrador and a review of the results of three years of archaeological reseach in Red Bay, Labrador.

The Basque Whalers

The Basques are known as the pioneers of large scale whaling, having established the trade as early as the twelth century and probably earlier (Markham, 1881: 969). The initial area of their whaling exploits was concentrated in the Bay of Biscay along the coastline of the Basque country, which extends from Bayonne in the north to Port Bilbao in the south. The Basque country, although mountaineous, is heavily forested and has for many years supplied timbers for use in ship construction. The Basques never lost their forests despite the shipbuilding boom of the sixteenth century. This was due to strict conservation laws which ensured that for every tree felled a new tree was planted. The effects of this law are most evident throughout the Basque country which today continues to supply timbers for ship construcion. The area around Onate is typical, and it was here at the University of Onate that Selma Barkham has conducted much of her archival research. The references to the San Juan come from the small port of Pasajes in the parish of San Juan. Throughout Pasajes there is evidence of the importance of shipbuilding with ship's timbers often used in house construction. Intriguing wood technology details such as triangular recesses, found on the Red Bay wreck, can be seen on some sixteenth and seventeenth century buildings. Many of the local shipyards still use traditional Basque shipbuilding practices, and it is not uncommon to find oak and eucalyptus timbers seasoning in nearby rivers.

The whalers come from all over the Basque country and literally emptied villages and small cities of all able bodied men. The Basques are generally believed to have hunted baleen whales, specifically the Right and Bowhead whales. These whales, averaging 12 to 18

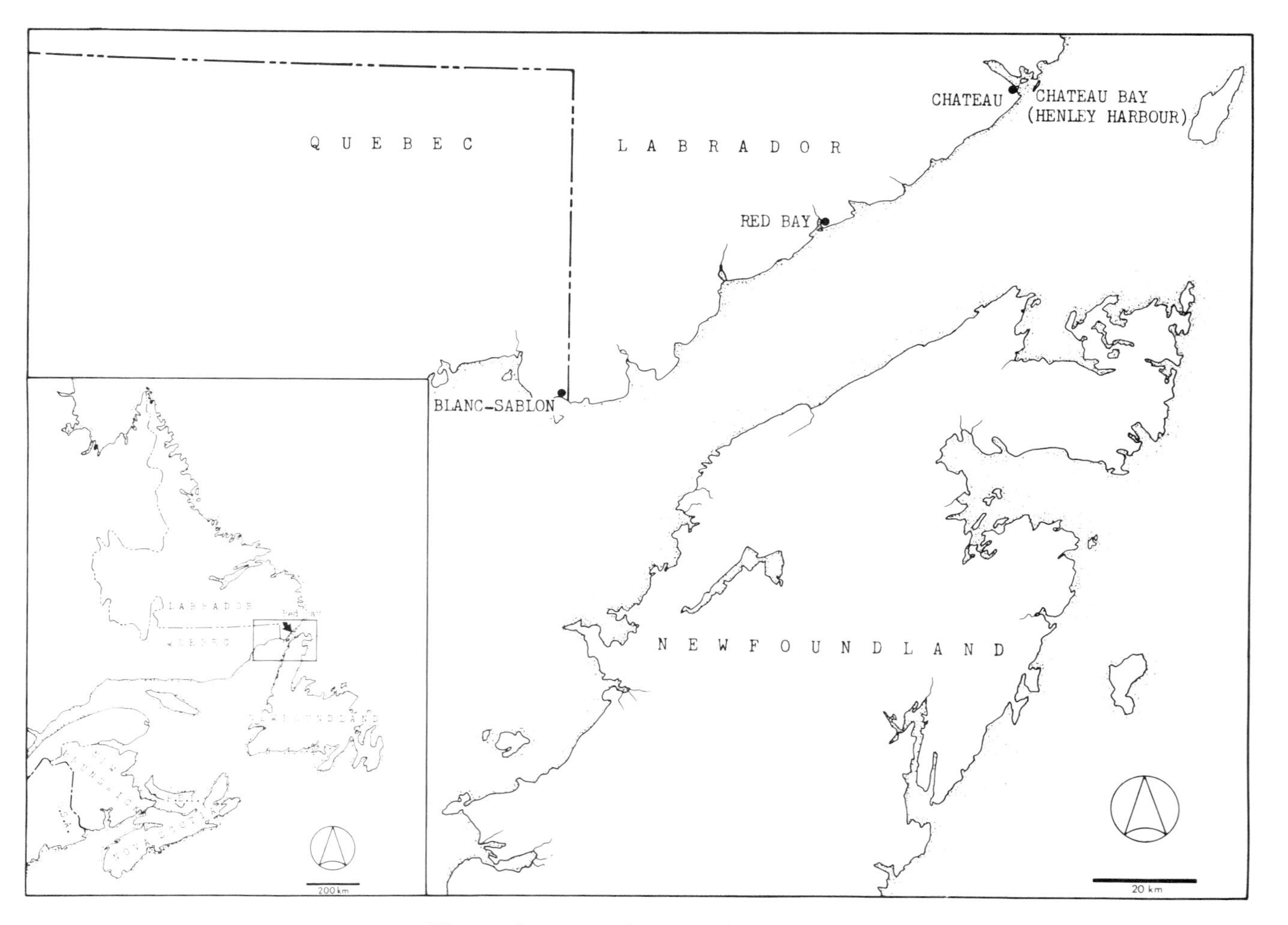

Figure 1. Map showing the locations of Red Bay and Chateau Bay.

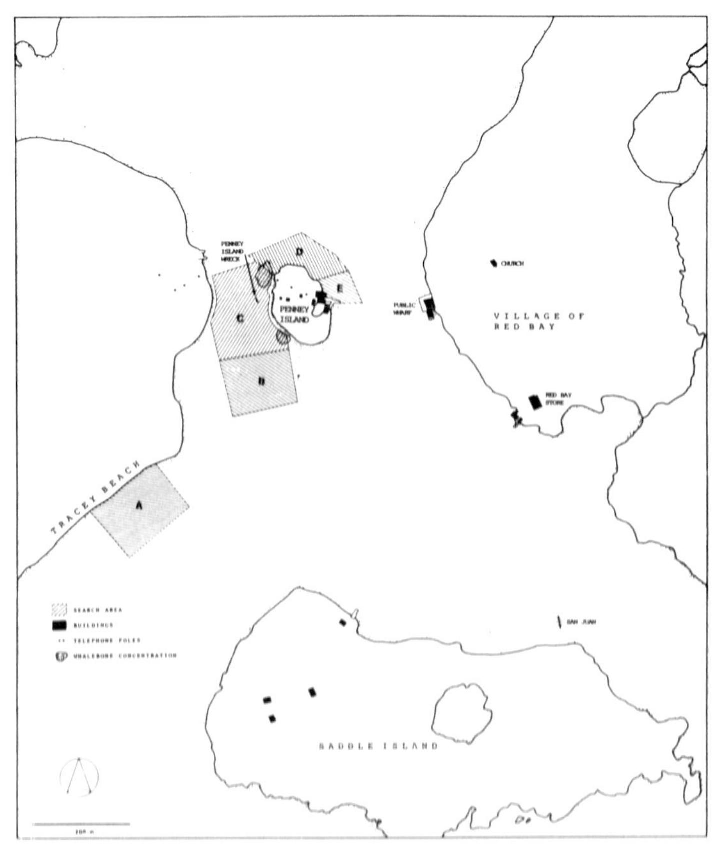

Figure 2. Map showing the location of the wreck site and of the harbour areas surveyed.

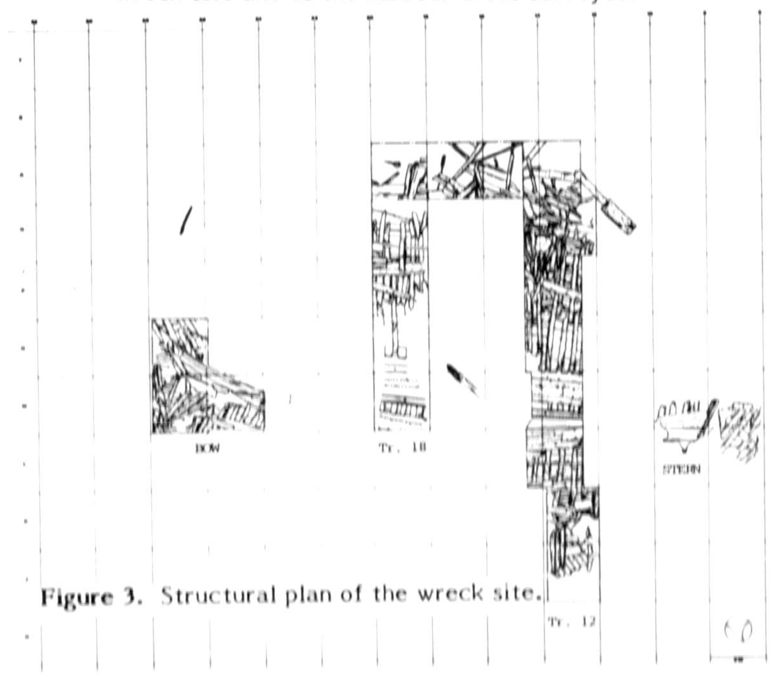

Figure 3. Structural plan of the wreck site.

meters in length, are characteristically slow swimmers, have a high oil yield and can be found near shore. They are also very buoyant after harpooning and lancing. Together these characteristics enabled the Basques to be highly successful at their trade, so much so that by the sixteenth century they had to sail as far away as Greenland and North America in search of the large whale herds. The importance of the whale fishery can be seen from the various documents now being studied and which refer to large amounts of money spent on ship construction and in particular insurance coverage.

The south coast of Labrador is not unlike the coastline of the Basque country. The high mountains overlooking the Straits of Belle Isle made for ideal natural watch towers from which to spot migrating herds of whales, while many of the well protected and deep harbours provided excellent locations from which to construct the shore facilities necessary for processing whales. The Basques were involved in whaling all along the south coast of Labrador and by the mid-sixteenth century as many as two thousand persons were engaged in the whale fishery, which extended from about July to late fall and oftentimes into January. The two main centers of activity were Chateau Bay and Red Bay (**Figure 1**).

Archaeological Research

Archaeological investigations began in 1979 with an initial strategy to test the site with a series of trenches through which a positive correlation could be made to sixteenth century ship architecture. This series of tests proved quite successful in locating and identifying the bow, the stern and the two sides of the solid portion of the well preserved structural remains of the ship, flattened on the bottom of the harbour in approximately ten meters of water (**Figure 3**). Subsequently, a full scale excavation project was initiated in 1980. Prior to excavation a grid was installed consisting of two meter by two meter squares. Additional grid sections were added when the need arose. Once in place the grid provided both horizontal and vertical control through its relationship to the site's main datum line. The principal means of excavation was the airlift while a water dredge was used in more shallow water. Recording techniques included underwater tracing, detailed mapping of all loose timbers and integral structure, as well as triangulation of whale bone and artifacts. Stero-photogrammetry has been used along with standard 35 mm photography, underwater video, 16 mm and super 8 mm cine.

Parks Canada Conservation Division is managing the conservation responsibilities for all artifacts and features, as well as experimenting on new means and ways to improve lifting techniques, packaging and recording of artifacts. All recovered timbers, following detailed suface recording, are reburied in deep trenches which are then silted in for protection. This reburial process is also being monitored by the Conservation Division.

Ship's Structure. The ship has a keel length of about 15 meters with a stem to stern length around 26 meters and a beam of 8.5 meters. These dimensions have been estimated from the 'as found' design characteristics which include a long overhanging stem and a strongly raked stern post which is linked to the upper extension of the keel by a fore-and-aft scarf, both typical of the sixteenth century. The transom was a square tuck design with the outside planking in a reversed v-shaped pattern similar to the Henry Grace à Dieu built in 1514. This planking revealed evidence of at least one gun port although no armaments have yet been found in this area of the site. A portion of the ship's rudder has been uncovered with complete excavation of this important artifact planned in the 1982 field season.

Several other interesting parts of the ship have already been raised and are presently undergoing conservation and analysis at Conservation Division. These include: a capstan, a swivel gun or verso, an anchor, and a bilge pump consisting of a square tube of beech, bevelled on the edes with the plunger mechanism still very much intact. The long range objectives of the ship study are to reconstruct the ship and its technologies in plan and ultimately as a model, with the interim step of constructing an 'as found' model of the wreck site.

Harbour Survey. Apart from the actual wreck excavation a harbour survey is also being undertaken. The survey, consisting of free swimming and towed searches, will encompass most of the underwater area of Red Bay harbour with the objective of providing additional information so as to better understand the overall context of the San Juan and to aid in the interpretation of Red Bay harbour as a whaling station. As part of the survey, a trench, referred to as the shore trench, was excavated between the wreck site and the whaling station on Saddle Island. The purpose of this trench was to gain a better understanding of the relationship of the wreck to the whaling station. A secondary aim was to determine the spatial and temporal distribution of whale bone with an outlook of providing the impetus for a comprehensive study of Basque whaling technology. The excavation has produced many interesting discoveries along with a variety of artifacts. The stratigraphic record was dominated by the presence of debris or discard material from Saddle Island. This included peat, woodchips, rock collapse and a series of articulated whale vertebrae within a deposit of codfish bone. This deposition of fish bone and whale vertebrae suggests that whaling and codfishing were contemporaneous operations, although the extent of the codfishery was not immediately evident.

Artifact and Other Related Research

The Red Bay Project is also deeply engaged in artifact research, carried out by the Material Culture Research Unit of Parks Canada. The San Juan was loaded with casks of train oil when the ship sank, and while much of the cargo was salvaged at the time of wrecking it now appears that as much as two-thirds of the cargo remained in the wreck. Thus the principal artifacts collected have been cask remains which include thousands of barrel staves, hoops, cants, and other miscellaneous pieces. Most of these casks have been badly damaged by the collapse of the wreck; however, complete casks are being found. The majority of casks were made of oak, although beech was also used. Makers assembly marks can be observed on most staves indicating manufacture, disassembly and reassembly at time of use. Four different sizes of containers have been identified: full barricas, one-third barricas, half barricas, and a small bever cask.

Various other research reports have been compiled on a wide variety of artifacts recovered from the site. These include: footwear, coarse earthenware, roofing tiles, ceramic and glass vessel sherds, weaponry (verso), and ship's fittings and rigging. A preliminary study of ballast stones has been completed with emphasis on stowage patterns, the geological and geographical provenience of the stones through lithic analysis, thus providing some indication as to the ship's origin, and finally to identify the cultural use of lithic materials. Other research includes the compilation of reports on sixteenth century shipbuilding and shipboard life, Selma Barkham's continued studies on the Basque whalers, investigations into the history of the whaling economy and technologies before and after the Basques, a study of faunal remains, literature searches in the Basque Studies Program in Reno, Nevada and in Ottawa, and finally various other researches in Basque whaling technology as the archaeological evidence becomes available.

Summary

The Red Bay Project is indeed an ambitious undertaking with many avenues of research as of yet untouched. However, in terms of overall project goals it is intended that we aim towards a reconstruction of sixteenth Basaque whaling technology, centered around the ship and shipbuilding technology, through which to gain a better understanding of the Basques of Red Bay.

References

Barkham, Selma and Robert Grenier, "Divers find sunken Basque Galleon in Labrador," Canadian Geographical Journal, Jan. 1978-79, pp. 60-65.

Markham, Clements, R. "On the whale-fishery of the Basque provnces of Spain," Proceedings of the Scientific Meetings of the Zoological Society of London, 1881. pp. 969-76.

UNDERWATER MOLDING TECHNIQUES ON WATERLOGGED SHIPS TIMBERS EMPLOYING VARIOUS PRODUCTS INCLUDING LIQUID POLYSULFIDE RUBBER

Tom Daley
Lorne D. Murdock
Charlotte Newton
Parks Canada, Conservation Division,
1570 Liverpool Court,
Ottawa, Ontario,
K1A 1G2

Abstract

When excavating a shipwreck marine archaeologists have at their disposal a variety of methods for recording various ship's features in-situ (underwater) e.g. photography, stereo photography, photogrammetry, drawing etc. The main disadvantages of these methods are that they have variable accuracy and rely upon two dimensional representations of the results. Parks Canada, Conservation Division is developing a new method of accurately reproducing ship's features in three dimensions, with retention of fine detail using a two component liquid polysulfide rubber. This paper described the preliminary work carried out in the lab and on site, in Red Bay, Labrador including methods of application and results.

A 23 minute 16 mm film of the molding technique was shown. It is our intention to produce a properly filmed and edited version of this movie, if and when the necessary funds become available. This film would be made available to interested individuals or organizations through Parks Canada. The article does not appear in full here, since the first stage of the experimental work has been published recently (1).

The second stage of the work was carried out in the 1981 field season and this resulted in several improvements to the technique. This work will be reported shortly (2).

Conclusion

Parks Canada Archaeologists working on the Red Bay site in Labrador, seemed pleased with the development of this new recording technique and have requested the authors to assist in a molding - recording project of the San Juan hull in the summer of 1982.

Other information on polysulphide molding compounds has been published elsewhere (3,4,5 & 6).

Supplier

The Polysulfide Rubber, FMC 200, is available at a cost of about $60.00 U.S. per gallon from Smooth-on Inc., 1000 Valley Rd., Gillette, New York, 07933, U.S.A.(6)

References

1. Murdock, Lorne D. & Tom Daley, "Polysulphide Rubber and its Application for Recording Archaeological Ship's Features in a Marine Environment," International Journal of Nautical Archaeology and Underwater Exploration, vol. 10, no. 4, 1981, pp. 337-341.
2. Murdock, Lorne D. & Tom Daley, "Progress Report on the Use of Polysulphide Rubber for Recording Archaeological Ship's Features in a Marine Environment," ibid., (to appear in the February 1982 issue).
3. J. Spriggs, this publication.
4. Hamilton, D.L., Conservation of Metal Objects from Underwater Sites: A Study in Methods, (Austin, Texas: The Texas Memorial Museum, and The Texas Antiquities Committee, 1976), Texas Memorial Museum, Misc. Papers no. 4, Texas Antiquities Committee, pub. no. 1).
5. Katzer, M.L. & F.H. van Doorninck, Jr., "Replicas of Iron Tools from a Byzantine Shipwreck," Studies in Conservation, vol. 11, no. 3, 1966, pp. 133-142.
6. Smooth-On Inc., "Polysulfide Flexible Mold Componds," Bull. of Information, no. 12, FMC Series, (Gillette, New York: Smooth-On Inc.).

Questions

Richard Clarke: In the blocklift did you think it worth trying accelerators with the plaster?

Tom Daley: No, because it only took one hour for the plaster to harden.

Jim Spriggs: There are two catalysts (a fast and a slow one) that can be used to cure polysulphide rubber. Which did you use?

Tom Daley: We tried both, but preferred the slower. Mixing on the surface and taking the fast curing mixture down didn't work well because curing was taking place as we tried to make the cast, whereas the slow curing mixture flowed in easily. We did however experiment with the thixotropic variety for horizontal surfaces but found that it wouldn't stick to itself underwater, nor spread to conform with the surfaces to be moulded.

Jim Spriggs: Did you mold on vertical surfaces and did you employ the thixotropic polysulphide rubber for this?

Tom Daley: We did mold on vertical surfaces, but did not use the thixotropic variety for this.

David Grattan: Do you plan to mold a complete timber, rather than just small features?

Tom Daley: Robert Grenier is planning to do this.

REVIEW OF THE CONSERVATION OF MACHAULT SHIPS TIMBERS: 1973-1981

Victoria Jenssen
Lorne Murdock

Conservation Division,
Parks Canada,
1570 Liverpool Court,
Ottawa, Ontario,
K1A 1G2

This case history is intended to familiarize colleagues with a large-scale treatment of waterlogged European white oak timbers which has been in progress here in Canada since 1972. Since the treatment has not been totally successfuly, we felt others would benefit from description of the decisions made during the wood's career. That much of this report had to be gleaned from various files and interviews, then edited together to make this mosaic, reflects a certain uneveness in the entire project. Somehow, the larger the artifact and the further it lies from the site of the decision making, the more likely the project can either miscarry or simply be deferred. A summary of the treatment chronology is shown in **Table 1.**

Introduction

From 1969 to 1972 an archaeological team from Parks Canada excavated the hull of the 18th Century French supply ship Machault and its buried contents. It is located in the north channel of the Restigouche River at the point where it flows in the Baie de Chaleur, separating Quebec and New Brunswick. The Machault was a 3-masted 5th rate frigate, about 145 feet (44) metres long with a displacement of 550 tons and armed with 22 heavy iron cannon.

The Machault is significant because it was involved in the last naval engagement of the Seven Years War in the New World between France and England; The Battle of Restigouche which the French lost.

An estimated 90% of the Machault's oak hull was preserved in the thick anoxic silt of this shallow brackish tidal area which has large mud flats at low tide. The site is also 3 km downstream from a paper mill, and previously a lumber mill operated nearby, adding much organic debris to the environment.

Decision to Retrieve and Preserve Hull Sections

At the end of the 1971 field season Parks Canada decided to retrieve in the following year a cross-section of the Machault along with its large rudder and other timbers which could be studied, recorded and ultimately displayed to the public (8).

Thirty-five years before, the Marquis de Malauze had been retrieved from the Restigouche in 1934 by local people under the direction of Father Pacific and left outdoors. Seen today under its shelter, the wood may be studied but has undergone damage due to climate and drying stresses.

The quantity was estimated to be about 90 cubic metres (3,244 cubic feet or 25 1/2 cords). A Parks Canada Visitor Reception Centre would be built near the site to display the numerous finds and a large cross section of the original hull.

Decision to Slow Dry by Burial in Sand

In the fall of 1971, conservators at Parks Canada were notified of the decision to retrieve and display portions of the hull and were asked to propose various conservation treatments for the massive oak sections including approximate costs and duration of treatment. Several approaches were considered, but really the only obvious examples in 1971 of treatments of massive well-preserved oak timbers were to be found in the Wasa and the Bremen Cog projects.

Two possible treatments were suggested which were thought to be feasible for such a large quantity of oak. Tank treatments in aqueous PEG solution using 18 m x 9 m x 2.4 m epoxy-coated concrete tank, at an estimated cost of $350,000, was recommended. As a cheap but unproven alternative, slow-drying by burial in sand was suggested at the estimated initial cost of $30,000. A committee of archaeologists, display people and a conservator chose the cheaper but untried method: slow drying by burial in sand (1).

In short, a major decison was made about the treatment of the timbers before excava-

TABLE I

TREATMENT CHRONOLOGY: MACHAULT TIMBERS

1971	Excavation	
1971-77	Slow-Drying	Burial in Sand
1977-79	Slow-Drying	Spray with 15% PEG 540 blend
1979	Slow-Drying	"Seal" with layer of 100% PEG 540 blend
1979-81±	Slow Drying	Spray with molten PEG 540 blend
1981-?	Slow-Drying	(to be determined)
1983-84	Reconstruction and Display	

Figure 1: Portion of midship section of Machault in August 1972 after excavation (Photo -Parks Canada).

tion and before any practical evaluations were made regarding the timbers' treatment requirements.

Retrieval of Timbers: Summer 1972

The ship's stern and the central hull sections were severed underwater with a submersible hydraulic chainsaw and were lifted to the excavation barge by its 30-ton hydraulic crane. During their documentation, transportation to the conservation facility, and temporary storage in a nearby field, the timbers were kept wet by spraying with water pumped from the river. **(Figure 1.)**

The pumps would sometimes run out of gasoline, or water when the tide fell below the intake hose, so that the wood was drying intermittently in the sunlit field. Surface checking was observed in some wood during that summer, suggesting that the timbers had started their own uncontrolled premature drying programme prior to burial in sand. Actually, conservators had favoured a large polyethylene lined pit filled with water and a suitable fungicide for interim storage. However, the field conservator who had been assigned to the excavation was transferred to Ottawa in early May. The field conservation facilities were being moved in the Fall of 1972 to headquarters in Ottawa.

Initial Conservation Treatment: Slow Drying by Burial in Sand (1)

While the timbers were being excavated, staff were constructing a double walled insulated building (18 m x 9 m x 2.4 m) with forced hot air heating to house the sand and timbers. The conservation building is still located on the Quebec shore of the river, 3.3 km from the wreck site.

The initial programme intended that the wood reach a moisture content of 13% over a two year period (1). The sand was meant to retard the drying of the wood and to provide support to retain the shape of the timber. The coarse gravel on the buildings base was to provide drainage for the sand and wood above. In this way, it was thought that the timbers would have a reasonably constant environment and that they would not freeze.

By October 1972, the timbers had been loaded in 200 cubic meters of washed sand. The initial 0.3 metre layer of sand was put on the coarse gravel floor. Then timbers or a hull section were put in position. The fungicide sodium pentachlorophenate (Dowicide G) crystals were sprinkled on the timbers. A large sandwich consisting of 0.3 m layers of sand, timbers, sodium pentachlorophenate and more sand were applied until the fill reached 0.6 m from the top of the walls. Finally the roof was constructed on the building with two 0.3 m square ventilation holes one at each end of the building.

Temperature and humidity monitors were installed 0.6 m above the sand at rafter level and an additional thermometer probe was buried 1 metre deep in the sand.

Thus the wood embarked on its first treatment programme, which was estimated to be complete in 2 years. Already the initial expense exceeded the predicted cost of $30,000 by $18,000.

From the outset, one could predict that the progress of the treatment would be difficult to evaluate since the timbers were buried and no studies had been undertaken before burial to establish moisture content, moisture gradient, or wood-drying characteristics.

During the five following years, from 1972 to 1977, four field trips were made from Ottawa to the Restigouche conservation building to inspect the project and to meet with the watchman who recorded the temperature and humidity. During this period, responsibility for the Machault project passed from the headquarters' laboratory in Ottawa to the new Quebec regional office of Parks Canada. Also in 1973, the Ottawa Conservation Division laboratory with its large professional staff was established.

Moisture content of the wood was checked yearly with a resistance meter which indicated a moisture content above 65% on each visit.

Second Conservation Treatment: Slow-drying with dilute PEG Spray (3)

In early 1977, the Conservation Division decided to remove the timbers from the sand since they didn't appear to be drying out at a fast enough rate. The sand was found to be extremely wet, apparently providing good wet-storage conditions. It was thought that ground water was being drawn up through the gravel at the same rate as the evaporation from the

TABLE II

PROPOSED SPRAYING PROGRAMME FOR THE TREATMENT OF TIMBERS FROM THE MACHAULT

Dates	Period	Spray Solution	Application	RH
September 1977 to January 1978	4 months	15% PEG 540 Blend	3 times per week	90%
January 1978 to January 1978	4 months	25% PEG 540 blend	"	70%
April 1978 to October	18 month	25% PEG 540 blend to 60% PEG 540 blend	"	70% to 55%
October 1979	6 months	100% PEG 540 blend	"	55%

Figure 2: Overall view of Machault timbers stacked for slow drying treatment in the convervation building in Matapedia, Quebec. Note the vapour-proofed walls; air-circulating fans; gravel floors (Photo - George Vandervlugt).

surface of the sand. Core samples taken in January 1977 had moisture contents between 77-105%, based on the dry weight. It was then proposed to spray the timbers with an aqueous PEG 540 Blend solution, and to reduce the humidity according to the programme shown in **Table II.**

PEG was intended to retard the rapid evaporation of water and to replace free water in the outer wood cells and cavities. PEG 540 Blend was chosen for its combination of two molecular weights, 1540 and 300, because of our successful use of it for treatment of large cedar and elm timbers.

To remove the wood from the sand and to place it on stickers (narrow pieces of wood to separate the wood during seasoning. ed.) for spraying meant dismantling part of the conservation building, disposing of the sodium-pentachlorophenate contaminated sand, hiring cranes, refitting the building with monitors and controls and incurring a cost of about $25,000 (**Figure 2**). The sand and wood had been found to contain sodium pentachlorophenate in concentrations from 60 - 8600 ppm. This continues to impose a considerable safety hazard to workers.

During the transfer from the sand in 1977 the timbers were exposed to hot humid summer conditions in Restigouche, and although well covered with polyethylene sheet they were in danger of partial drying. (The conditions were similar to those of the excavation in 1972.) The effects of this exposure are difficult to assess.

The PEG spray treatment required more coordination between the conservation laboratory in Ottawa and the drying operation at Restigouche than the previous sand burial. Furthermore, a technician had to be hired to carry out the treatment. In practice the project was beset by non-technical problems: keeping or attracting good staff to work, ensuring the building was providing a proper environment and protection from vandalism. Indeed, there was at least one period of 2 months in whch we are certain no spraying was carried out. Contrary to schedule, the 15% solution was not increased because a trial with 25% PEG didn't appear to soak into the wood. In May 1979, the technician left the project, precipitating a crisis which had been building for some months: whether or not to rehire a technician, whether or not to continue spraying with the dilute PEG solution at all.

Third Conservation Treatment: Slow-Drying after Applying a Thick PEG Surface Coating (6)

In May 1979, the Conservation Division decided not to replace the technician and to discontinue spraying with dilute PEG in favour of applying a thick layer of PEG 540 Blend directly to the timbers. The PEG was intended to retard drying while the RH in the building would be lowered from 90% to 50% over the next two years. Completion was predicted for June 1981. This action was welcomed by Parks Canada during a period of fiscal restraint when the priorities of projects were being changed. Moreover, in 1979 the Visitor Reception Center seemed a distant possibility. In effect, the Machault was to be "mothballed."

However, this "mothballing" project did not go as planned. When experimenting with methods of applying the molten 100% PEG, we quickly discovered that the wood absorbed the sprayed layer overnight. The room temperature was 18°C, the RH at 80%. Three spray applications were required to coat the wood, using pressurized fire extinguishers, modified for this purpose (**Figure 3**). The PEG was quickly absorbed we believe for several reasons. Since the wood had not been sprayed for at least two months, it was highly absorbant. However, we also had previous experience surface-treating waterlogged cedar (Artillary Park, Quebec) with molten PEG 540 Blend and knew its strong penetrating power.

Was it not likely that very dilute solutions tend to flush from the surface as much PEG as they are meant to put in? In 1979 we entered a period of reflection and "brainstorming." On the one hand we submitted a proposal to try the outdoor freeze-drying method of Grattan et al. (4,5) on Machault timbers, and on the other we re-read the PEG bibliography to re-examine our theories of treating archaeological wood. We also asked our Analytical Section for an evaluation of the Machault oak.

One concept that emerged is that regardless of whether wood has a high or low moisture content, drying of wood is a diffusion process, controlled by the diffusion of moisture through that part or zone of the wood which is below

Figure 3: Conservator Bob Marion sprays Machault planking with molten PEG 540 Blend using a modified pressurized stainless steel fire extinguisher. Above note one of the four electric space heaters installed to supplement furnace heat to maintain 18°C in the building (Photo - George Vandervlugt).

the fibre saturation point (FSP). The depth of this zone in water-saturated wood can be small (at the surface) or up to 1-2 cm depending on ambient RH, wood species, the available hydroxyl groups, directions of moisture movement in relation to wood structure (7), as well as degree of degradation which may deepen this zone.

If there is PEG in the wood, it must affect the diffusion of moisture. Firstly because it is so hygroscopic it would tend to reduce the fibre saturation point of the PEG/wood composite, and because PEG replaces part of the water. Secondly the vapour pressure of a PEG solution is much lower than that of pure water because of the low vapour pressure of PEG. Both effects will lower the rate of outward diffusion of water by (a) the attraction of PEG to the water at the surface of the wood and (b) reducing the amount of diffusing moisture vapour. For the wood in question, however, PEG only occupies the outermost layer and only in this part of the wood will the nature of diffusion be affected although the rate of diffusion for the entire piece will be considerably reduced.

The excellent penetrating ability of the pure PEG when applied to the surface was no doubt due to osmotic pressure effects arising from large concentration gradients, and also possibly from the often overlooked and remarkable solvent ability of PEG.

Since May 1979, we have continued the applications of molten PEG at approximately 6 month intervals; ie. on each field inspection trip. The wood still absorbs the PEG, but takes increasingly longer periods. We have been reviewing data from limited tests to attempt to predict the wood's drying charactristics. The environment in the conservation building has been maintained at 80% RH using water bubblers, fans and exhaust fans, and at a temperature of around 18°C.

Proposals for the Fourth Conservation Treatment

In May 1981, we learned that the Machault display will be built and opened in 1984, and that portions of the timbers are to be displayed in the Visitor Reception Center. This has changed the complexion of our discussions, for we now know when the timbers are required. The actual reconstruction and use of the timbers are still open to discussion.

The problem for us to decide is simply, how should we approach the stabilization of the wood? The conservation must produce wood fully equilibrated to the 35% RH environment imposed by the center's design, within the two year deadline. The two other factors to be considered are the wood's present moisture content (MC), and the reconstruction requirement that the wood be stable at 35% RH both during the carpentry work and the subsequent display.

The following properties of the oak have been studied: moisture content, moisture gradient, desorption-resorption ability, and general behaviour in slow drying trials given to selected pieces brought back to Ottawa for study. We did not study the deterioration although it is worth noting that the ship seems to have suffered a certain amount of rot during its service (**Figure 4**). One timber actually supported a rot during burial in sodium pentachlorophenate; this was identified by Environment Canada as a basidiomycete similar to Merulius lacrimans. We have not considered the few iron wood composite pieces: the few iron bolts are corroding from the high humidity and the presence of PEG. We have not considered the effect of iron corrosion products on the behaviour of the wood, although we have noticed less cracking etc. in zones rich in iron corrosion products.

We are not able to discuss here the various methods and problems of measuring moisture content and gradient. It is a difficult subject and it would be a mistake in this context to look for anything more than trends.

Moisture contents (see **Table III**) taken over the 9 year period show drying in the wood from over 100% moisture in 1973 down to 35 $\pm$ 5% by 1981.

By examining **Figure 5** the moisture desorption-absorption curve which was obtained from the Machault oak in 1980, we can study the drying characteristics and moisture content. Nineteen large core samples 3.8 cm in diameter (taken in the tangential direction) were conditioned at the same humidities until they reached constant weight.

Accepting the wood cores were totally dried by the freeze-drying, we learn the following:

- when conditioned at 35% RH, the average MC was 6%;
- if the oak were freeze-dried to constant

Figure 4: Portion of Machault midship section in November 1979 after two years of slow drying. Note evidence of rot during service in rib; radial cracking in ribs and planking (Photo - George Vandervlugt).

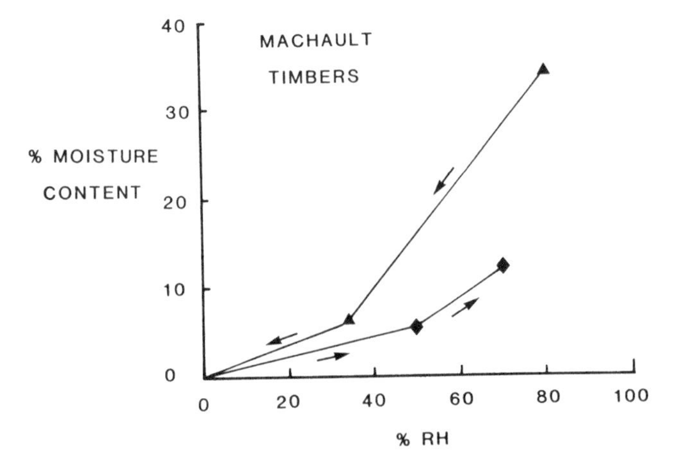

Figure 5: Water absorption/desorption curves for PEG treated oak timbers from the Machault. At 35% RH for 42 days; freeze-dried 39 days; 55% RH for 78 days; at 80% RH for 30 days.

weight, upon rehumidification to 70% the wood would only attain an 11% moisutre content;
- finally, the average MC based on the freeze-dried weight was 35%, tending to corroborate the MCs found by oven drying.

In 1979, attempts were made to determine moisture gradients by measuring the MC at 1 cm increments taken from 1.3 cm diameter core samples 8 - 10 cm long. The data of 1979 and 1980 show few trends in common except that the MC at the surface is lower than that in the interior. This problem probably illustrates the uneven drying behaviour of the wood, and effects caused by sampling difficulties.

Likewise, attempts have been made to determine the extent of the penetration of PEG into the timbers by methanol extraction of 1 cm incremental samples ca. 9 cm x 0.6 cm core samples. The "involatile extractives" (which can be interpreted as PEG) seem not to occur significantly beyond the first 2 centimeters below the surface of the oak.

Finally, some practical drying trials of selected Machault ribs and planks in the Ottawa lab have shown that when conditioned at 55% RH, the ribs seem to lose their moisture slowly and do not exhibit much dimensional change. For instance, a rib of weight 29 kg and MC 40% took 7 months to achieve a MC of 22%; it lost approximately 100 g per week. At 55% RH, 25°C modern oak would have a MC of 10.5%. (see **Tables IV & V**).

With these general results in mind, we ask:
- should we keep spraying PEG?
- should we start lowering the relative humidity levels?
- if so, at what rate?
- do we seek a stable average MC in the wood at say, 35% RH?
- are we looking for wood which no longer moves (i.e. dimensionally stable), when kept at stable temperature and humidity?
- are visual examinations, photographs, pin measurements and weights monitored every 6 months adequate for evaluation of the progress of treatment?

Naturally, each of us attached to the project has his own opinions about the timbers: Some think the timbers will move during drying. Others think that the bulk of the distortion has already happened, particularly at the time of excavation, and that they won't open up further during drying.

The list of people associated with the Machault is a long one and we would like to mention the following in alphabetical order:

Conservators: Brian V. Arthur, Tom Bryce, Tom Daley, Sten Holm, Victoria Jenssen, Derek Jones, Eric Lawson, Bob Marion, Lisa Mibach, and Lorne Murdock.

Scientists: Andy Douglas, Jim Moore, and Judy Schlieman.

References

1. Arthur, Brian V., "The Conservation of Artifacts from the wreck of the Machault," in: preprints of the ICOM Committee for Conservation: 3rd Trienniel Meeting (Madrid: 1971).
2. Cech, M.Y. and Pfaff, F., Kiln Operator's Manual for Eastern Canada. Eastern Forest Products Laboratory Report OPX192E (Ottawa: 1977) Table 3, p. 165.
3. Daley, Tom, "Progress Report: Slow Drying of Waterlogged Ships Timbers from the Machault," in: abstracts of the 4th Annual IIC-CG Conference, (Fredericton, New Brunswick: 1978).
4. Grattan, D.W. and J.C. McCawley, "The Potential of the Canadian Winter Climate for the Freeze-Drying of Degraded Waterlogged Wood," Studies in Conservation, vol. 23, no. 4, November 1978, pp. 157-167.
5. Grattan, D.W., J.C. McCawley and C. Cook, "The Potential of the Canadian Winter Climate for the Freeze-Drying of Degraded Waterlogged Wood: Part II," Studies in Conservation, vol. 25, no. 3, August 1980, pp. 118-136.
6. Jenssen, V., "Giving Waterlogged Wood the Brush," in: abstracts of the IIC-CG Sixth Annual Conference, (Ottawa: 1980), p. 29.
7. Kollman, F.F.P and W.A. Côté Jr., Principles of Wood Science and Technology Vol. I: Solid Wood. (New York: Springer Verlag) 1968, p. 220.
8. Waddell, Peter, "The Restigouche Underwater Archaeology Project: 1972," Research Bulletin No. 10, Dept. of Indian Affairs and Northern Development, National Historic Sites Service, (Ottawa: December 1972).

ON-SITE PACKING AND PROTECTION FOR WET AND WATERLOGGED WOOD

Deirdre A. Lucas
Conservation Division,
Parks Canada, 1570 Liverpool Court,
Ottawa, Ontario,
K1A 1G2

Introduction

The Conservation Division of Parks Canada has been involved in field work at a number of marine sites, most recently in Red Bay, Labrador (1979-81). Previously we have worked in Bay Bulls, Newfoundland (1977), Dingwall, Nova Scotia (1977-78) and Restigouche, Quebec (1969-73). Each site presents various wet-wood problems. In this paper I intend to describe each situation and how they were dealt with. This paper will refer principally to the 1979 and 1980 excavations in Red Bay.

Packaging and storage are our main responsibilities, occasionally however, artifacts are superficially cleaned if the researcher or archaeologist needs to make a more detailed examination. In the more fragile pieces the attached seabed silt can be used as cushioning.

Prior to our field work we meet with the staff archaeologist to discuss the kinds and quantities of artifacts expected. This enables us to prepare well in advance for the packing and storage problems that may arise. For instance, before the 1979 season in Red Bay the archaeologists were able to give us sketches, dimensions and approximate weights of a large oak capstan and iron anchor which they intended to excavate and transport to Ottawa.

Field Storage Methods

We received information that a "great number" of wooden barrel pieces would be excavated, the largest measuring approximately two meters in length. However, by the end of the 1979 season approximately 10,000 barrel pieces had been removed from the site. A further problem was that all laboratory materials had to be shipped from Ottawa approximately 2700 kilometers by land and coastal boat. Thus to send sufficient tanks (estimated 10 x 2 meters in length) was very expensive and hence not feasible. The solution adopted was to use flexible PVC swimming pool liners, supplied to our specifications by a local swimming pool manufacturer. Suitable tank liners were made from stock 0.76 mm (30 mil) PVC sheet. The supporting frames were easily constructed of plywood and lumber purchased in Red Bay while the laboratory was being set up.

Locally available containers were also used and included plastic herring barrels, squid and fish tanks. All were capable of holding a variety of sizes of artifacts.

Polyethylene containers intended for storing food in refrigerators were used for small artifacts because they are light to transport, and also serve as shipping containers at the end of the season. Polyethylene garbage buckets which have the advantage of tight fitting lids, handles and strength (from the cylindrical shape) are very useful for storage.

For storage, the solutions we generally employ are: fresh water for holding an artifact before further study or shipment; sea water for temporary holding on the Dive Support Vessel; 1% phenol in fresh water for shipment of objects to Ottawa. Fortunately, in Red Bay we work in cool conditions so there have been minimal problems with slime development. The acidic bog water may be a factor in this regard. The water in the tanks is regularly changed to limit the obnoxious hydrogen sulphide fumes usually produced by wet archaeological wood.

For packing artifacts for holding and storage we use "Whirl pak bags" (a small polyethylene bag with a built in twist tie), larger polyethylene bags either purchased commercially, or bags made on site from polyethylene sheet with a heat sealer.

Problems Related to Tank Storage

On site, we are sometimes faced with security problems, either from theft or simple misplacement. Theft can be discouraged by the obvious use of burgler systems, by never leaving an open lab unattended and by displaying signs which say: "Danger, Corrosive Fluid", etc. These signs will usually deter even the most stout-hearted tinkerer.

An important requirement is that most artifacts in tank storage must be readily available to the archaeologists and researchers. Tanks are therefore built with low sides to allow one to fish around with gloves for an elusive piece.

We maintain for the Field Registrar a card inventory of all artifacts. These are filed by the number corresponding to their storage tank number. These cards are also used to record movements of the artifacts from tank to tank, the storage solutions used, further conservation procedures and any noticeable condition problems.

Providing tours of the facilities to local residents and visiting dignitaries (major and minor) is often part of our job. This means that some of the most interesting pieces are displayed and thus available to people bent on touching them, and are also open to the danger of air drying. We have dealt with this by spraying the wood with atomizers containing water and by not leaving the artifacts unattended; placing a few artifacts in the center of a large layout area deters all but the most persisent guest from touching them. Tours provide time for the conservation staff to explain conservation to the curious and to demonstrate a few of the more simple techniques.

Labelling Materials

"Teflon" (polyytetrafluoroethylene) has been our standard material for tagging archaeological artifacts since 1973. The label is numbered with a hand held "Dymo" embossing machine. The problems with the embossed "Teflon" label so produced are its poor readability, (white numbers against a white background), its tendency to wear out the embossing guns and the relatively high cost. It is almost indestructible and withstands most solutions including phenol and acetone. As an alternative, for artifacts returning to Ottawa, we use a tag made from a spun polyethylene paper known as "Nalgene Polypaper" or "Tyvek". This can be written on by pen, pencil or waterproof felt tip pens.

Once a label material has been chosen, we then decide on a method of affixing it to the artifact. We have found PVC coated florist's wire quite adequate since it is water resistant, is supplied on large spools, and is flexible enough to be twisted or knotted. Its disadvantage is that it can break in the process of knotting. We also use tacking guns with "Monel" staples (a nickel-copper corrosion-resistant alloy) to attach labels onto very large pieces.

Storage by Reburial

In the 1979 Red Bay field season a decision was made by the Site Director to rebury a substantial quantity of waterlogged wooden barrel staves no longer required for study.

Reburial takes place after a complete inventory is done. The material is taken down by divers and left under a weighted fish net, and at the end of the season it is placed in a trench which is then back-filled.

Requirements for packaging and labelling the material for reburial are as follows: Bundles of staves have to be sturdy to withstand handling and compact for ease in moving. The material for reburial has to be labelled in a permanent way to aid possible recovery of the artifact in the future. We opted to tag the artiacts with the embossed "Teflon" tied on with the florists's wire. Bundles of 10 to 15 staves are then bound together with strips of fiberglass mesh (window insect screening). Layers of "Micofoam" (expanded polyethylene sheeting) are inserted between staves to protect wood edges. The bundle was then placed in a perforated plastic garbage bag or polyethylene tubing heat sealed at both ends to make a bag and the whole unit bound again with the screening strips. Each bag was numbered with a "Teflon" tag. In the 1980 season we began to use monel staples for tag attachment. The divers found the sealed tubing which we also began to use in the 1980 field season much more manageable underwater than the garbage bags, and much less prone to tearing.

Packing Materials

Crates:

We were fortunate in Red Bay 1979 to have had forewarning of the size of some of the artifacts. We already knew of the intended excavation of the capstan and the anchor. Before leaving Ottawa for the field season, wooden crates were prefabricated and shipped as flat lumber, thus saving valuable time on site.

Polyurethane Foam:

We have found definite advantages in using expanding polyurethane foam for packing both large heavy pieces and very light fragile artifacts.

Its curing (or foaming) time can be adversely affected by cold and damp. We found that by storing the cans in a heated building and by carrying out the initial processing in a warm area, we could then take the measured components (in disposable containers) outside for their final mixing and pouring without much ill effect. It can be used to encapsulate or to simply support an artifact on one side. Encapsulations can be used when an important artifact arrives in the lab in a mud matrix; when an artifact cannot support its own weight; or if an artifact is so heavy that other conventional packing methods do not protect it. For a small artifact, a box is lined with plastic, the artifact placed on the bottom and covered with aluminum foil as a separator. The expanding foam is then poured in small batches until it rises just over the top of the object. The foam sets fairly quickly and the unit can be removed from the box. The plastic box lining can be pulled away and the top surface of the foam can be sawn flat. The object is then inverted and the process is repeated to fully encapsulate the object. The encapsulated artifact can be maintained mould free by spraying it with "Lysol" (0.1% Dowicide no.1) and spraying with water mist. Polyurethane foam, though expensive is worthwhile.

Styrofoam:

A less satisfactory alternative to polyurethane foam is the use of expanded polystyrene, "styrofoam". The artifact is wrapped in water-soaked cotton flannel, and then is padded with bubble pak. "Styrofoam" panels are cut into layers to fill a crate. The center of the block of "styrofoam" so formed is curved out so that the wrapped artifact accurately fits the hole so created.

Bubble-Pak

"Bubble-Pak" (air entrapped polyethylene sheeting) has proven invaluable in the field, although bulky, it is light to ship. It can be used to cover tanks (to prevent evaporation and dust deposits) as well as to pad and to protect artifacts for shipping.

Microfoam:

"Microfoam" (expanded polyethylene foam) is used for protecting the edges of the wood. This material can be easily stapled and therefore can be used to make envelopes or sleeves for smaller more fragile pieces. Galvanized staples are used in "clipperstyle" staplers to join the various plastics. Its main disadvantage is that it can cause smaller pieces to float.

Fibreglass Screening:

Fibreglass screening (non-corroding window screening) is used to make bags by stapling rolled seams. It can also be cut into strips and used as a binding material providing care is taken to protect the edges and surfaces of the wood. Screening can leave an imprint if the wood is soft. It is necessary to wear gloves since strands of the fiberglass are freed at cut edges.

Cotton Gauze:

Cotton gauze was found to be very useful for reinforcing loose material. The gauze will hold fragile objects leaving little or no imprint on the artifact. Its main disadvantage is that it will disintegrate after more than 2 months in water.

Vinyl Tape:

In the past few years we have been using 5 cm wide vinyl tape with vinyl adhesive made by the "3M" company. We use it to seal bags, mounts, encapsulations etc. It can be used on material to be taken underwater. Applied to a bag in 1979 it was still attached quite firmly when re-excavated in 1980, although the tape had darkened, the writing on it was still legible.

Polyethylene Tubing:

Polyethylene tubing can be purchased in a number of widths and thicknesses and is very useful for making specific or odd size bags. The ends are heat sealed to ensure a good joint and the bag is perforated to allow water exchange. The tubing can also be cut into strips and used for binding. Artifacts prepared for shipment back to headquarters can be placed in heat-sealed tubing (without perforations) and shipped in a padded container. The advantage of this method is that it gives a very light shipping weight and prevents immediate water evaporation.

Sphagnum Moss:

Materials used in padding artifacts both in storage and in shipment are usually synthetic materials but sometimes are quite literally locally grown. In Red Bay we were fortunate to have an unlimited supply of living sphagnum moss. We used the wet moss to pack the artifacts, to maintain a damp environment and to provide a natural biocide. Wood shavings and seaweed have been used with success as well.

Artifact Shipment

Artifacts excavated by Parks Canada have been shipped back to Ottawa by almost every mode of transport available. Each method has its own problems. Shipping by air, although quick, is very expensive and therefore we only use this when time is limited. Rail (when possible) has been used with moderate success. It is slower but there is more control in the loading and unloading of crates. Very large artifacts from Red Bay have been shipped by coastal freighter. This method is very slow because it is actually a combination of boat and truck, but it does have the capacity to move very large pieces relatively cheaply. The bulk of our artifacts are usually returned in a van driven by field staff at the end of the season. In Red Bay 1979 a trailer, hitched to the van, was used as an enormous packing crate. Artifacts heat sealed in polyethylene tubes were placed in the trailer, which was padded with "styrofoam" and foam rubber. In 1978 when potentially valuable artifacts had to be shipped, we used a security courier sevice. The problem of this was the complete loss of control over the shipping, timing and the conditions.

Examples of Packing Methods

The capstan from Red Bay was approximately 4 metres long. Packing was started underwater by covering it with cotton flannel, fiberglass screening strips, and polyethylene sheeting. The capstan was then raised to the deck of the barge where it was generously soaked with a 1% phenol solution and then tightly wrapped with vinyl tape.

The crate with a lining of polyethylene sheeting taped inside and an initial layer polyurethane foam, had been prepared on deck. Preliminary shaping of the foam had been carried out with an axe. The capstan was lowered into the crate and enough foam to completely encapsulate it poured in. The crate lid was screwed on and had steel bands around it for added security.

In 1979 the large quantity of barrel parts presented us with a problem. Staves were packed for transit by stacking them in small piles of comparable lengths and curvature. Staves were interleaved with 1 cm diameter bubble-pak and handfulls of soaked sphagnum moss. Each bundle was wrapped in bubble-pak and tied with strips of fiberglass. The wrapped unit was put in a polyethylene tube sealed at both ends. Normally a bundle would consist of 7-10 barrel parts. In 1979 a pulley block with rope attached to it was excavated. Most of the silt was removed from it at the time to view the artifact, so that extra padding was needed to ensure that the rope would not move during transportation. The artifact, was wrapped first in cotton gauze and then polyethylene film to provide a closely bound layer. Bubble-pak was used to wrap the piece and to pad the small plastic container.

Barrel hooping is delicate when not in place on the cask. Hoops of degraded alder have little or no strength and willow caning unravels easily. It is necessary to use gauze bindings and the pieces are then encased in microfoam envelopes. For transport to headquartrs, each unit was wrapped in bubble-pak and placed in polyethylene tubing heat sealed at both ends.

During the field season small wooden artifacts are packed like most small finds; i.e. in individual polyethylene bags to prevent abrasion and loss of label. These small artifacts are stored in polyethylene tanks which have snap-on lids. In preparation for transport the storage solution is poured out, the box lined with bubble-pak or microfoam and generous amounts of sphagnum moss soaked in water, placed in layers throughout. Finally, padding is placed on top of the artifacts to ensure that there is stability inside the box.

Conclusion

In Canada the role of the field conservator is determined by many people of different disciplines such as registrars, researchers, and archaeologists. They can have a direct effect on our methods of packing and storage. The care for the wood is sometimes secondary to their immediate need for information. It is imperative for these reasons that our conservation techniques be flexible. Our challenge is that we must be able to adapt our equipment and time to the various requirements of others. We have to be imaginative, inventive and above all jacks of all trades to support our share of the workload.

Material Suppliers

Mermaid Pools, Methot Sales, 1373 Ogilvie Road, Ottawa, Ontario K1J 7P5
- custom built polyethylene liners (30-40 mil in thickness)

Frig-O-Seal, Modern Plastics, Ottawa, Ontario
- small polyethylene tanks with snap on lids (bulk orders)

Fisher Scientific, 112 Colonnade Rd., Nepean, Ontario K2E 7L6
- Phenol -Whirlpak bags (various sizes)

Canus Plastics, 340 Gladstone Ave., Ottawa, Ontario, K2P 0Y8
- Foaming polyurethane (1 & 5 gal. cont.)
- Teflon (can be cut by the supplier to required widths)
- Nalgene Polypaper
- Polyethylene tubing and sheeting
- Polyethylene buckets
- Vinyl coated florist wire
- Heat sealers

Arrow Staples, R.W. Alexander and Co., 1275 Wellington St., Ottawa, Ontario, K1A 3A6
- Monel staples (variety of sizes)

Most Department Stores
- "Lysol"
- "Frig-O-Seal"
- fiberglass screening (variety of widths)

Dorfi Two, Packing Supplies, 2491 Kaladar Ave., Nepean, Ontario, K1V 8B9
- 6 litre disposable buckets
- Air cap (Bubble Pak)

Bibliography

Dowman, Elizabeth A., Conservation in Field Archaeology. (London: Methuen and Co., 1970).

Grosso, Gerald H., ed., Pacific Northwest Wet Site Wood Conservation Conference, 2 vols., proc. of the conference, 19-22 September 1976 (Neah Bay, Washington, 1976).

Lawson, Eric, "In Between: The care of artifacts from the Seabed to the Conservation Laboratory and Reasons Why it is Necessary," in: Beneath the Waters of Time, proc. of the Ninth Conference on Underwater Archaeology, ed., J. Barto Arnold III, (Austin, Texas: Texas Antiquities Committee Publication 6, 1978), pp. 69-91.

Leigh, David, First Aid for Finds, (Southampton, England: Department of Archaeology, University of Southampton, 1972).

Questions

Colin Pearson: Does the exothermic reaction during the formation of the polyurethane foam cause any problems?

Deirdre Lucas: The heat output is not very great, and with waterlogged wood this does not pose a real threat. The objects are also well padded and wrapped in two layers of plastic and thus a long way from the heat source.

Herman Heikkenen: Why do you pefer rigid polyurethane foam to flexible foam?

Deirdre Lucas: Because it is important that the objects, which are sometimes very bulky and heavy do not move during handling in transit. Movement within the packing cases would be very damaging.

John Dawson: What phenol did you use?

Deirdre Lucas: Phenol i.e. carbolic acid.

Figure 1. The main section of the hull of the Brown's Ferry Vessel, supported by its cradle that is welded to a truck for transport.

THE RECOVERY AND CONSERVATION OF THE BROWN'S FERRY VESSEL

Katherine R. Singley
Institute of Archeology and Anthropology,
University of South Carolina,
Columbia, South Carolina,
U.S.A., 29208

Introduction

The Brown's Ferry Vessel, a small merchant ship dating from 1735 to 1740, was recovered from the Black River in eastern South Carolina in 1976. This river is dark in colour since it is tannated, i.e. rich in tannins. The boat is approximately 48 feet (14.6 metres) long and 12 feet (3.66 metres) wide at midships. Its cargo capacity, estimatd at about 22.7 tonnes, was maximized by the shallow draft and compound curves of the hull. Double-ended and flat-bottomed, the Brown's Ferry Vessel was designed for manoeuverability in various kinds of coastal waters. A boat of this type, probably plantation built, would have been the backbone of the colonial system of local transportation and movement of cash crops, supplies, and people along the waters of the Carolina low country.

More information concerning the boat's construction has been published (Albright and Steffey 1979). While exact parallels are unknown at present, general parallels may be found in the flat-bottomed, sprit-sailed ferries and cargo vessels of Holland and England in the 17th and 18th centuries.

Initial wood identification in 1976 by the Forest Products Laboratory, Madison, Wisconsin, showed that the woods used in the vessel are of species that would have been locally available and must point to a Southern tradition of shipbuilding which is otherwise poorly documented. Live and white oak (Quercus virginiana and Leucobalanus) were used in the frames and bow stem; baldcypress (Taxodium distichum) was used in the king plank, wales, and 36 foot (11 metre) keelson. A variety of yellow pine was used in the planking. Patches of pitch exist on the inside surface of the hull. Teredo damage in some of these bottom planks indicates that the Brown's Ferry Vessel sailed in saltwater bays and estuaries as well as freshwater rivers.

Recovery and Storage

Both the size of the vessel and financial constraints placed limitations on the lifting and storing of the Brown's Ferry Vessel before conservation could begin. Although one side of the wreck had collapsed under the weight of the cargo (10,000 bricks), more than half the boat was lifted in a single section of hull. A special sling system, made from air cargo straps and ratchets hanging from railroad rails, was constructed in the field for removing the hull. By tunnelling with water jets, the straps were passed under the hull at 60 cm intervals. Boards 25 cm wide, placed between the staps and the wood, distributed the weight in order to minimize damage to the timbers. Even tension and support along the length of the hull was achieved by adjusting the ratchets, and then the 2.5 cm steel cables connecting the frame to the lifting hook were adjusted so that the twisted hull became staightened during lifting.

The network of cargo straps and railroad rails was then welded to the back of a flatbed truck for transport to Columbia, over one hundred miles away (**Figure 1**). The cradle continued to support the Brown's Ferry Vessel during four and a half years of storage. The boat was kept first outdoors under sprinklers for a year, and then was reimmersed in a spring-fed pond for three and a half years. As a precaution, new straps were substituted for the old ones before the vessel's final move to its conservation tank in May, 1981.

The relatively good condition of the Brown's Ferry Vessel was due partially to the depth at which it was found (7.6 metres) and partially to the fine silt (up to 30 cm of overburden) which covered it. Furthermore, the chemistry of the river itself may have been a factor. The Black River contains humates and organic acids, with a probable molecular weight range of 1,0000-10,000, in a concentration of about 10 ppm. More information has been furnished by the South Carolina state health department, which for seven years has been taking monthly readings at a field station 100 metres above the site. In one year (1976) the pH of the water varied from 5.5 to 7.7, and the water temperature ranged from 2.8°C in winter to 31°C in summer. Recorded also are levels of dissolved oxygen and carbon. Free metals, which complex freely with organic acids, and carbon both reflect the dissolved organic matter.

While the overall chemistry of the river is thus well documented, whether this information can be related to the chemistry and micro-environment of the sediment remains to be investigated. In any event, it is evident that the Black River was more conducive to preservation than the farm pond. Although the seasonal temperature and pH levels of the waters were about the same, the light levels were much higher at less depth (3 metres) in the pond. Algae abounded on the boat during immersion there, and intensified deterioration was noticed.

Documentation

During storage, all pieces were tagged and numbered. Over 122 loose timbers were photographed in 0.46 metre sections. Details (tool marks, treenails, grain irregularities, etc.) were sketched 1:1 on Mylar. Eventually this documentation will aid in monitoring the wood during "curing" and in reconstructing the vessel for display in a museum. Moreover, the creation of a 1:10 scale model of the boat (**Figure 2**) helped both to clarify details of the construction, as well as to create more of an impression in the public eye while funds for conservation were sought.

Analysis of the Wood

Wood samples were removed from frames and planks and analyzed by Dr. Robert Harris of the Department of Forestry, Clemson University. Photographs taken through a scanning electron microscrope show highly degraded cell walls. In one sample of pine for example, there is little difference in the thicknesses of cell walls in spring and summer wood.

Results for one sample of live oak and two samples of pine, with comparative values for new wood in parentheses, are shown in the **Table**. Samples were extracted using alcohol/-benzene. Since yields were low, lignin and holocellulose determinations were carried out on dry, unextracted meal. Lignin was isolated as "Klason Lignin" with sulfuric acid. It was not corrected for ash. Holocellulose was determined using the sodium chlorite method. Samples were ashed at 575°C for three hours.

The low specific gravities and values for holocellulose in both the pine and oak suggest weak, highly hydrolyzed wood. Lignin constitutes an overwhelming percentage by weight of the remaining wood. The inorganic contents of the woods, represented by ashing, are more than ten times greater than the values for new woods. This is probably due to the absorption of sedimentary materials by the wood. The red color of the residues of all three samples may indicate a high content of iron oxide.

Other small samples tested for moisture content (expressed as a percentage of dry wood) give results of 593% and 415% for pine, 323% for oak, and 184% for cypress. On the other hand, one centimeter samples removed by increment borer show much lower moisture contents for oak and pine than the larger samples. Samples average 222% for pine, 91% for oak, and 153% for cypress. Also, there is not an extreme gradation of moisture levels from the inside to the outside surface, but more of a uniformity of moisture content. This tendency may be due to the heat generated by hand boring.

Conservation

Because of the range of species, conditions, and moisture contents of the woods in the Brown's Ferry Vessel, a polyethylene glycol of medium range in molecular weight was selected for conservation. Chosen was, Union Carbide's Carbowax 1450, with a molecular weight of 1300-1500. The live oak, with its tight pore structure and tyloses, will be difficult to penetrate. Pockets of compression and tension wood exist in frames cut from limbs that were probably selected for their blended shape. In contrast, the pine and cypress should be more easily impregnated. PEG 1450 is regarded as a compromise for treating these widely different woods.

In May, 1981 the Brown's Ferry Vessel was moved into its conservation facility, constructed for $300,000 with matching federal and state funds. The tank, the largest in North America, has a capacity of 131,164 litres. Despite the greater expense incurred by treatment in one piece, it is felt that dismantling the boat, held together by treenails and some large spikes, would be more damaging. The gradual addition of PEG will begin after two months of initial hot washing to remove as

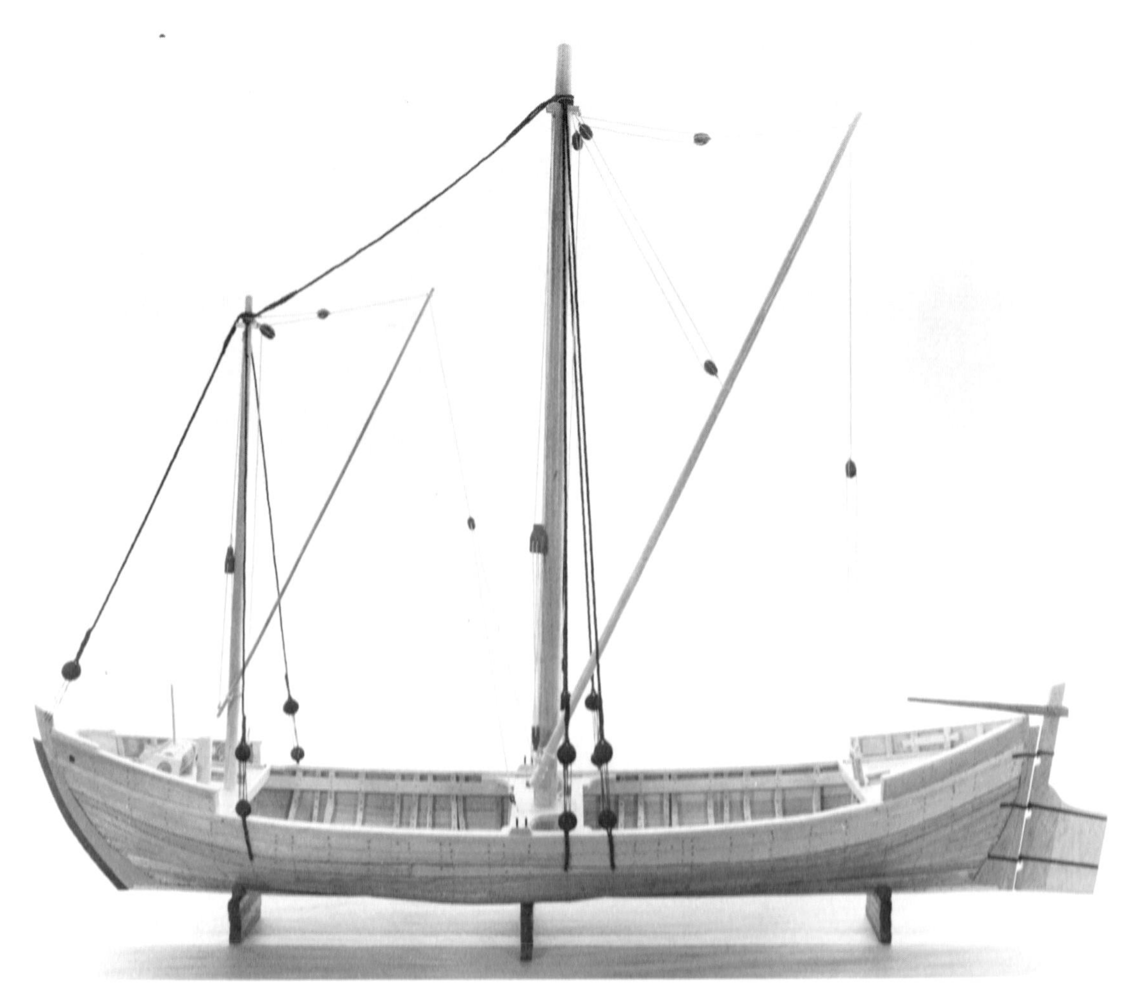

Figure 2. The model of the Brown's Ferry Vessel constructed in 1978 by J. Richard Steffey of the Institute of Nautical Archaeology, Texas A & M University.

THE CHEMICAL ANALYSIS OF WOOD FROM THE BROWN'S FERRY VESSEL

Sample	Sp. Gr.	% MC	% Holocell.	% Lignin	% Ext	% Ash
frame 15B (live oak)	.17 (.81)	496	33 (75)	67 (25)	.74	5.8 (.4)
plank S6SBS no.656 (pine)	.26 (.45-.56)	318	52 (70)	49 (28)	.86	2.1 (.2)
plank S3 no.505 (pine)	.14 (.45-.56)	648	29 (70)	71 (28)	.46	2.1 (.2)

$$\text{Sp. Gr.} = \frac{\text{Oven dry wt.}}{\text{Green vol.}} \qquad \text{MC} = \frac{\text{Green wt.} - \text{Dry wt.}}{\text{Dry wt.}} \times 100$$

much extraneous material as possible from the wood. The solution will be kept throughout at 60°C. The concentration will be increased slowly by about 1% every two weeks, to a final concentration of about 60% over a two year period. Because of the volume of treatment solution, this means that about 208 litres (55 gallons) of PEG 1450 must be added about every three days.

"Curing" the wood in an environment of gradually decreasing relative humidity (from 95% to 65%) will take about another two or three years. A 30% solution of PEG 1450 or 4000 will be applied by hand where more consolidation of timbers is needed. A local application of a phenolic fungicide will be made where needed. The Brown's Ferry Vessel probably will be dried under tension in another cradle system.

Conclusion

With a length of about 48 feet (14.6 metres), the vessel is just within a manageable size for recovery and conservation in a bath of PEG. Throughout the project, during excavation, recovery, documentation, and moving, efforts have been made to keep the costs within reason and to keep the logistics of handling within the capabilities of staff and volunteer help. Many of the early stages of the project, until the construction of the conservation facility, were carried out cheaply. Reasonable levels of documentation and analysis have been carried out. Much of the credit for the success of the project so far is due to the Brown's Ferry Vessel itself which, overbuilt originally, seems to have withstood much of our punishment.

The conservation of the Brown's Ferry Vessel provides an excellent opportunity for evaluating PEG as a preservative for large-scale wooden objects. Moreover, long-term research projects are anticipated, and these include an examination of the sediment and water in the Black River and their roles in preservation.

Acknowledgement

I would like to thank the following people for their help in the preparation of this paper: Mr. Alan Albright, Institute of Archaeology, University of South Carolina; Dr. Terry Bidleman, Department of Chemistry, University of South Carolina; and Dr. Robert Harris, Department of Forestry, Clemson University.

Reference

Albright, Alan B. and J. Richard Steffey. "The Brown's Ferry Vessel, South Carolina, Preliminary Report," The International Journal of Nautical Archaeology and Underwater Exploration, vol. 8, no. 2, 1979, pp. 121-142.

Questions

Richard Clarke: We have found that the ash content of wood from the Mary Rose is high, like that from the Brown's Ferry wreck. It seems to be typical of waterlogged wood. Research at Portsmouth Polytechnic employing electron microprobe analysis showed the presence of much iron and other metals in the wood. Metals and siliceous materials apparently enter the wood and add to the ash content.

Mary-Lou Florian: Was any soft rot found in the late wood tracheids in the pine from the Brown's Ferry wreck? Do you think that the heavily tannated river may have played a role in the preservation of the wood? Tannic acid is used as a biocide in the lumber industry.

Katherine Singley: I am not able to answer these questions yet since I do not yet have the results of analysis. The tannates may play a part in the wood preservation as you suggest. Studies on tannated water have been carried out on swampy rivers in South Carolina and in South America. From the former it has been fund that the level of the tannins and humates in the first 2.5 cm of sediment is much lower than the water, and that as you go through the sediment layer there is an exponential decrease in humates and tannates.

GENERAL DISCUSSION PERIOD - THE CONSERVATION OF SHIPWRECKS

Questions addressed to Willis Stevens

John Dawson: How deep are the trenches for the re-burial of the artifacts?

Mary-Lou Florian: Can you tell us about the sand bagging method in the re-burial, and what are the direct effects of sand bags as opposed to silt on the re-buried timbers?

Willis Stevens: There are really two parts to this and it needs to be clarified: (i) The method of closing the excavation down for the winter and (ii) the permanent re-burial of excavated material in a location close and similar to that of the shipwreck. When the site is closed down at the end of each field season, the excavated trenches are filled in with sandbags. However, if structure or timber is to be covered, a layer of silt or sand is placed on these first, before sand bags are used. Trenches are backfilled to prevent storm or ice damage. The re-burial operation takes place off the wreck site itself in an area just to the west of the wreck, on the lee side of the incoming tide. On the lee side we find that there are large accumulations of organic materials such as natural harbour silts, peat from the island etc. On that side a heavy compact silt has accumulated to a depth of up to 2 1/2 metres. This is, we believe, a good environment for the preservation of timber. Our re-burial trench is therefore up to 2 1/2 metres deep and at the minimum 1 metre deep. For reburial, a 4 x 4 metre pit is excavated to a depth of say 2 metres, lined with plastic sheet and silt allowed to accumulate for a while. Then the timbers are installed and loose sand put in by cutting open sand bags. Sometimes silt is introduced by re-directing an airlift. When full of sand and silt, the surface of the trench is packed with a layer of sand bags.

John Dawson: Is there any damage from freezing?

Willis Stevens: Although the harbour freezes over, reburial takes place at a depth of 7-10 metres where the water remains permanently unfrozen. Icebergs however, pose a threat but they have not caused any damage yet.

John Dawson: Will the re-burial material be examined in future to determine the success of this method?

Willis Stevens: Not yet but it is in the planning stage. Timbers reburied in a like manner for the winter season, show no physical damage but some discolouration.

Victoria Jenssen: There has been some visual documentation of selected timbers. We are preparing a monitoring programme and selected timbers will be examined annually.

Questions to Victoria Jenssen and Lorne Murdock

Jim Spriggs: Are you worried about the danger of excessive hygroscopicity of wood treated with PEG 540 Blend? The PEG may weep out of the surface and hence attract dirt. If so, much after-care would be needed for objects to be displayed. Did you consider these points in choosing PEG 540 Blend.

Lorne Murdock: The primary reason that we use PEG 540 Blend for surface treatments was that earlier work with PEG 4000 had proved unsuccessful; as it solidified almost immediately, became very brittle and eventually flaked off the surface of the wood. Despite the hygroscopicity of the PEG 540 Blend we have not had the problems you describe; in fact we achieve better penetration with 540 Blend and it does not harden or creep; but we do remove the excess PEG from the surface of the wood after treatment. Migration has not been observed. We consider that a bigger problem is the presence in the wood of the fungicide sodium pentachlorophenate (Dowicide G) which had concentrations as high as 9000 ppm. This potential health hazard will be a major concern if and when the Machault ships timbers go on display to the public.

David Grattan: I have found that wood treated in this way is less hygroscopic than might be supposed from the hygroscopicity of pure PEG. This is because there is an interaction between the wood and the PEG, probably involving hydrogen bonding, which lowers the hygroscopicity of the PEG/wood composite. At high humidities above 75% RH there could be serious problems of weeping, but below 60% RH there is no problem.

Allen Brownstein: How does the molecular weight of the PEG effect the depth of penetration?
Lorne Murdock: The lower the molecular weight the better the penetration.
Victoria Jenssen: Our main aim in using PEG is not to penetate the wood, but simply to slow the drying of the wood. We are basically trying to avoid steep moisture gradients which create large stresses within the wood.
Lorne Murdock: In our experience it is not possible to penetrate sound white oak with PEG 540 Blend to any considerable depth with any spraying method.

Allen Brownstein: I want to reinforce the point that when PEG is incorporated into wood, you have in fact formed a new composite material and the properties of pure PEG are no longer applicable. PEG very readily forms complexes with many materials. The polyether oxygens may complex with various wood components and some metal ions present. Such complexes will distort the properties of PEG making it difficult to try and relate the properties of pure PEG with that found in treated wood.
Lorne Murdock: I'd like to make a comment about the reversibility. I feel that no large scale PEG treatments are truly reversible, in the sense that the practicalities of removing it are, at present for many reasons out of the question.

John Dawson: Why did you use the toxic biocide sodium pentachlorophenate?
Lorne Murdock: The decision to use this biocide took place before our involvement in the project. We feel that it was probably made after seeking advice from a wood technologist, or from wood technology literature. This fungicide is very effective and obviously useful - but no thought seems to have been given to the possibility of public contact with the objects when they eventually go on display. It illustates the problems of communication that sometimes exist.

Lorne Murdock: I'd like to add one further comment about the disposal of the fungicide. We had great difficulty in finding somebody at Environment Canada who would show any interest in advising us in how and where to dispose of the contaminated sand. The sand was finally removed in the late fall to a dump site as selected by Environment Canada. However, no one in authority took any steps to indicate that at the site x tons of contaminated sand had been buried. As far as Environment Canada was concerned, once the site was found, their job was over. The warning sign and pickets that I installed in 1977 have probably fallen down by now and there remains all that toxic waste of which the location is only known to a few people.

Richard Clarke: If you wish to spray a large structure with biocide, there is a real problem because of the atomized spray; particularly if you wish the public to view it. One of the biggest problems with the Mary Rose timbers is not fungi but slime. The most effective "slime-cides" are quaternary ammonium salts. The attitude of the wood technologist is that these do not present a serious health hazard because the spray operators only have to wear rubber gloves, face masks, rubber suits, and eye protectors! Our difficulty has been to get over to the wood technologist that here is a slightly different problem from the treatment of commercial timber.

Anon Question: Were any measurements of dimensional change or splitting made?
Victoria Jenssen: Fourteen timbers have been measured and monitored since 1977. In addition some measurements on selected pieces brought back to Ottawa have been made over a one year period. (Sept. 80 to Sept. 81); mostly in the tangential dimension. Over the same period, relative humidity decreased from 80% to 55%, and moisture content from 40% to 20%. Shrinkages were between 1% to 4%. I find shrinkage measurements a little difficult to interpret because of the appearance or disappearance of cracks. For this reason I do not trust their validity.
Kirsten Jespersen: You can make a 1:1 scale drawing very accurately, and have all of the initial cracks marked on this. You can also use the drawing to calculate shrinkages. We make these drawings on plastic sheets

mounted over the artifact, and held on a transparent table.

Per Hoffmann: To measure shrinkages all you have to do is to subtract the width of the crack from say a measurement between pins, and then the determination becomes perfectly valid.

John Dawson: A photo-copying machine is also useful for getting an accurate 1:1 picture of an artifact.

Victoria Jenssen: I worry about the drawing method of Jespersen because it relies on having a very skilled person to do it.

Questions to Howard Murray

John Dawson: In storing the wood in your passive system did you use a fungicide?

Howard Murray: Yes, 5% Borax/Boric acid. On the subject of fungicides, I think that good housekeeping is the best fungicide. The best approach is to try and curtail the invasion of fungus, because once fungus grows on something, it is very difficult to get rid of. Once fungus is present this fungicide is not very effective.

John Dawson: How did you assess the deterioration of stored wood?

Howard Murray: Visually. In plastic bags shrinkage of the wood took place, whereas submerged in a tank of water, the wood was inclined to swell and actually become partially dissolved in the water.

David Grattan: Have you tried PEG 400 solution for pre-treating the wood before freeze-drying?

Howard Murray: A few pieces of wood were tried, but the results differed little from wood freeze-dried without pre-treatment. With PEG 400 pre-treatment acceptable objects were produced in one step, this therefore is a more efficient process.

David Grattan: What was the concentration of the PEG 400 pre-treatment solution that you used?

Howard Murray: I used a 10% solution.

David Grattan: Our results indicate that it is necessary to emloy a much higher concentration of up to 25% v/v PEG 400 to control shrinkage satisfactorily.

Anon: What was the water content of the wood?

Howard Murray: It was up to 500% on an oven-dry weight basis, but varied from object to object.

Richard Clarke: The hull of the Mary Rose is very variable in water content. In the centre of an oak block it can be as low as 80-90% on an oven-dry weight basis. So that you can assume that the wood is quite sound. On the outside it can be from 200 to 500%. For pine the water content was usually found to be quite uniform from the centre of blocks to the outside. It was variable from about 100% up to levels of 200% again on an oven-dry weight basis.

Per Hoffmann: The problem in the measurement of water content is that it neglects the possibility of the presence of trapped pockets of air, and even in deteriorated wood they can occur. What we really want is a measure of the wood substance remaining and maximum water content is the most accurate method of providing this information. If for instance you find the maximum water content is 400% then from this you can get a measure of void volume, which in this instance is about 80%, and this will tell you something about the wood which is missing. To carry out the method you meed only put a sample in a flask under water and evacuate. Then the vacuum is released and the air pressure forces the water into the wood. If this cycle is repeated several times, all of the pockets of trapped air will be removed. The wood is then oven-dried in the conventional fashion.

Richard Clarke: We sometimes find that there is a technical problem with this procedure. Fragmentation can occur with really degraded wood; it only works with fairly non-degraded wood.

Questions to Daniel Nelson

Mary-Lou Florian: During the geotechnical part of the site survey of the sediments and core analysis would you get enough infomration to reveal the presence of a reducing/non-reducing layer just below the interface between the water and sediment? This area is where there is a high level of biological and chemical activity and where maximum degradation can be expected.

Daniel Nelson: The concern here is that the presence of a reducing/non-reducing layer in the first 10 cm of sediment would cause a higher than average level of deterioration of the ships' timbers in a ring, where the hull rests in the sediment. We don't have this information, but I feel that even if this type of deterioration has taken place, that it would be unlikely to penetrate through the planking to the frame or to the keel, so I don't see that it would cause structural weakness. In connection with this, the oxygen content of the water has been found to be uniform throughout the water column, therefore it is likely that the oxygen content of the sediment would be similar to that in shallower water.

One other factor I would like to mention is the cost of working at a depth of 88 metres. Costs increase exponentially with depth of water, and the estimated expense for working on the site of the Hamilton and Scourge wrecks runs at around ten to eleven thousand dollars per day. Thus we must plan our sampling very carefully, and the idea of sampling of the hull below the sediment level is certainly being considered.

Lorne Murdock: In my experience, the usual places where the wood is most deteriorated are in the bilges (this is usually caused by biological attack when the ship was in service), rather than the wood buried in the silt. We have found that silt buried (anaerobic) wood is usually in better condition than any exposed part of the hull.

Richard Jagels: What were the ages of the Hamilton and Scourge when they sank?

Daniel Nelson: The Hamilton was 4 1/2 years old at the time of sinking.

Richard Jagels: The Indiana was about ten years old when she sank in Lake Superior in the mid nineteenth century. Wood from it was brought up by the Smithsonian Museum in Washington and I examined it and found it to be in sound condition.

Daniel Nelson: The Alfred Clark which sank in Green Bay, Lake Michigan in 1864 and was recovered in 1968 was found to be in excellent condition and floated on its own bottom after pumping out.

The Hamilton and Scourge were both originally merchant ships and therefore built from properly seasoned wood; probably with oak frames and black walnut or cedar planking, unlike the hastily built naval vessels of that time.

David Grattan: Last November at the meeting to discuss the possibilities for the conservation of the Hamilton and Scourge a stepwise approach was discussed. The idea was to firstly collect wood samples which would be analysed etc. and then would be used to test various possible submersion conditions, if a successful environment could be achieved, the proposed storage/display tank could be tested on an intermediate scale by using the ship's boat which is presently lying off the stern of the Hamilton. If this is successful then one of the schooners could be raised and placed in one tank. Then you would have a schooner in each location and their relative conditions could be monitored. If it was felt that the schooner in the tank was faring better, the second vessel could be raised and would join it.

Daniel Nelson: Conservation decisions must be made on a compromise basis. They depend on the funding available. If you build a whole conservation complex for two ships it would be very expensive to install one and run the programme for a number of years and then go and collect the other. The ship that remains would be in increased jeopardy if only one vessel is raised. Its location will inevitably become better known after the raising operation. I am not ruling out the stepwise programme, but before decisions can be made the site survey must be completed.

Questions to Richard Clarke

David Grattan: What can you tell us about the atmospheric freeze-drying method of Drocourt in Marseilles? Does he employ PEG pre-treatment before freeze-drying? In the pre-freezing stage which uses liquid nitrogen, doesn't the wood tend to crack?

Richard Clarke: As far as I am aware Drocourt did not pre-treat his wood, which is apparently quite badly deteriorated. I don't have direct knowledge of the results of his method, but some wood from the Mary Rose treated by his process was not satisfactory. It fragmented badly, had very heavy cracks and was badly distorted. Perhaps some of this was caused by too rapid freezing in liquid nitrogen.

David Grattan: I would like to make a comment about the natural freeze-drying method It is not a complete freeze-drying process, since all of the ice is not removed by it. If you consider a frozen artifact as essentially a block of ice containing wood, then what is achieved is that the ice recedes between 1 and 2 cm below the surface. This may not sound very much but for many small artifacts can represent a considerable percentage of the volume.

Howard Murray: In our work we find it unnecessary to continue to freeze-dry once the ice has receded below the surface layer. At that point we remove the artifacts and allow them to air dry.

Mary-Lou Florian: As Richard Jagels mentioned the surface zone is where most of the strength loss of the wood has occurred. When the drying stresses are greater than the compressive strength of the wood, then collapse occurrs. This is the situation for wood with a degraded surface. If you can freeze-dry this outer zone and prevent collapse then perhaps this is all that is necessary. The cells which are inside and not deteriorated are strong enough to withstand drying stresses.

Victoria Jenssen: What sort of treatment trials have you carried out, and what are your results with the timbers from the Mary Rose?

Richard Clarke: We have not been able to test several methods because of limitations placed upon us. Fire regulations do not permit us to use large volumes of volatile solvents, large-scale freeze-drying facilities are not available to us, gamma ray sources are also not available so resin polymerisation is not possible. We have tried to be practical and consider those techniques which can be used on a large scale. Much work has been done with spraying with different grades of PEG, controlled air-drying, and small scale freeze-drying.

As for the results so far: For small pieces freeze-drying has been found to be successful when PEG pre-treatment is carried out. For the oak, it seems to be a matter of stabilizing the outer decayed layer, because the majority have a substantial inner core. Our aims in using PEG can therefore be different according to the condition of the wood and the techniques available to us whether we want to:

(a) pentrate the wood to the middle.
(b) form a bridging layer between the less decayed inner material and the highly decayed outer material.
(c) simply slow down the rate of drying.

Freeze-drying and spraying with PEG 1500 solutions have given good results except that in the latter, slime development has been a problem.

PEG 1500 concentration in the spray method begins at 2-5% and is increased by 1% per week until a concentration of 50% is achieved. This avoids too much drying or swelling stress on the wood.

Victoria Jenssen: I don't think that high concentrations of PEG cause shocks or stresses to wood when applied as surface treatments.

Richard clarke: High concentrations of PEG applied to the surface of waterlogged wood causes dehydration of the interior.

David Grattan: I think that you have to take account of the grade of PEG. If there are mobile low molecular weight fractions in the PEG these will penetrate rapidly and counterbalance the outward flow of water. Thus I expect PEG 540 Blend and PEG 400 to be less harmful in this respect than PEG 4000.

John Dawson: Can you suggest any guidelines as to whether or not a wreck should be raised?

Richard Clark: Political and financial pressures, questions of prestige all play a part. I think that given the choice, my advice to the archaeologist would be that they do what they can by excavating on the sea bed, bring up a representative selection of artifacts and leave the hull behind. I question the need to bring up whole hulls, because according to one eminent marine archaeologist, you are able to learn far more about marine architecture by dismantling the structure. What this means is that it isn't really a decision that can be made by conservators alone, but by archaeologists and conservators in consultation with one another.

Lorne Murdock: It seems to me however, that the voice of conservation is far too weak in this field, we are there simply to provide a service.

Colin Pearson: An so we meekly say yes sir?

Lorne Murdock: No we don't, but I do like my job!

(Laughter!)

Victoria Jenssen: Just so, but I often feel that we in conservation act as the conscience for archaeologists. I feel that they should take more responsibility for their actions in disturbing historical material.

Colin Pearson to Victoria Jenssen and Lorne Murdock: You mentioned that with the Machault timbers, you were required to supply them for display so that they would fit in with display conditions chosen for reasons not connected with the stability of the artifacts. Isn't this putting the cart before the horse?

Victoria Jenssen and Lorne Murdock: Yes.

Eric Lawson: If you return the treated wood with your recommendations and say that if it is not kept within the specified conditions of RH and temperature you can not assume any further responsibility for it, wouldn't this solve the problem?

Victoria Jenssen and Lorne Murdock: Yes that is what we are doing, but the response is that we are told to develop better methods!

Willis Stevens: In Parks Canada, raising a ship has never been recommended by a marine archaeologist. As for myself, this action would be the last thing I would ever recommend. I find however, that we are under pressure from the conservator to supply more wood.

Richard Clarke: The reason why we have to have so much wood is simply to test and develop processes.

Victoria Jenssen: Decisions to raise ships grow like Topsy. Just who is responsible in the final analysis for making the decision to raise or not to raise shipwrecks seems very hard to identify. Shouldn't we therefore be trying to frame some guidelines to develop a process of decision-making acceptable to all parties!

SESSION II
THE ANALYSIS AND CLASSIFICATION OF WOOD

Ian Wainwright
Session Chairman

Introductory Remarks

The subject of today's session is the microscopical and chemical analysis of wood, and like the treatment of the artifacts themselves this is labour intensive and time consuming. I think that what we should attempt to do here this morning is to try at the beginning to develop a common language so that we can all learn from the experiences of other workers elsewhere who are doing the kinds of analysis we are doing. I have just returned from a conference on Rock Art where a lack of communication was displayed. Here in Canada we have done a lot of work in this area, but to some extent it is lost research unless it can be communicated to people, for example in South America or Africa where so much Rock Art exists. Similarly in waterlogged wood we must learn to share our experiments and to develop a common language so there isn't a repetition of research which we simply cannot afford in the conservation world.

A DETERIORATION EVALUATION PROCEDURE FOR WATERLOGGED WOOD

Richard Jagels
University of Maine,
School of Forest Resources,
Orono, Maine, U.S.A.

The extent and kind of deterioration in waterlogged wood can significantly affect the results of conservation procedures. A survey of conservtion techniques applied to deteriorated wood indicates that seemingly identical conservation procedures, when utilized with woods having different deterioration histories, can yield very different results.

I would like to propose that a reasonably uniform procedure for evaluating deterioration in waterlogged wood be adopted so that conservation treatments could be assessed by reproducible comparative methods. This would allow conservators from various labortories to adopt procedures best suited to their particular cases. In addition, such an evaluation wuld have value in historical documentation.

The evaluation procedure which I will describe should be considered a tentative, preliminary procedure. As conservators use the procedure, adaptations and improvements can be incorporated, but as the procedures are revised, these changes should be well-publicized so that all laboratories can follow similar protocol. Hopefully many of the features of this initial assessment can be omitted from subsequent evaluations. The eventual goal is a simplified and easily conducted assessment.

Section I of the procedure considers the factors which would be found in any sound piece of wood; those which could affect conservation treatments. Species identification, when matched to species characteristics provides considerable information relating to permeability, strength, durability, etc. (3, 19, 20, 22). Identification of deteriorated wood may pose particular problems. Botanical microtechnique methods for embedding may be required (1,11).

For many species, sapwood and heartwood differ significantly, especially with regard to permeability (16). For instance, the sapwood of the Southern hard pines can be impregnated readily with wood preservatives, but the heartwood, which contains more latewood and resins, is much more refractory (15).

Separating heartwood from sapwood is relatively easy in the sound wood of some species, but difficult in others. Color, presence or absence of tyloses, crystals, gums and other inclusions may be of some help, but can be caused by other factors such as disease. Once the wood has been waterlogged the differences become further obscured. Chemical staining procedures can differentiate sapwood from heartwood on fresh green wood of certain species (17), but these procedures will, for the most part, be of little value for waterlogged wood. In many cases the distinction between sapwood and heartwood will not be possible.

Growth rate (as measured in some convenient way like rings/cm) can be of considerable value. For ring-porous hardwoods, rapid growth indicates increased specific gravity and fewer earlywood pores per unit volume (10). Permeability, therefore, is reduced in rapid growth, compared to slow growth, ring-porous hardwoods. Conversely, in many coniferous species rapid growth wood is more permeable than slow growth wood. Permeability is just one of several properties (including strength) related to growth rate (10).

Natural defects can cause considerable problems in conserving wood. Reaction wood (tension wood and compression wood) shrinks differently from normal wood as it dries; thus, combined with normal wood it produces distortion and checking. Compression wood is less permeable and more resistant to invasion by brown rot fungi (12,13). Knots, pitch pockets and distorted grain can cause drying distortion, and can also impede the penetration of PEG or other consolidating solutions.

Section II of the evaluation procedure assesses wood deterioration prior to immersion. It is not always possible to decide whether damage took place before or after immersion, but where possible it can be of value, especially in historical documentation. Clearly, Item A, carbonization, would precede immersion. Carbonized wood will often be very refractory to conservation techniques, and therefore is sometimes discarded. However, I would like to point out that, where biological deterioration may be extensive in neighboring uncarbonized wood, it is often absent in carbonized wood which has no food source. In these cases, carbonized wood may be the most useful for identification purposes since the carbonizing process does not significantly alter the anatomical features (8).

Of all the biological deteriorating agents, fungi are the most destructive to wood (5,9). Yet wood immersed in water lacks one essential ingredient for fungal attack; a source of oxygen. Therefore, most fungal activity will precede immersion. Exceptions are found for wood exposed to wet conditons, but not fully immersed, such as in mud flats or in well-aerated tidal waters.

The ascomycete fungi, which lack clamp connections, cause the least amount of damage. Mold and sap-staining fungi, for instance, primarily digest stored starch in ray cells. However, spiral checking (not to be confused with spiral thickening) of tracheid and fiber cell walls is often associated with staining fungi. Soft rotting fungi will digest cell walls, but they work primarily on wood surfaces, so damage is generally shallow. Presence of ascomycete fungi can enhance wood permeability (5).

The basidiomycetes, on the other hand, can extensively degrade lignin or cellulose (or both) and cause degradation which affects both wood strength and the effectiveness of conservation treatment (18). Among these basidiomycete fungi it may be possible to determine whether the decay occurred in the living tree or at a later time. For instance, several white rotting fungi, which degrade primarily lignin, cause a "pocket" rot only in living trees (a good example is pecky-cypress). Determining the presence of a pocket rot can be of value both for the conservation treatment procedure and for historical documentation. In most instances adequate assessment of fungal penetration and type of decay will require microscopic analysis (23).

Insect attack can be analyzed in analogous ways to fungal decay. For instance, some insects, like ambrosia beetles, are restricted to living trees. Shape, size and extent of insect galleries as well as presence or absence of fungi associated with the galleries can be diagnostic, but areas of confusion exist. For example, carpenter bee nest galleries are not unlike some of the pocket rots, and in waterlogged wood, fine distinctions, such as the presence of white fungal mats may be lost.

Bird attack is relatively minor and needs no extensive discussion, except to point out that it may be an indication of insect damage, since woodpeckers attack wood to extract insects.

Surface additives such as paint are often not a problem in water deteriorated wood since they are usually rapidly lost. Asphaltic products and Chinese lacquer, however, may remain and can interfere with conservation treatments. Mechanical or physical damage may either be highly visible or difficult to see (particularly compression failure). Hidden fractures may present logistic problems in large objects. X-ray or sonic analysis could be of assistance in this diagnosis.

Analysis of chemical deterioration is the subject of succeeding papers, so I will omit a discussion of this topic here, but will point out that chemical and structural degradation are interrelated (4).

Bacterial activity is difficult to assess, since positive identification generally requires elaborate culture techniques or electron microscopic analysis. Pre-immersion bacterial activity will be minimal, primarily associated with wetwood which is common in oaks and ubiquitous in elm. Bacterial activity can be greatly accentuated in anaerobic waterlogged conditions (6).

Bacterial activity can cause strength loss and greatly affect permeability (1,21). At least some of this permeability change seems to be related to bacterial preference for pit membranes (7).

In addition to enhanced bacterial activity, wood immersed in marine or brackish water can be subject to shipworm damage (14). Two basic groups are involved: (B) Molluscs, which typically produce tiny surface entry holes, but penetrate deep within the wood and produce a myriad of internal galleries; and (C) Crustacea which produce readily observable surface damage, but the galleries only penetrate short distances into the wood.

Clearly, in such a brief discussion only a few highlights of a deterioration evaluation can be emphasized. However, I should point out that the evaluation procedure need not be quite so awesome a chore as this discussion might indicate. For most waterlogged wooden objects the array of deterioration factors will be quite limited. But the assessment of those factors in a systematic evaluation could prove most valuable in standardizing conservation techniques.

Waterlogged Wood Deterioration Evaluation

I. **Assessment of Non-Deterioration Factors**
  A. Identification of wood to species.
  B. Presence of heartwood or sapwood.
  C. Growth rate (rings/inch).
  D. Natural defects (tension wood, compression wood, spiral or interlocked grain, knots, heartshake, pitch pockets).

II. **Pre-Immersion Damage** (*indicates exclusivity to this category)
  *A. Carbonization.
  *B. Fungal Activity.
    - Ascomycetes (lack clamp connections): molds, stains, or soft rot.
    - Basidiomycetes (clamp connections): brown rot, white rot.
  C. Insect Attack.
    - wood nesting insects (carpenter ants, carpenter bees).
    - termites.
    - wood boring beetles.
  *D. Bird Attack (woodpeckers).
  *E. Additives (paint, tar, pitch).
  F. Mechanical or Physical Damage.
    - abrasion, fractures, physical signs of weathering.
  G. Chemical Deterioration.
    - contact with corrosive materials (copper, iron, marine concretions).
    - loss of chemical constituents due to weathering or dissolution (lignin, cellulose, hemicellulose).
  H. Bacterial Activity (presence and analysis of bacteria).
  I. Contaminants (salt or mineral infiltration).

III. **Post-Immersion Damage** (*indicates exclusivity to this category)
  A. Site (freshwater, brackish water, seawater).
  *B. Molluscan borers (Teredo, Bankia, Martesia, etc.)
  *C. Crustacean borers (Limnoria, Sphaeroma, Chelura, etc.).
  D. Mechanical or Physical Damage (see above).
  E. Chemical Deterioration (see above).
  F. Bacterial Activity (see above).
  G. Contaminants (see above).

References

1. Berlyn, G.P. and J.P. Miksche, Botanical Microtechnique and Cytochemistry, (Ames, Iowa, Iowa State Univ. Press, 1976).
2. Boutelje, J.B. and A.F. Bravery., "Observations on the bacterial attack of piles supporting a Stockholm building," J. Institute of Wood Science, vol. 20, no. 4, part 2, 1968, pp. 47-57.
3. Côté, W.A., Jr., "Structural factors affecting the permeability of wood," J. Polymer Science, part C, vol. 2, 1963, pp. 231-242.
4. Côté, W.A., Jr., "Wood ultrastructure in relation to chemical composition," Recent Advances in Phytochemistry, vol. 11, 1977, pp. 1-44.
5. Darrel, D.N., ed, Wood Deterioration and its Prevention by Preservative Treatments, vol. 1, (Syracuse, N.Y.: Syracuse Univ. Press, 1973.).
6. Ellwood, E.L. and B.A. Ecklund, "Bacterial attack on pine logs in pond storage," Forest Products J., vol. 9, no. 9, 1959, pp. 283-291.
7. Goldstein, I.S., Wood Technoloogy: Chemical Aspects. ACS symposium series 43, (Washington, D.C.: American Chemical Society, 1977).
8. Hanna, R.B. and W.A. Côté Jr., "Current applications of electron microscopy in wood research," in: 37th Annual Proc. Electron Microscopy Society of America, ed. G.W. Gailey (San Antonio, Texas: 1979), pp. 176-179.
9. Hunt, G.M. and G. Garratt, Wood Preservation, (New York: McGraw-Hill 1953).
10. Jagels, R., "Growth rate and wood properties," Woodenboat, vol. 39, 1981, pp. 115-116.
11. Jensen, W.A., Botanical Histochemistry, (San Francisco: W.H. Freeman & Co., 1962).
12. Jutte, S.M., W.L. Jongebloed and I.B. Sachs, "Influence of water encironment on normal and compression wood of a Picea species observed by scanning electron microscopy (SEM)." in: proc. of the Workshop on Other Biological Applications of the SEM/STEM. IIT Research Institute, Chicago, March 1977. Scanning Electron Miscroscopy, vol. 2, 1977.

13. Jutte, S.M. and I.B. Sachs, "SEM observations of brown-rot fungus Poria placenta in normal and compression wood of Picea abies in: proc. of the Workshop on Plant Science Applications of the SEM. IIT Research Institute, Chicago, April 1976. Scanning Electron Microscopy, vol. 2, 1976.
14. Jutte, S.M. and I.B. Sachs, "SEM observations of teredo attack of tropical hardwoods in brackish water," in: proc. of the Workshop on Plant Science Application of the SEM. IIT Research Institute, Chicago, April 1976. Scanning Electron Microscopy, vol. 7, 1976.
15. Koch, P., "Utilization of the Southern Pines," in vol. II Processing. Agriculture Handbook no. 420. USDA Forestry Service.
16. Krahmer, R.L. and W.A. Côté, "Changes in coniferous wood cells associated with heartwood formation," TAPPI, vol. 46, no. 1, 1963, pp. 42-49.
17. Kutscha, N.P. and I.B. Sachs, Color tests for differentiating heartwood and sapwood in certain softwood tree species, Report No. 2246, (Madison, Wis.: U.S. Forest Products Lab., 1962).
18. Liese, W., ed., Biological transformation of wood by microorganisms. Proc. Sess. Wood Prod. Pathol., Second Int. Cong. Plant Path., Minneapolis. (New York: Springer-Verlag, 1975).
19. Panshin, A.J. and C. deZeeuw., Textbook of Wood Technology, 4th ed., (New York: McGraw-Hill, 1980).
20. Suolahti, O., "Dependence of the resistance to decay on the quality of Scots pine," Papper Tra, vol. 23, 1948, pp. 421-425.
21. Unligl, H.H., "Penetrability and strength of white spruce after ponding," Forest Products J., vol. 22, no. 9, 1972, pp. 92-100.
22. U.S. Forest Products Laboratory., Wood Handbook: Wood as an Engineering Material. USDA Agriculture Handbook 72, rev., (Washington, D.C.: U.S. Gov't. Printing Office, 1974).
23. Wilcox, W.W., Preparation of decayed wood for microscopical examination, U.S.F.S. res. note FPL-056. (Madison, Wisc.: Forest Products Laboratory, 1964).
24. deVries-Zuiderbann, L.H., ed. Conservation of Waterlogged Wood. International Symposium on the Conservation of Large Objects of Waterlogged Wood, proc. of the symp., 24-28 Sept. 1979. (The Hague: Government Printing and Publishing Office, 1981).

Questions

Mary-Lou Florian: Would you agree that if you observe a higher or lower than normal specific gravity of wood, it means that something is seriously wrong with the wood; either the ash content is high or the wood is very degraded (respectively)?

Richard Jagels: It is possible to have a lower than normal specific gravity and yet still have a very high ash content. It is therefore important to measure both specific gravity and moisture content.

CHEMICAL WOOD ANALYSIS AS A MEANS OF CHARACTERIZING ARCHAEOLOGICAL WOOD

Per Hoffmann
Deutsches Schiffahrtsmuseum, D - 2850 Bremerhaven, Germany.

Abstract

An introduction is given into the ultrastructural organisation of the wood cell wall. Then a standardized chemical analysis of waterlogged wood is proposed. It comprises the quantitative determination of water content and maximum water content; lignin, holocellulose and ash content; hot water and alkali extractives. From these values an idea of the state of the wood on the ultrastructural level can be developed. Results are given for the chemical analysis of twelve samples of waterlogged wood aging up to 2000 years. Degradation mainly affects the carbohydrate fraction; up to 95% of which have been removed, whereas nearly all the original lignin remains. All archaeological samples contain 10-60 times more mineral compounds than does fresh oak wood. Micrographs and electron micrographs of the woods analysed are shown.

Introduction

More than a hundred years ago the routine conservation of waterlogged archaeological wooden finds started in the National Museum in Copenhagen. And for one hundred years conservators have been irritated, because while one piece of wood might be well stabilized another may not. This frustration is still alive, despite the replacement of the original alum treatment by more modern conservation methods.

Obviously, waterlogged wooden objects have to be graded according to their properties; they have to be classified somehow into groups showing the same behaviour. Then all objects of the same group can be treated appropriately. We all dream of the simple test, the result of which will tell us what treatment to use by applying it to a set of tabulated data! The most evident difference between finds is their state of degradation. Waterlogged wood, thousands of years old, can look as if newly cut, or may resemble black cottage cheese.

Based on this, the Nestor of waterlogged wood conservation B. Brorson Christensen, introducted the needle test: with a needle the scientist sensitively pricks the wood and decides: "Alas! very soft," or, "still a bit hard," or "well, strong as fresh wood!" or even "strong heart in a soft shell!" To the conservator these are important statements, but they are quite subjective; and also, the needle needs to be standardized!

The Structure of the Cell Wall

The properties of a piece of wood, its shrinking and swelling, as well as its behaviour after an attack by chemicals or organisms, have their origin in its chemical and physical structures. These structures are visible to a certain degree, but some cannot be resolved even with the electron microscope. They can be derived only with the help of chemical methods. In **Figure 1** I would like to show you what idea wood scientists have of the ultrastructure of wood. If you look at the cross-section of a piece of oak with a lens, you will see that the annular growth increments, known as annual rings, consist of a zone with wide pores and a zone of very fine pores. Running radially you see the so-called "rays." Under the light microscope the "pores" are identified as cross-cut very wide cells. They are called "vessels." The tissue between the vessels is made up of cells with thin walls, tracheids, and cells with very thick walls, the fibres. The rays are made up of thin-walled cells called parenchyma which typically are filled with cytoplasmic debris. The cavities of all these cells, called "lumina," are connected to each other by small pores through the cell walls, called "pits." In the living tree the pit acts as valves, since they contain a membrane that can close the pit openings. In wood that has been allowed to dry out, most of the pits may have become closed, much to the conservators distress. In the electron microscope you can see that the walls of woody cells consist of several layers. Onto the middle lamella between two neighbouring cells four layers have been built up; the primary wall and the outer, middle, and inner secondary wall. The electron microscope reveals that these layers are composed of fibrils; as if the whole cell were a sort of cocoon spun of filaments. If for example you expose the fibrils to ultrasound they disintegrate into thinner micro-

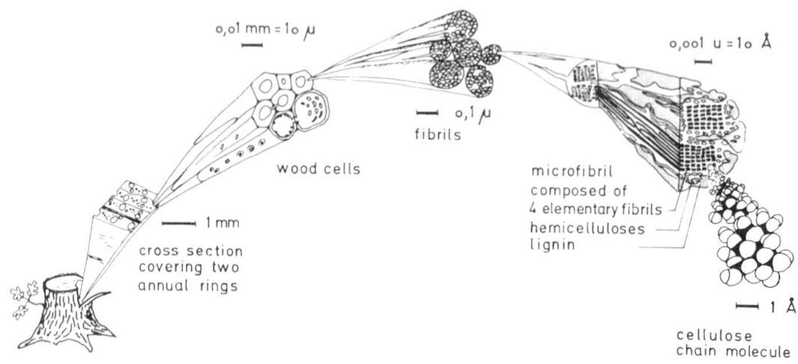

Figure 1

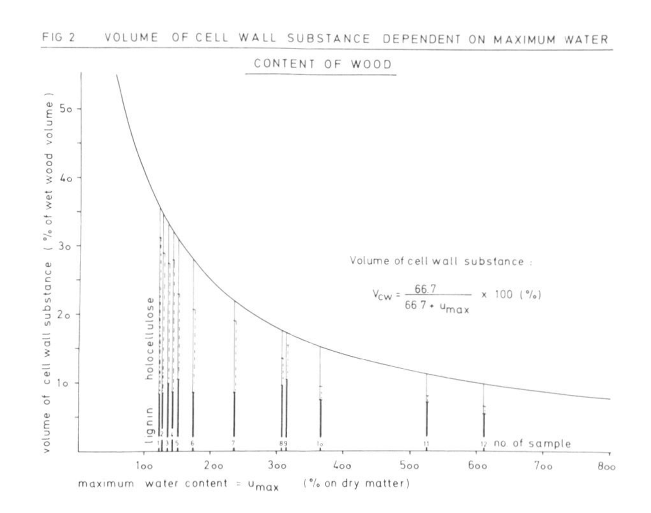

TABLE 1

The Capillary System of the Cell Wall

fissures within elementary fibrils	1 nm
capillaries between elementary fibrils	10 nm
capillaries between fibrils	up to 80 nm
pores in the pit membranes	up to 150 nm

Dimensions of some Molecules

water	0.2 nm
PEG 400 length x width	2 x 0.25 nm
PEG 1000 " "	4.5 x 0.35 nm
PEG 4000 " "	18 x 0.35 nm

fibrils, which in turn are composed of elementary fibrils. These cannot be further subdivided without destruction, and this is the limit of what can be made visible.

By means of chemical and physical methods it has been shown that elementary fibrils consist of about 40 very orderly packed cellulose molecules forming a quite crystalline arrangement. Cellulose molecules are extremely long chains of 5,000 to 15,000 glucose units. The number of monomeric units is called the degree of polymersiation (DP) of the macromolecule.

The elementary fibrils are surrounded by shorter chain molecules called "hemicelluloses" which are composed of a number of different monomeric sugars. They have DP's of some hundreds, and are adapted to the crystalline order of the cellulose to a certain degree only. Onto and into this loosely packed layer of hemicelluloses, lignin is deposited after the formation of fibrils has ended. Lignin bonds chemically to hemicelluloses, but not to cellulose.

Lignin is an amorphous substance, which gives pressure resistance and rigidity to wood. It consists of large three dimensionally cross-linked molecules. How large these are is not known, as it is not possible to separate the lignin from the carbohydrates without degrading it.

When lignin and other components, such as colouring matter, tannins, fats or waxes, are deposited into the cell wall it is not filled completely. A manifold system of open capillaries and cavities is left between the fibrils, the microfibrils, and even within the microfibrils. This capillary system makes up about 40% of the volume of the cell wall. It has a surface of 100 to 200 m^2 for each cubic centimeter of cell wall. This is the so-called "inner surface" of the cell wall. In the living tree the inter- and intrafibrillary capillary system is filled with water and the wood is in a swollen state, but if this water evaporates the wood shrinks.

The capillary system of the cell wall is where the conservator has to work if he wants to stabilise his object by means of a "bulking method." The water of the cell wall has to be substituted totally or in part by some other material that can prevent the cell wall from shrinking. In discussing what molecules can be used, it is helpful to know about the dimensions of this capillary system and some are shown in **Table 1.** Nevertheless, the problem of space is not the only factor affecting the complex problem of the permeability of wood and the infiltration of it with chemicals.

The Chemical Analysis of Archaeological Wood

The ultrastructure of the wood, i.e. the structure of the cell wall and its chemical composition of cellulose, hemicelluloses, lignin, extractives, varies remarkably little within a tree and from tree to tree within the same species. For every species there are approximate "standard values." This is why it seems reasonable to investigate, if the degradation that most archaological woods have undergone can be traced in a change of their chemical constitution. If so, perhaps we can learn to draw conclusions from this information about the sort and degree of modification of the wood, and hence might be able to predict the response of the wood to different conservation treatments.

The actual and the maximum, water contents give the first impression of the conditon of a piece of wood. After weighing, drying, and reweighing, one can calculate what percentage of the volume is occupied by cell wall substance, and what is made up of the void system of the lumina and the capillary system of the cell walls. A difference between the actual and the maximum water contents is a measure of gas in the sample. If there is a lot of gas in the object, vacuum/pressure impregnation of some kind might be considered.

Figure 2 shows the curve of the volume percentage of cell wall substance plotted against maximum water content. For the calculation of volume from weight the density of the cell wall substance can be taken as 1.5 g/cm^3. The main part of the chemical wood analysis is the quantitative determination of cellulose, hemicelluloses, lignin, extractives, and minerals.

As mentioned before, the quantitative composition of wood is relatively constant within a species. If nevertheless quite substantial differences appear in the literature, this can in most cases be traced to the fact that different methods of investigation have been used. Many procedures have been published to make the impossible possible: to separate the wood components quantitatively, and none is 100% successful. Either the fraction to be determined is not pure, or losses have to be accepted to produce an absolutely pure component. For comparative investigations one method only must be used and, most important, it must be clearly specified. The scheme in **Figure 3** demonstrates the sequence of analysis used in this work. With about 40 grams of dry matter one can get the most important values in double determinations. Although in many cases 60-80 grams are needed as the necessary milling yields a lot of dust which is not suitable for the procedure. In **Table 2** are arranged the results of the analyses of 11 archaeological finds of oak wood up to 2000 years old. These are arranged in order of increasing maximum water contents. (This parameter is often employed as a measure of degradation.) As water content increases, lignin content increases and carbohydrate content decreases in a remarkably regular fashion. (Holocellulose comprises cellulose and hemicelluloses.) In the degraded woods relatively more carbohydrates than lignin have been lost. If the relative percentages are converted into absolute amounts in most samples all the original lignin is still present. Only the heavily degraded samples nos. 11 and 12 seem to have lost some lignin along with the carbohydrates.

Indeed lignin is a material quite resistant to chemical and microbial attack. Only a few organisms can digest it, and then only to a certain percentage. The degradation of wood is therefore mainly a degradation of the carbohydrates. A detailed study of the carbohydrate fractions of the woods (not part of the routine analysis shown here) revealed something very interesting: from all samples, both cellulose and hemicelluloses had been removed in more or less the same ratio as they were present in the wood. This result is astonishing, since hemicelluloses can be chemically extracted or degraded much more easily than crystalline cellulose.

The important question is whether one can develop a picture of the ultrastructure of the wood from chemical analysis of it, because it is the ultrastructure that governs the behaviour of the wood on which the selection of a suitable conservation method depends. When carbohydrates are removed from the fibrils of the cell wall further cavities are formed, until at a certain point the residual lignin-carbohydrate skeleton will break and collapse.

The solubility of the wood in 1% hot sodium hydroxide solution can tell us something about this. The alkali extracts short chain carbohydrates, hemicelluloses and degraded cellulose, and fragments of lignin. If the alkali extract is high and the value for holocellulose low, as for the last 3 samples, in **Table 2** it means that a large part of the wood substance has lost its structural coherence.

Archaeological woods have a much higher mineral content than fresh wood, as shown by the high ash contents. The minerals are determined as their oxides after the organic content has been completely burnt away. Sometimes it is important to know what metals are present and the ash can be further analyzed. For example it is known that iron compounds will depolymerize PEG at high temperatures. The "conserved" wood may become sticky mess instead of hard and dry.

Examination

In order to examine how the changes of the chemical composition of degraded wood are accompanied by visible modifications of the cell wall structure a series of micrographs and electron micrographs were made. For micrography the sections were stained with safranine and astra blue for better contrast.

Fresh wood is easily distinguished from archaeological wood by the double staining technique: Lignified cell walls of fresh wood retain only the red colour of the safranine, especially intensely in the middle lamellae and the primary walls. In the archaeological woods so far examined the cell walls retain the blue colour of the astra stain, too. This seems to indicate a swollen state of the cell walls, which means they contain more and larger capillaries than fresh wood, into which the staining particles can enter. The more the cells appear to be swollen, the more intensely the wood becomes stained. (ed. note - Hoffmann has a series of colour photographs illustrating this, which can only be displayed in black and white).

Figure 4 shows fresh oak heartwood with thick-walled fibres and thinner walled tracheids. The original is stained uniformly light red with the compound middle lamellae darker red. In the electron micrograph **Figure 5** all

TABLE II

Chemical Analysis of Archaeological Oak Woods (european oak)*

Sample	Water Content	Max. Water Cont.	Sol. in 1% NaOH**	Hot Water Ext.	Lignin (Klason)	Holocell.+	Ash	Total**
1 recent	12	122	30.2	12.9	23.3	64.6	0.2	101.0
2 Zuider Zee	126	127	34.0	3.3	24.5	59.2	2.2	89.2
3 Drakenbg. centre	131	134	26.4	14.0	30.2	52.8	4.0	101.0
4 Vejby	140	142	22.6	6.2	26.9	60.3	2.8	96.2
5 Bremen 1 centre	147	150	34.4	4.3	34.2	40.1	2.7	81.4
6 Bremen 2 centre	9	173	41.7	12.5	30.4	43.4	4.5	90.8
7 Cologne 1 centre	215	236	29.7	4.3	38.8	48.7	2.6	94.4
8 Bremen 1 outer	307	307	41.1	4.2	54.4	22.6	7.7	88.9
9 Cologne 1 outer	313	313	35.5	6.5	60.3	28.6	7.0	102.4
10 Drakenbg. outer	349	365	54.9	0.4	49.0	12.9	12.4	74.7
11 Cap Vol	523	525	58.3	4.7	63.7	9.1	12.6	90.1
12 Cologne 2	527	611	59.0	3.5	56.9	11.2	12.9	84.5

*All percentages are based on the dry matter
** % on nonextracted wood.
+ Corrected for residual lignin.

Fig. 3 Proposed scheme for the chemical analysis of waterlogged wood

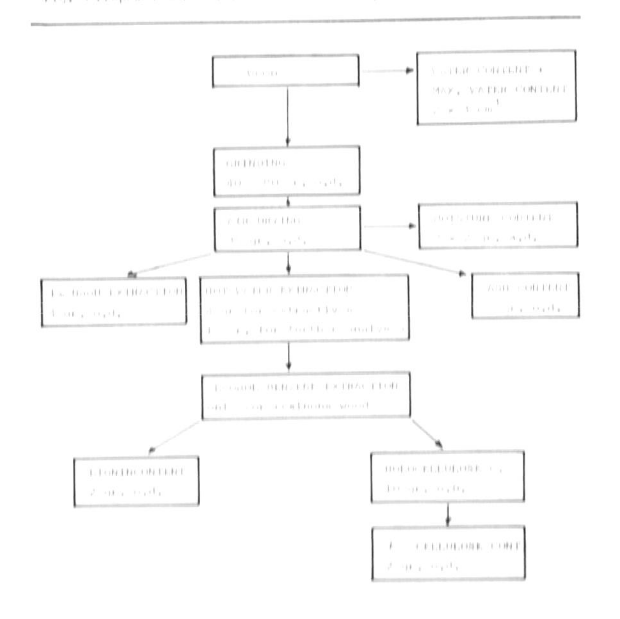

LEGEND TO THE MICROGRAPHS

Figure 4, 5: fresh oak heartwood (sample 1)
All micrographs are cross sections. ML = middle lamella, P = primary wall, S_1 = outer secondary wall, S_2 = middle secondary wall, S_3 = inner secondary wall, E = extractives, L = lumen, A = artefact, Ch = pit chamber. The distance bar means 10 microns in the micrographs and 1 micron in the electron micrographs.
Figure 6, 7: sample 4.
Figure 8,9,10: sample 5.
Figure 11,12: sample 7.
Figure 13: sample 9.
Figure 14: sample 10.
Figure 15,16: sample 12.

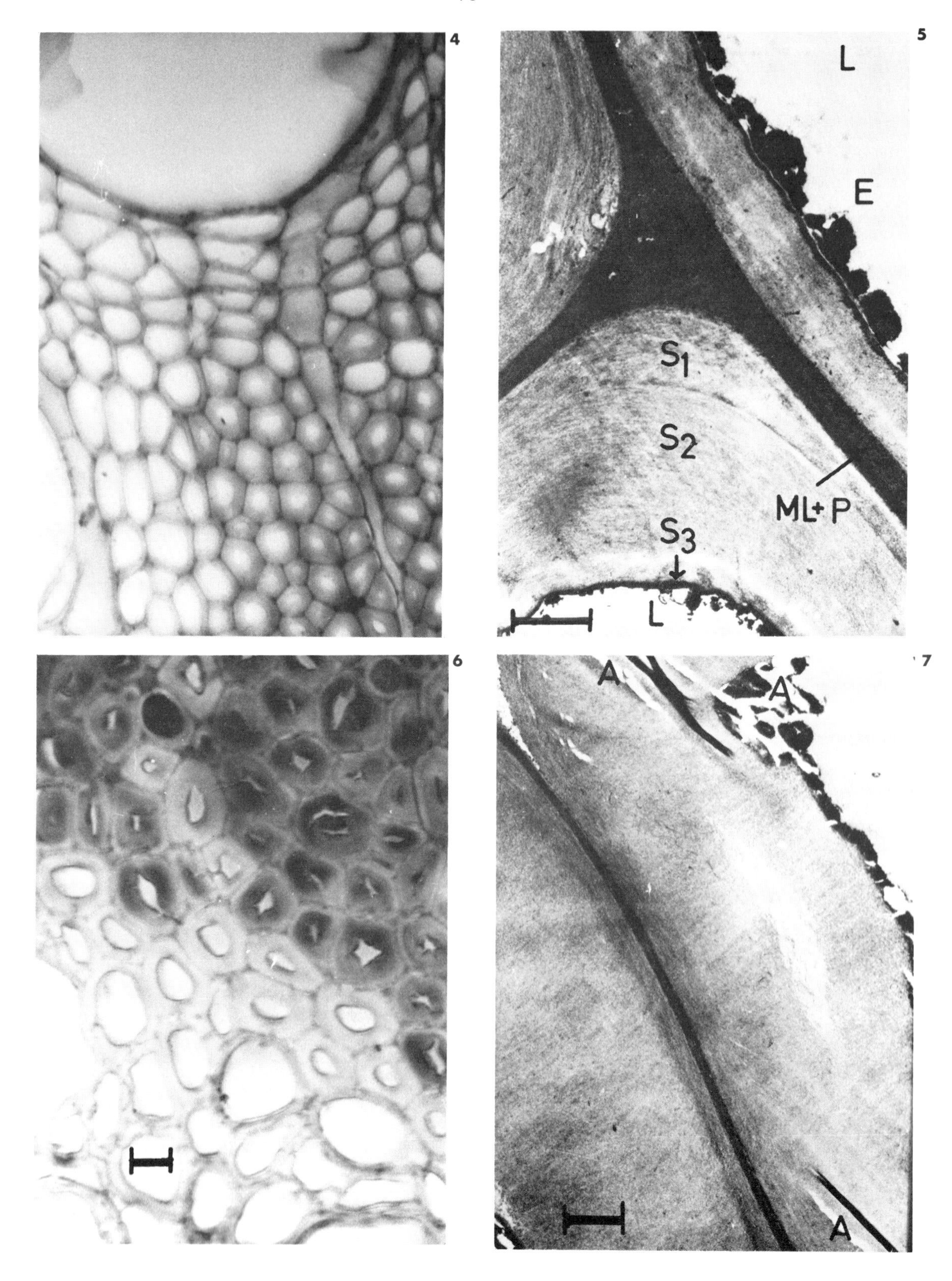
4
5
L
E
S_1
S_2
ML+P
S_3
L
6
7
A
A
A

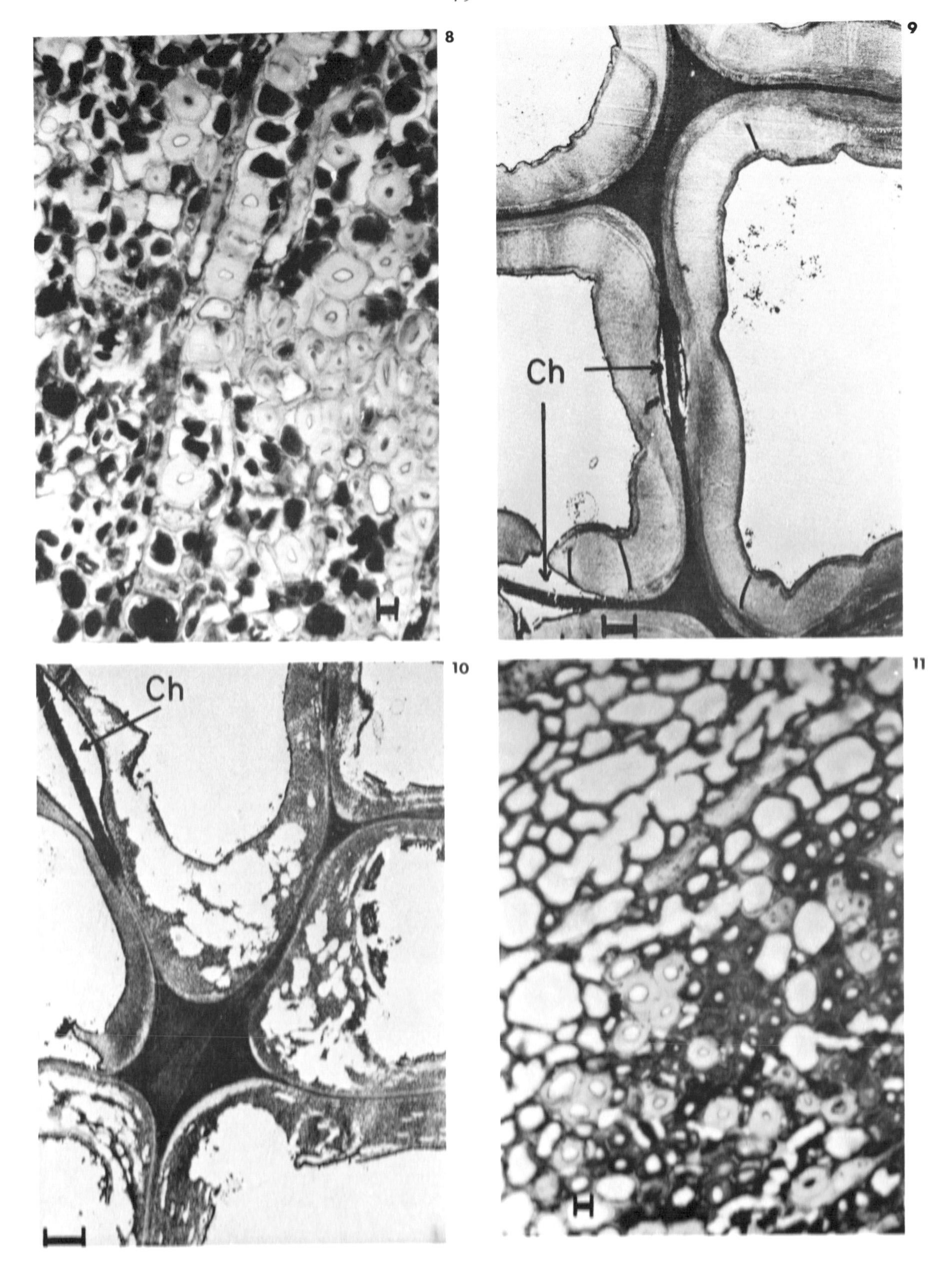
8
9
Ch
10
Ch
11

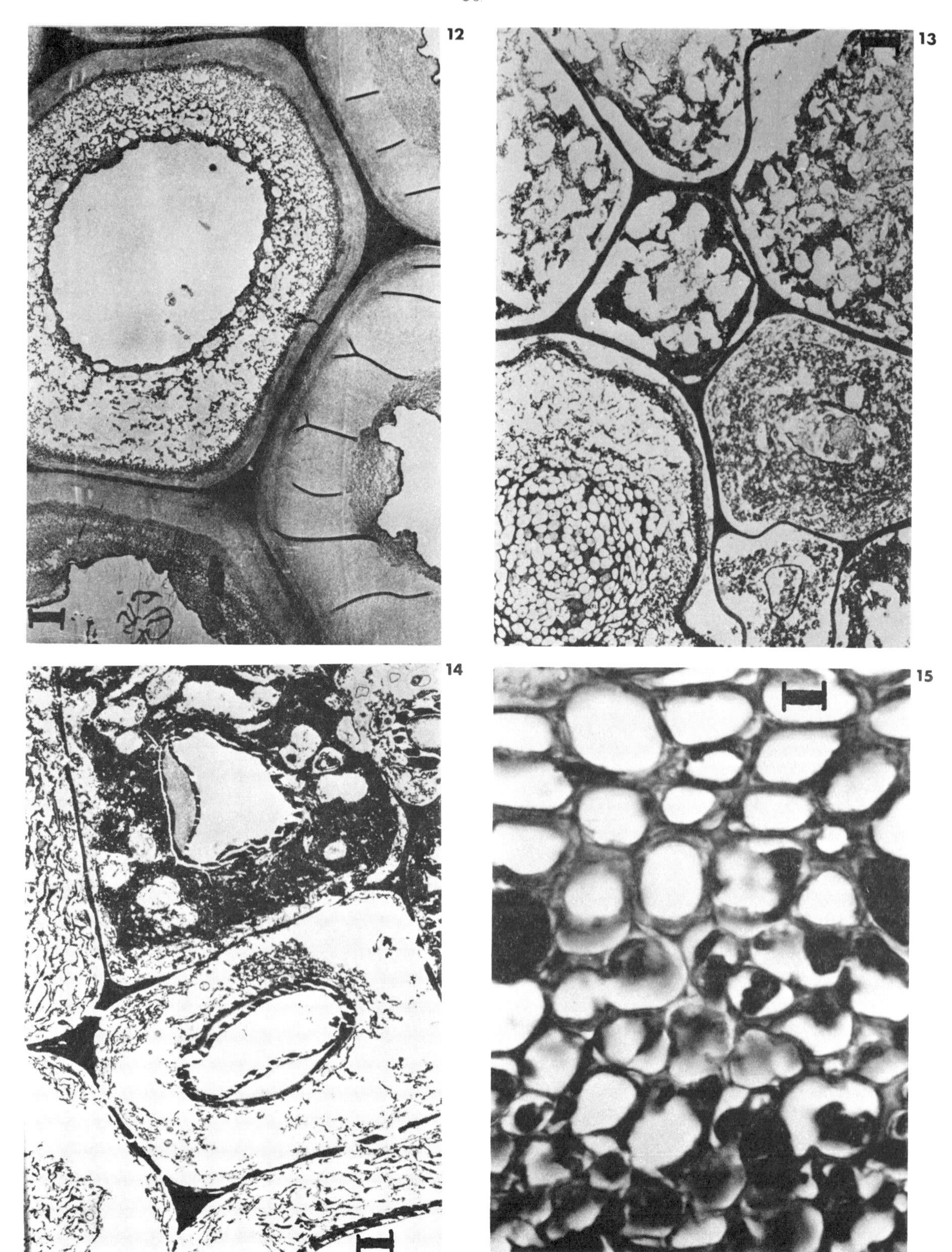
12
13
14
15

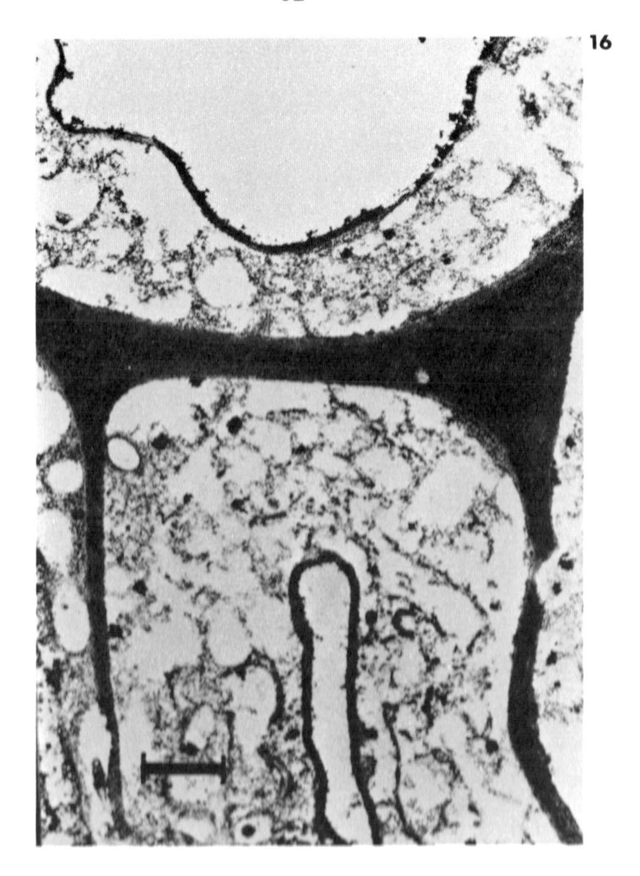
16

four cell wall layers can be seen. The black "blobs" on the S_3 might be extractives.

Figure 6 shows wood from Sample No. 4. The micrographic section is light blue with darker blue areas (the darker grey areas in this picture). Notice that the areas retaining more stain are in cells with swollen cell walls, discernable through their "bulging" surface towards the lumen. The closer view at these cell walls reveal a loosening of the cell wall structure, even fissures within the S_2 may occur, **Figure 7**. This is accompanied by slight carbohydrate losses of the wood.

In the sample No. 5 carbohydrate losses have led to heavy shrinking in many secondary cell walls resulting in their splitting off the compound middle lamellae (= middle lamella +primary wall). **Figure 8.** The residue of these shrunken cell walls is heavily stained and appear as "blobs" in the centre of the lumen. The degradation of cells seems to take place quite irregularly, heavily degraded cells neighbouring quite intact ones. Electron micrographs of this sample show not only pure chemical degradation, indicated by swollen cell walls but physical degradation also. (What appears to be the beginning of the splitting off of secondary walls in **Figure 9** is in reality a cross cut pit chamber). There are typical break-down patterns of the cell walls caused by fungi, too (**Figure 10**).

According to the chemical analysis, sample no. 7 is more degraded than sample no. 5. It has lost about one third of its original cell wall substance. At first glance this is not evident from the micrograph **Figure 11.** The fibres seem more intact than those with shrunken cell walls in sample 5 (**Figure 8**). But the thin walls of tracheids start to lose their shape due to failing compression strength, and cracks develop in the tissue. The electron micrograph of this sample, **Figure 12,** shows very clearly how chemical degradation of the S_2 starts from the lumen and progresses into the cell wall, accompanied by swelling of the affected areas (the radially oriented black lines in the S_2 of the cells to the right are artefacts, folds produced during the preparation of the ultrathin sections for electron microscopy).

Figure 13 shows that in this sample (no. 9) only the compound middle lamellae form a coherent structure filled with cell wall debris. The egg shaped caverns of different sizes are channels produced by fungal hyphae. This sample has lost about 80% of its carbohydrates. In the final state of degradation even the system of middle lamellae is corroded and dissolved, as seen in **Figure 14.** One third of the wood substance is left, and only 5-10% of the carbohydrates remain. A structure like this is only held in shape by the water that fills it completely, when it is excavated as waterlogged wood.

Figures 15 and 16 are added as a warning: They show that the strong retention of the astra blue in the extremely loose residual cell wall debris might give the false impression of relatively intact cell walls (**Figure 15**). Only the electron micrograph (**Figure 16**), and the chemical analysis (sample no. 12) reveal the real state of affairs: It is an extremely degraded piece of wood.

Conclusion

The wood analyses presented in this paper are initial results from work only recently started. They are not confirmed. Nor was it intended to examine all sorts of possible wood degradation. The material analysed is a random collection of archaeological wood as it is presented to the conservator. Analysis of what kind of biodeterioration might have affected the wood is not the objective of this paper; this was treated by Richard Jagels in his paper at this conference. However I wanted to show what wood analysis can reveal and how the results can vary. And, most important, I want to show that interpretation of results can be a bit difficult.

You have probably realized, that often for archaeological wood the sum of the components determined does not reach 100%, whereas for fresh wood one will usually arrive at 97 to 103%. In degraded wood evidently larger amounts of the material have been modified in such a way that they are not measured by the analysis. As the measurements of the lignin content showed here, in most cases the original amount is still present, therefore the losses must have been carbohydrates. It is known that even with fresh wood 2 to 5% of the hemicelluloses are lost in the determination of holocellulose. In the degraded woods, sometimes a large amount of carbohydrates must have been modified, depolymerized, or transformed to other compounds so that they are not determined as holocellulose. At present I do think that the chemical analysis can yet give an adequate characterization of archaeological wood. Synthesising all of the results one can construct an idea of the ultrastructural state of the wood. But which conclusions the conservator has to draw from this still remain to be found out. For this, experiments are required: The state of the wood has to be documented as well as the conservation method applied and the results obtained with it. To get as many results as possible in a reasonable time we can all undertake these experiments. We can arrange our daily work so that the necessary investigations are standardized and applied to all the objects that have to be treated. A number of laboratories are already testing the methods of analysis described here; perhaps we will soon hear what their experiences and opinions are and what proposals they can make to optimize the methods. The "Newsletter" would be a good forum for this discussion. In the course of time enough material could be treated so that we can look for confirmed relationships between the state of a wood as defined by analysis and its reaction towards defined conservation treatments. And from this one could conclude which form of conservation will give the best result for a wood of defined chemical and physical constitution.

Author's Note

The standard procedures of the Technical Association of the Pulp and Paper Industry (TAPPI-standards) can be obtained from "Technical Association of the Pulp and Paper Industry, 1 Dunwoody Park, Atlanta, Georgia 30 341, USA." On request, I will send a set of the relevant standards along with some comments and recommendations as to their handling.

Appendix 1: Material and Methods

The oak samples (european oak) analysed:

1. - Recent wood, air dried;
2. - vessel from the Lake Jyssel/Holland, 17th century;
3. - dugout from Drakenburg/river Weser, undated, inner part;
4. - Cog from Vejby Strand/Denmark, 14th century;

5&6 - Bremen Cog: 14th century, inner part of plank;

7. - boat plank, Cologne/river Rhine, 1st century, inner part;
8. - as no. 5, surface parts of plank;
9. - as no. 7, surface parts of plank;
10. - as no. 3, surface parts;
11. - Roman wreck off Cap Vol/Mediterranean, 1st century;
12. - dugout, Cologne/river Rhine, 1st century

Appendix 2: Chemical Analyses

1. Water content - TAPPI standard - T12
2. 1% NaOH solubility - " T212
3. hot water extraction - " T207
4. ash content - " T15
5. alcohol benzene extraction - " T204
6. lignin - " T222
7. alpha - cellulose - " T203
8. holocellulose - L.E. Wise et al., Paper Trade Journal, vol. 122, no. 2, 1946, p. 35.
9. maximum water content - the sample is submerged in distilled water in a suction flask. The flask is closed and evacuated for 30 min. Then the vacuum is released for 15 min. This cycle is repeated two times. The sample is then weighed, oven dried at 105°C and weighed again. The maximum water content (Umax) is calculated as 1% of the dry sample.

For light microscopy wet wood was embedded in PEG 1500. Sections of about 10 microns were double stained with saffranine and astra blue, an acetic acid-copper phthalocyanin compound.

For tansmission electron microscopy small wood blocks were treated with 1% aqueous $KMnO_4$ as a fixataive, dehydrated in acetone, and embedded in Spurr's medium (epoxy resin). Thin sections of 60-80 nm were contrasted with Reynold's lead citrate and examined in a Siemens Elmiscope 101.

Acknowledgements

The electron microphotographs were most kindly prepared by Professor Narayan Parameswaran at the Federal Research Institute on Wood and Wood Technology, Hamburg.

The investigations of this paper were supported with financial help by the Stiftung Volkswagenwerk. For this I wish to express my thanks.

ANOMALOUS WOOD STRUCTURE: A REASON FOR FAILURE OF PEG IN FREEZE-DRYING TREATMENTS OF SOME WATER-LOGGED WOOD FROM THE OZETTE SITE

Mary-Lou Florian
Richard Renshaw-Beauchamp
British Columbia Provincial Museum,
685 Belleville Street,
Victoria, British Columbia,
V8V 1X4

Abstract

Eleven waterlogged wood samples from the Ozette site, Washington State, were given to the British Columbia Provincial Museum Conservation Laboratory to determine if the samples could be conserved by a freeze-drying treatment.

Each sample was divided into three parts for the treatment: 1) freeze-drying; 2) PEG pre-treatment/freeze-drying; 3) control. Analysis of the samples was undertaken before treatment to determine if the analyses could be used to predict the success of the treatment. The analysis included: 1) species identification; 2) assessment of inherent impermeability due to anatomical features; 3) assessment of inherent dimensional instability due to normal and anomalous growth; 4) deterioration of the tissue; structural features; chemical and physical changes; and biodeterioration.

A prediction on the basis of the analysis was made as to the success of the conservation treatments.

During treatment, weight and dimensional changes were recorded.

An explanation of the shrinkage or distortion, brash breaks or cracks, radial, tangential and cuboidal cracks can be given in reference to the analyses.

In most cases it was possible to predict the response of the samples to the treatment.

The most significant finding was that inherent dimensional instability due to anomalous growth could not be overcome by the treatment.

I. Introduction

This project was not intended as a research project. It was the result of a service request from Ms. Milfie Howell, Conservator at Neah Bay in 1979 to test freeze-drying treatments on some waterlogged non-artifact wood samples from the Ozette site to determine if these treatments could be used for artifacts of comparable material. A standard polyethylene glycol (PEG) treatment had not been successful for some small artifacts and it was thought that freeze-drying might be a good alternative.

The project presented an opportunity to do some analysis which would enable us to try "to predict" the results of freeze-drying.

Predictions of response to treatment were made from a compilation of the observed condition of the wood samples and the occurrence of various anatomical features for wood species which are reported in the literature (Panshin and de Zeeuw 1970).

2. Material

Non artifact samples of waterlogged wood (NBL-1, NBL-3 to 12 (5.V.25)) excavated in the summer of 1974 from the Ozette site had undergone routine field treatment at Cape Alava. This consisted of soaking in 50% aqueous PEG 540 Blend. NBL-1 was soaked in water 6-8 months prior to this project. NBL-3 to 12 blend. NBL-1 was soaked in water 6-8 months prior to this project. NBL-3 to 12 were removed from PEG treatment on the 12th April 1978 and then placed in a water bath with 1.4% w/w boric acid and 0.6 w/w borax. On the 15th March 1979 all the samples were placed in a water bath in the British Columbia Provincial Museum Conservation Laboratory and remained there until the freeze-drying trials commenced on the 7th June 1979.

3. Analysis and Results

3.1 Identification of the Sample

Species Origin and Plant Parts:
Microscope slides of the thin sections from the samples, prepared by standard histological techniques were examined under the microscope to observe the salient characteristics for identification. The results are summarized

Number	Species	Common Name	Growth Pattern of Species	Plant Part
Hardwoods				
NBL-7	Menziesia ferruginea Smith	Fool's Huckleberry	small shrub	3 year old branch (Phloem, xylem and pith)
NBL-6	Acer circinatum Pursh.	Vine Maple	small shrub	1 year old branch (xylem and pith)
NBL-1	Acer macrophyllum Pursh.	Broadleaf Maple	tree	xylem-root or knot tension wood
Softwoods				
NBL-5	Picea sp.	Spruce	tree	compression wood (branch)
NBL-8	Picea sp.	Spruce	tree	wood
NBL-12	Picea sitchensis (Bong.) Carr.	Sitka spruce	tree	wood
NBL-11	Picea sitchensis (Bong.) Carr.	Sitka spruce	tree	wood
NBL-9	Thuja plicata Donn.	Western Red Cedar	tree	wood
NBL-10	Thuja plicata Donn.	Western Red Cedar	tree	wood (root)
NBL-3	Thuja plicata Donn.	Western Red Cedar	tree	wood
NBL-4	Thuja plicata Donn.	Western Red Cedar	tree	wood

Table I - The Species Origin Growth Pattern of Species and Plant Part of the Samples from the Ozette Archaeological Site.

Normal growth and growth related defects	Hardwoods			Picea sp.					Thuja plicata		
	NBL-7	NBL-6	NBL-1	NBL-5	NBL-8	NBL-12	NBL-11	NBL-9	NBL-10	NBL-3	NBL-4
Mixed tissues (branch or root)	x	x	?	?					?		
Differential shrinkage of late and early wood				x	x	x	x				
Inherent anomalous shrinkage above fiber saturation point								x	x	x	x
Sapwood	x	x							x		
Knot			?								
Tension wood	?	?	?								
Compression wood				x							
Expected volumetric shrinkage											
Recorded normal Volumetric shrinkage*	-	13-15%	13-15%	11.5%	11.5%	11.5%	11.5%	6.8%	6.8%	6.8%	6.8%
Significant deterioration											
Charred wood						x	x				
Soft rot				x				x			
Pectin loss	x	x	?	x	x	x	x	x	x		x
Cellulose change	x	x	x	x	x	x	x	x	x	x	x
Predicted dimensional changes											
Excessive longitudinal shrinkage	x	x	x	x					x		
Distortion			x	x							
Collapse	x	x	x	x	x	x	x	x	x	x	x

Table II - Normal growth, growth related defects, expected volumetric shrinkage, and significant deterioration of the samples NBL-1, NBL 3-12 (5.V.25) from the Ozette Archaeological Site. From these features, dimensional changes on drying are predicted.

* From green to oven dry wood. (Panshin & de Zeeuw, 1970, pp. 504, 627)

in **Table 1.**

3.2 Assessment of inherent impermeability due to normal anatomical features:

The following mainly anatomical characteristics are considered to reduce permeability.

(i) few and small hardwood vessels
(ii) inclusions (gum and tyloses) in hardwood vessels
(iii) abundant hardwood thick walled fiber cells
(iv) short soft wood longitudinal tracheids
(v) absence of ray tracheids
(vi) absence of radial and longitudinal resin canals
(vii) aspirated and encrusted bordered pits
(viii) blind simple pits
(ix) high specific gravity
(x) ray parenchyma containing resin or extraneous material
(xi) reduced tangential wall pitting
(xii) extractives in heartwood

3.2.1 Results

The analysis of the above features in each species represented in the samples, suggested that they should not be impermeable, but the degree of permeability should be variable. Obviously, it is the extent and number of these features present, that is important. For example, both Picea sp. and Thuja plicata are reported to have no tangential permeability due to: the small amount of tangential wall pitting; blind pits; and resin in ray parenchyma cells. Also, Picea sp. has resin canals and ray tracheids which are not present in Thuja plicata and thus it will be more permeable than Thuja plicata.

Longitudinal permeability should be excellent in both these woods, unless the bordered pits are aspirated or encrusted. It will be apparent later that due to deterioration, these pits are open and are therefore not a limiting factor.

3.3 Assesssment of inherent dimensional instability due to normal or anomalous growth:

The following list summarizes the normal growth and growth related defects of the samples, which may cause unusual dimensional instability, and the normal volumetric shrinkage (Kelsey, 1963; Wardrop, 1965), and is reported in **Table II.** The following list is based on information given by Panshin and de Zeeuw (1970).

(i) branches or roots of mixed tissue (bark, phloem, xylem and pith) of different shrinkage rates cause distortions such as warping (e.g. NBL-7, NBL-1, NBL-10).
(ii) normal volumetric shrinkage from the green state to air dried wood will naturally occur. In **Table II** the normal shrinkages for the wood species in question are listed. These are for wood going from the green to the oven dry condition and are intended as reference values . This shrinkage occurs below the fibre saturation point and normally does not result in warping or serious distortions.
(iii) collapse of large thin walled early wood tracheids adjacent to thick walled late wood tracheids has been observed in some woods. In waterlogged wood it is possible this could occur and cause uneven surface grain.
(iv) collapse of the radial walls of tracheids, which takes place above the fiber saturation point is an inherent weakness in Thuja plicata (NBL-3, 4, 9 & 10).
(v) abundance of parenchyma (longitudinal or ray) will cause excessive longitudinal shrinkage, e.g. roots and hardwoods (NBL-7, 6 and 10).
(vi) knots may exhibit checking and splits with moisture loss. These are due to abnormal tissue relationship resulting in differences in shrinkage parallel and across the growth rings within the knot (NBL-1).
(vii) reaction wood (tension wood in hardwoods and compression wood in softwoods) has an increase in longitudinal shrinkage. In hardwoods the longitudinal shrinkage may not exceed 1% but in softwoods it can be as much as 6-7% (NBL-1 & 5).

3.4 Deterioration of tissue prior to treatment:

3.4.1 Methods of analysis

For examination by light and polarized light microscopy, freehand radial, tangential and transverse sections were made with a single-edged razor blade, air dried and then cleared in xylene.

Histo-chemical tests were observed using the following standard staining methods:

(i) Toluidine Blue O for polychromatic staining of primary and secondary cell walls (O'Brien et al, 1964)

(ii) Weismer reaction for lignin (phloroglucinol / HCl) (Johansen, 1940)
(iii) Sudan III for fatty acids (Johansen, 1940)
(iv) Ruthenium Red for pectin (Johansen, 1940)
(v) Ferrous sulphate for tannin (Johansen, 1940)
(vi) Tannic acid for ferric iron (Florian et al., 1978)
(vii) Aniline or Cotton Blue for fungal hyphae (Johansen, 1940)
(viii) Mercuric Brompohenol Blue for protein (Mazia et al, 1953)
(ix) Iodine/Potassium Iodide for starch

Crystalline cellulose was determined by birefringence using polarized light.

The following structural features were examined for deterioration: ray parenchyma, vessels, fibers, tracheids, pith and pits.

The presence of bacteria, soft-rot, mould fungi and decay fungi were recorded.

Other features such as carbonization wood, texture changes and abnormal colour were recorded.

Chemical analysis for lignin, extractives and ash on a few selected samples was undertaken at Forintek Canada Corporation, Vancouver, British Columbia. The results are presented in **Table III.**

3.4.2 Results

The deterioration of structural features, chemical and physical changes and biodeterioration were recorded (available on request).

The pertinent results and implications of these results are presented in summary here.

(i) All the samples showed a loss in crystalline cellulose expressed by reduced birefringence and lack of dichromatic reaction to Toluidine Blue stain. This indicated chemical leaching or biodeterioration.
(ii) All samples showed the presence of material (with some normal staining reactions) in ray parenchyma cells. This suggests the wood rays have become or were naturally impermeable. Because these radially oriented cells are less deteriorated than longitudinal cells abnormal drying tensions may occur.
(iii) All the samples except NBL-1, Acer macrophyllum sapwood (probably root or knot) and NBL-3, Thuja plicata show loss of pectin which normally glues cells together.
This suggests the possibility of separation of longitudinal cells from stronger ray parenchyma cells, resulting in radial cracks on drying.
(iv) All the softwood samples showed open enlarged tracheid bordered pits which will make woods abnormally porous.
(v) All softwood samples contained bacteria which are probably responsible for the enlarged pits.
(vi) Two Picea samples, NBL-12 and NBL-11, are partially carbonized (burnt). The difference in the degree of physical deterioration within a sample can cause dimensional instability.
(vii) Two samples Picea sp. NBL-5 and Thuja plicata NBL-9 have soft-rot which will cause surface cuboidal cracking on drying due to excessive longitudinal and radial shrinkage in the cell walls.
(viii) The sample NBL-1 Acer macrophyllum is compact, soft and waxy on cutting. Also it is dark in colour, shows collapsed compacted fiber cells and complete impregnation. This could be due to previous treatment with PEG 540 Blend, and a solution of borax/boric acid.
(ix) All the Thuja plicata samples contain iron from the environment. This may be species specific or chance.
(x) The analysis (**Table III**) shows an elevated lignin content which indicates the loss of cellulose. This is shown by the increase in lignin content compared to the normal amount as given in the literature.
(xi) The extractives (**Table III**) do not show significant differences, despite the fact that some of the samles had been treated with PEG.
(xii) The abnormally high amount of ash (**Table III**) in NBL-1 (Acer macrophyllum) could be the result of treatment prior to analysis.
(xiii) The branch, NBL-6, showed significantly different amounts of lignin in the sapwood and heartwood. This reflects not only the different chemical natures of these two tissues but the different degree of degradation.

Sample No.	Species	% Moisture	% Extractives	% Lignin (Moisture Free)	%Lignin (Moisture free extractive free)	% Ash
NBL-1	Acer macrophyllum		7.9 (7.04*)	57.6 (23.4-24.7*)	62.5	46.9 (0.24-0.26*)
NBL-6 Sapwood	Acer circinatum	8.4	9.4	48.3	53.3	--
Heartwood	Acer circinatum	7.8	4.7	35.1	36.8	--
NBL-11	Picea sitchensis		13.6 (5.2-10.3*)	60.1 (27.4-29.6*)	69.6	5.1 (2.21-2.22*)
NBL-8	Picea sitchensis	7.9	12.3	61.1	69.7	3.3
NBL-10	Thuja plicata	7.7	11.5 (10.7-27.6*)	36.0 (31.8-32.5*)	40.6	3.3 (0.24-0.32*)

Table III - Chemical analysis of selected samples after freeze-drying treatment.
Analysis done at Forintek, Vancouver, British Columbia.
*Values for normal wood (Isenberg, 1951).

Treatment		Hardwoods			Picea sp.			Thuja plicata			
	NBL-7	NBL-6	NBL-1	NBL-5	NBL-8	NBL-12	NBL-11	NBL-9	NBL-10	NBL-3	NBL-4
FD[1]	83.6	82.8	76.2	60.1	82.5	78.2	79.6	73.4	73	69.9	77.8
PEG/FD[2]	42	18	16.9	24.8	16.1	19.2	15.2	25	15.7	52.2	-

Table IV - The percent weight loss after freeze-drying of pieces of waterlogged wood samples NBL-1 and NBL-3 to 10 (5.V.215) excavated from the Ozette site. The PEG/FD weight change was calculated using original weight as weight after PEG pretreatment.

1. FD = freeze-dried
2. PEG/FD = pretreatment of PEG 600 followed by FD (see 5.1)

3.4.3 Summary

In **Table II** all of the pertinent information is presented in summary. From it one can see that all samples have potential dimensional instability on air drying.

4. Prediction of the response of the samples to freeze-drying treatment:

From the information on the inherent impermeability, the inherent dimensional instability and the deterioration of the samples, it was hoped that one could predict the reaction of the samples to freeze-drying and freeze-drying after pre-treatment with PEG 600 solution. It is precarious but the following were predicted:

(i) excessive longitudinal shrinkage resulting from loss of crystalline cellulose which causes surface cross grain cracks, probably cannot be prevented by freeze-drying, but pre-treatment by PEG 600 should help to overcome it by bulking the voids in the cell wall.

(ii) anomalous growth, abnormal deterioration and high ash content of NBL-1 (Acer macrophyllum) will probably prevent successful treatment, because the treatments act by overcoming drying stresses, and would not therefore affect the stresses caused by differential tissue strength.

(iii) NBL-5 may or may not respond to treatment because compression wood is present.

(The proportion of the compression wood present is a key factor and this was not measured.)

(iv) collapse and hence distortion of the sapwood in NBL-6,7 & 10 probably cannot be prevented; because of the extensive deterioration of this tissue. A consolidation treatment would be more appropriate for wood in this condition.

(v) In NBL-5, & 9 cuboidal cracks due to surface soft-rot probably will occur with simple freeze-drying. Pre-treatment with PEG 600 will probably prevent this by bulking the voids in the cell wall.

(vi) All the Thuja plicata (NBL-2,3,4,9 & 10) will probably respond well to either treatment, because only the drying stresses need to be overcome. Some cell collapse due to an inherent weakness of this wood species may have occurred prior to treatment and may not be overcome.

5. Conservation Treatment and Results

5.1 Conservation Treatments

All the samples except NBL-4 were divided into three pieces, one for each of the following treatments: air dried (AD); freeze-dried (FD); and PEG pretreatment plus freeze-drying (PEG/FD).

The pieces for AD treatment were placed in a polyethylene covered chamber for the purpose of slow drying. Constant weight was reached in thirty days.

The pieces for FD treatment were removed from water storage, weighed then frozen at -10°C in a deep freezer for two days. The frozen pieces were placed in the precooled (-40°C) vacuum chamber of an Edwards High Vacuum Freeze-drier and subjected to a vacuum of 0.1 torr. A constant weight was reached in 20 days.

The pieces for PEG/FD treatment were removed from water storage, weighed and photographed. They were then immersed in a cold solution of 10% PEG 600 in water in a heated tank.

The temperature was slowly increased to 40°C (Ambrose, 1976) and then allowed to cool overnight. This cycle was repeated for nine days. After this pretreatment the samples were weighed and frozen at -10°C in a deep freezer. The frozen samples were placed in the precooled (-40°C) vacuum chamber and freeze-dried at 0.1 torr for only one day due to a break down of the equipment. The samples were maintained at -10°C until the equipment was repaired. After three months the freeze-drying treatment was continued. After six days the weight was constant.

(It is important at this point to reiterate that this is not a research project, thus the realities of the technical service are presented.)

5.2 Results

5.2.1 Weight changes

The percent weight loss of the FD and PEG/FD pieces due to the freeze-drying treatment is summarized in **Table IV**. Weights of the AD pieces were not recorded.

From **Table IV** it is apparent that the percent weight change of the PEG/AD pieces is much less than FD pieces. This may be due to the replacement of water by PEG. The percent weight loss (**Table IV**) of FD pieces indicates percent moisture content of the waterlogged wood prior to treatment.

The moisture content (MC) of some pieces after FD treatment range from 7.7-8.4% (see **Table III**).

5.3 Explanation of Results

In air drying of the samples the dimensional changes and surface defects that occur can be explained in terms of inherent dimensional instability due to normal or anomalous growth which is assessed in (3.3). Deterioration (**Table II**) also enhances the dimensional changes and surface defects.

5.3.1 Hardwood samples NBL-7, NBL-6 are branches of mixed tissue which have different shrinkage rates resulting in distortions such as warping and excessive dimensional changes on air drying. Moreover, the tension wood in these branches has an abnormal degree of longitudinal shrinkage which results in abnormal bowing and irreversible collapse.

The FD and PEG/FD treatments overcame the major dimensional changes; against predictions; but not the surface cracks; as predicted. However PEG/FD did eliminate more of these cracks than FD alone.

5.3.2 The hardwood sample NBL-1 is a knot. Because of this, on drying, checking and splitting due to abnormal tissue relationships occurred. Also the presence of tension wood may have caused excessive longitudinal shrinkage.

These results were predicted.

5.3.3 The samples of Picea sp. all showed collapse of thin walled early wood cells adjacent to thick walled late wood. This was not predicted for this species, but was suggested as a possibility in 3.3, section (iii).

5.3.4 None of the samples of Thuja plicata showed dimensional instability. This result was predicted.

Some cell collapse due to an inherent weakness of the wood was expected but did not occur.

The surface cracks on the tangential surface of sample NBL-3 are probably due to soft-rot areas which were not observed in the sample observed for deterioration.

Conclusion

From the explanation of the results it is apparent that it is possible to predict the response of the waterlogged wood to the different drying treatments by using an assessment of the inherent impermeability, inherent dimensional instability and deterioration of the samples.

The most important observation is that inherent dimensional instability due to anomalous growth could not be overcome by the treatment.

References

Ambrose, W.R., "Sublimation drying of degraded wet wood," in: proc. of the Pacific Northwest Wet Site Wood Conservation Conference, ed. G.H. Grosso (Neah Bay: 1976), pp. 7-16.

Florian, M.L.E., Seccombe-Hett, C.E., and McCawley, J.C., "The Physical, Chemical and Morphological Condition of Marine Archaeological Wood Should Dictate the Conservation Process," in: Papers from the First Southern Hemisphere Conference on Maritime Archaeology. (Melbourne: Oceans Society of Australia, 1978), pp. 128-144.

Isenberg, I.H., Pulpwood of U.S.A. and Canada. (Appleton, Wisconsin: Institute of Paper Chemistry, 1951).

Johansen, D.A., Plant microtechnique. (New York: McGraw Hill, 1940).

Mazia, D., Brewer, D.A., and Alfert, M., "The cytochemical staining and measurement of protein with mercuric bromphenol blue," Biological Bulletin, vol. 1, no. 104, 1953, pp. 57-67.

O'Brien, T.P., Feder, N. and McCulley, M.E., "Polychromatic staining of plant cell walls by Toluidine Blue O," Protoplasma Bk, no. 59, 1964, pp. 367-373.

Panshin, A.J. and de Zeeuw, C., Textbook of Wood Technology, vol. 1, 3rd ed. (New York: McGraw-Hill, 1970).

Wardrop, A.B., "The Formation and Function of Reaction Wood," in: Cellular Ultrastructure of Woody Plants, ed. W.A. Côté, (Syracuse: Syracuse University Press, 1965), pp. 371-418.

Shrinkages

Sample	Water* Content	Air Dried	Freeze-Dried	Freeze-Dried/PEG
NBL 1	320%	61% r	10%	30%
NBL 3	232%	NC	NC	
NBL 4	350%	NC	NC	
NBL 5	151%	NC	NC	NC
NBL 6	481%	30% d	12% d	NC
NBL 7	510%	32.5±2.5% d	25% d	15%
NBL 8	474%	20% r	NC	NC
NBL 8	474%	30% t	NC	NC
NBL 9	276%	NC	NC	NC
NBL 10	270%	NC	NC	NC
NBL 11	380%	30% t	NC	NC
NBL 12	359%	10% d	NC	NC

* - Based on Freeze-Dried weight.
r - radial
t - tangential
d - diameter (orientation unknown))
NC - No change.

Table V

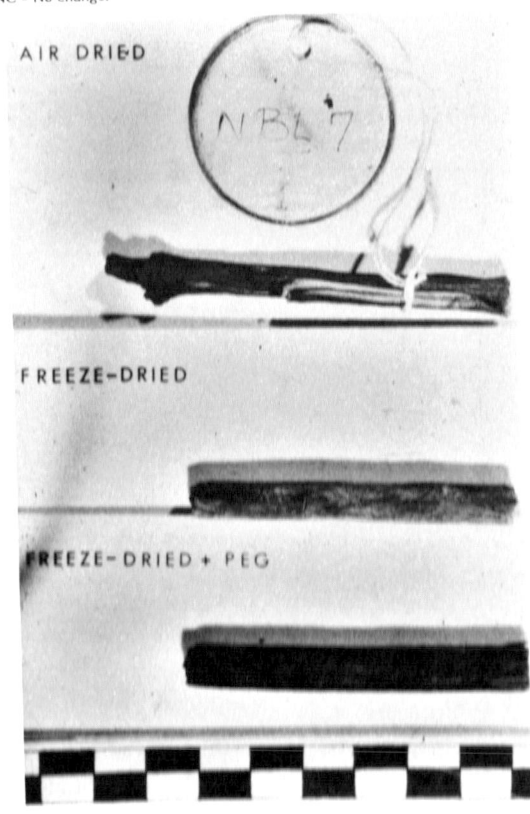

Figure 1:
NBL-7 Menziesia ferruginea Smith, 3 year old branch (xylem, phloem and pith)
(top) Air Dried: deep radial cracks due to extensive cell collapse.
(middle) Freeze-Dried: transverse and radial cracks
(bottom) Freeze-Dried/PEG: fine radial cracks occurred, but are not apparent in the photograph.

Figure 2:

NBL-6	Acer circinatum Pursh, 1 year old branch (xylem and pith)	
(top)	Air Dried:	extensive radial cracks and dimensional changes due to shrinkage and cell collapse (Some transverse cracks are present but are not apparent in the photograph)
(middle)	Freeze-Dried:	transverse brash breaks and radial cracks
(bottom)	Freeze-Dried/PEG:	radial cracks; no transverse cracks were present.

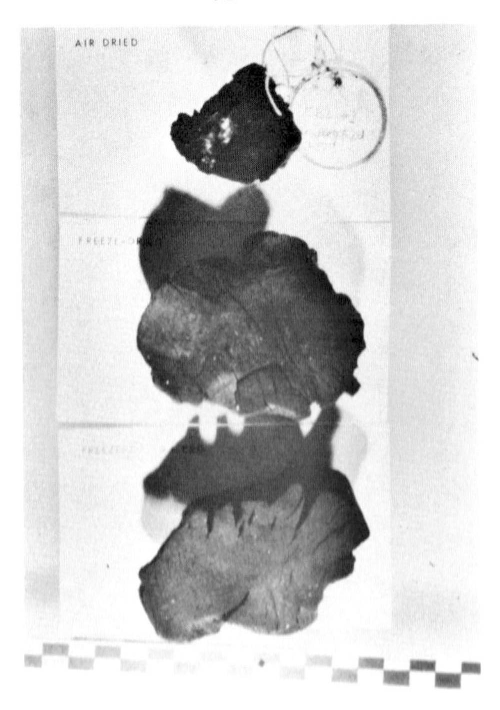

Figure 3a:

NBL 1	Acer macrophyllum Pursh, root or knot (xylem - tension wood)	
(top)	Air Dried:	reduced dimensions due to complete cell collapse and shrinkage.
(middle)	Freeze-Dried:	brash breaks and radial cracks
(bottom)	Freeze-Dried/PEG:	some cell collapse and radial cracks.

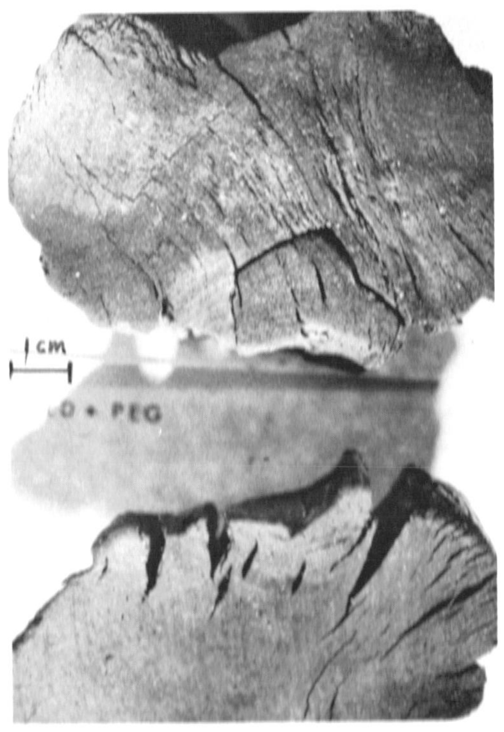

Figure 3b:

NBL-1	enlarged
(top)	Freeze-Dried:
(bottom)	Freeze-Dried/PEG:

Figure 4a:

	Picea sp., branch (compression wood) (mainly tangential surface view)	
(top)	Air Dried:	brash breaks and radial cracks, some cell collapse
(middle)	Freeze-Dried:	transverse and radial cracks on tangential surface.
(bottom)	Freeze-Dried/PEG:	some radial cracks on tangential surface

Figure 4b: (As **Figure 4a**) - (mainly radial surface view)
All three treatments show exaggeration of growth rings due to excessive early wood shrinkage or collapse.

Figure 5:

NBL-12	Picea sitchensis (Bong.) Carr., wood	
(top)	Air-Dried:	charred area on top and unburnt wood below shows cell collapse, shrinkage & radial cracks.
(middle)	Freeze-Dried	extensive brash breaks and radial cracks in unburnt wood.
(bottom)	Freeze-Dried/PEG:	minor radial cracks.

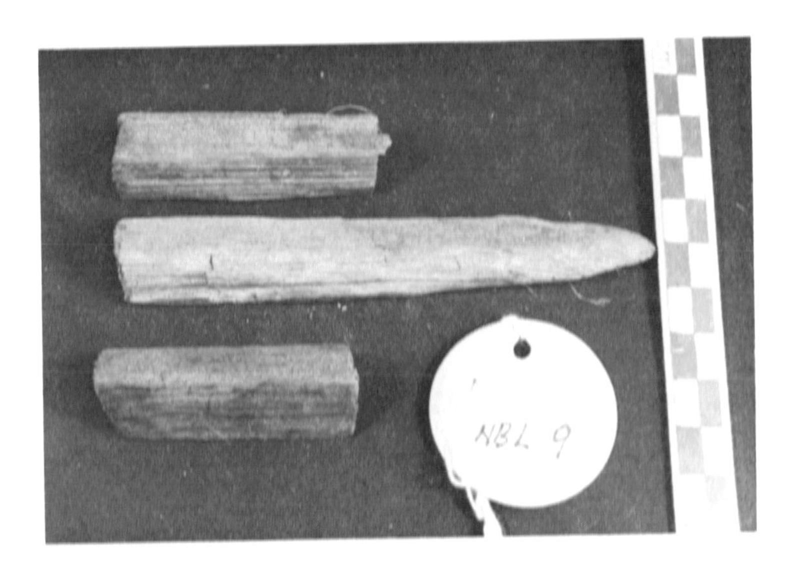

Figure 6:

	Thuja plicata Donn., wood
	For all three treatments no surface defects occurred.
(top)	Air Dried:
(middle)	Freeze-Dried:
(bottom)	Freeze-Dried/PEG:

Question

Richard Jagels: There has never been any evidence found for the existence of compression wood in true roots, although in exposed roots some has been found; tension wood has been found in hardwood roots.

Mary-Lou Florian: I did mention that tension wood is not characteristic in roots.

CONSERVATION OF WATERLOGGED WOOD IN SWITZERLAND AND SAVOY

Fritz H. Schweingruber
Swiss Federal Institute of Forestry Research,
CH-8903 Birmensdorf,
Switzerland

Introduction

For more than a hundred years archaeologists have been excavating wooden objects from submerged lakeside settlements. For eighty years the alum-linseed oil process was used to conserve the finds, although it did not give satisfactory results (Rathgen, 1924). Around 1950 laboratories in Switzerland developed the alcohol-ether method (Kramer and Muhlethaler, 1967/68) and the Arigal C process (Haas et al., 1961/62). The freeze-drying and PEG methods were also often used. In Grenoble, the radiation polymerisation process has been in use for some time.

By 1975 there were hundreds of preserved artifacts in museums, but a neutral assessment of the success of preservation had never been made and all discussions were based on incomplete comparisons.

Therefore, in 1976 representatives from five laboratories agreed to carry out a joint experiment in which different methods of conservation were applied to samples of the same type, taken from subfossil posts, and the results evaluated by a number of independent observers. The methods and results are published in detail (Braeker and Brill, 1979).

The subfossil and modern test samples, shown in **Figure 1**, were conserved with the following methods: lyofix DML, Arigal C, alcohol/ether/resin, freeze-drying, PEG/freeze-drying, PEG, radiation polymerisation of polystyrene.

After conservation, all samples were examined independently by those involved in the experiment and by neutral observers. We felt it important to consider criteria of concern for museum display/storage as well as to evaluate the wood scientifically; aesthetic criteria were not included, because these are evaluated differently by individual observers. In spite of some variation in the assessments (**Figure 2**), certain advantages and disadvantages of individual methods stand out clearly (**Figure 3**). With the exception of freeze-drying, none of the processes show fundamental faults.

Microscopic Identification of Wood Species

This is possible after application of any of the methods. However, there are technical difficulties in microscopic work with samples conserved by radiation polymerisation, since specially designed microtomes are necessary for the preparation of microsections. For material conserved with any of the other methods, the usual microscopic techniques can be used.

Depth of Penetration

The depth of penetration of the conserving agents is variable. The penetration of lyofix can be considerably increased by ultrasound (personal communication). Freeze-drying without previous treatment with PEG produces wide cracks due to shrinkage in the interior of large samples, and the subsequently applied stabilising resin does not penetrate deeply.

Shrinkage, Collapse & Cracking

Fractures appearing with the different methods practically always have a simple form and broken pieces can easily be repaired.

The compression and tensile strengths of the samples vary greatly with method of preservation (**Figure 4**). They are extremely high after radiation-plymerisation, high after conservation with lyofix, and low to very low after freeze-drying and after treatment with alcohol/ether/resin.

Stability of form may be regarded as satisfactory with all methods (**Figure 5**). Our experiments with PEG were prematurely broken-off, which explains the poor results.

Cracks appear more or less frequently with all methods (**Figure 6**), although the small transverse cracks produced by freeze-drying are a disadvantage. The formation of cracks can be avoided by immersing the samples in an 8% aqueous solution of PEG 1000 at 60°C for two to three months before freeze-drying (Elmer Zuerich verbal communication). The pure freeze-drying process is no longer used in Switzerland.

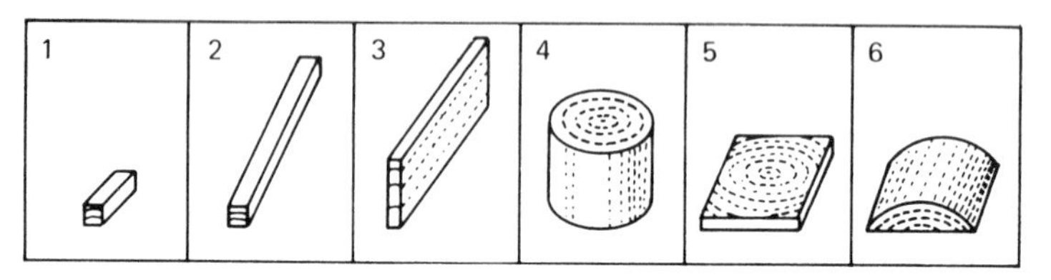

Figure 1 - Forms of the tested samples.

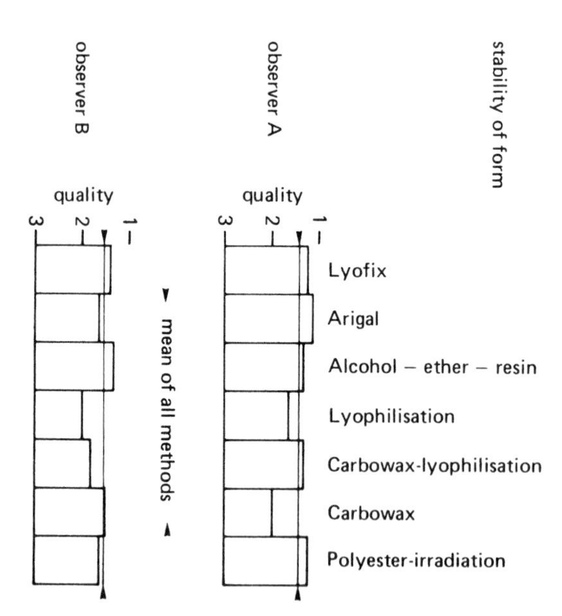

Figure 2 - Example showing variation of assessment between two observers.

	Lyofix	Arigal	Alcohol – ether – resin	Lyophilisation	Carbowax-lyophilisation	Carbowax	Polyester-irradiation	Scale
Structural properties microscopic identification	1 ✪	1 ✪	●	2 ✪	●	✪	3 ●	A
penetration depth	○	○	●	○	○	✪	●	A
breakage easy to repair	●	●	●	○	●	●	●	A
Strength properties compression strength	●	✪	✪	✪	○/✪	○/✪	●	A/B
tensile strength	●	○/✪	○/✪	○	○/✪	○/✪	●	A/B
Form properties stability	●	●	●	✪	✪	○	●	A/B
transverse crack	●	●	●	○	✪	●	●	A
radial crack	✪	✪	●	✪	✪	✪	✪	A
Dimensional properties volume shrinkage	✪	✪	✪	●	●	○	✪	B
weight	✪	✪	○	○	○	●	●	B

Evaluation:

	evaluation A	evaluation B
●	good	high
✪	satisfactory	medium
○	unsuitable	low

Figure 3 - Summarised results and evaluation of all observations on conserved wood samples.

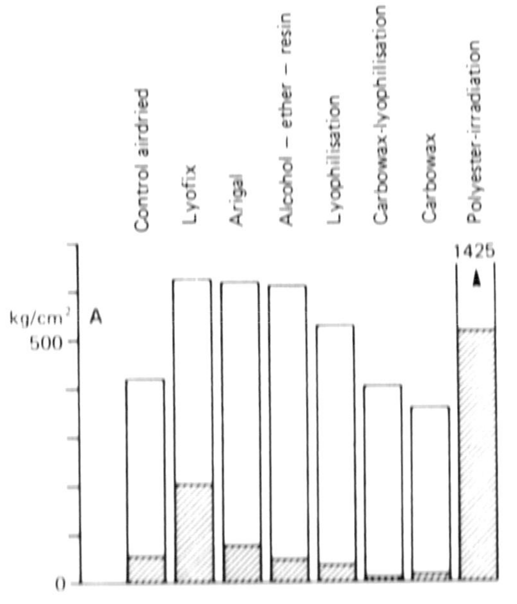

Figure 4 - Average compression strength of soft neolithic wood (ash, willow and elm).

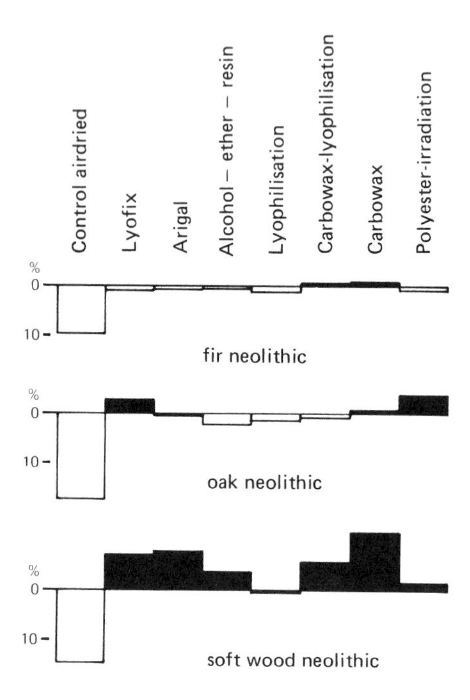

Figure 5 - Longitudinal shrinkage of neolithic samples.

Figure 6 - Formation of cracks and type of breakage in samples conserved with lyofix (above) and freeze-drying (below).

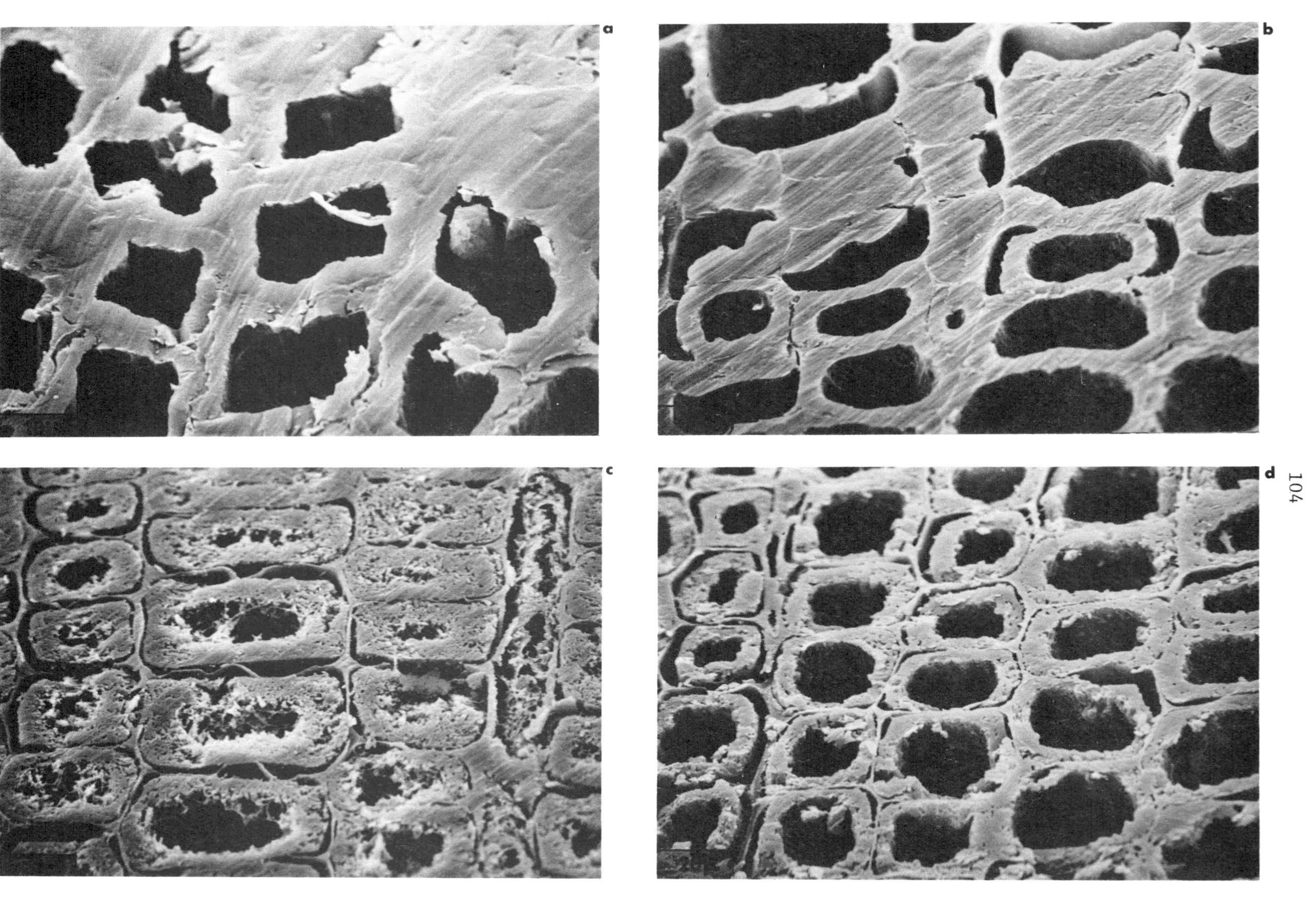

Figure 7. Cell wall structure after conservation by different methods (Magnification x 2000).
a. untreated
b. freeze-drying
c. lyofix
d. alcohol/ether/resin

Weight

The weight of the conserved samples shows great variation. Pieces conserved by PEG and radiation-polymerisation are very heavy; with lyofix or Arigal C of average density and after freeze-drying and treatment with alcohol/ether/resin are very light.

Colour

Opinions differ as to whether the colour of the conserved samples is natural, i.e. aesthethically. Material is often whitish when conserved with lyofix; yellowish when alcohol/ether/resin is used; dark after treatment with PEG and often slightly shiny after radiation-polymerisation. In no case does its appearance correspond to that of modern or freshly excavated subfossil samples.

Influence of Wood Species & Condition

Most of the qualitative features are determined by the anatomy of the wood. Pieces which are very much softened and degraded are easier to conserve than hard, only slightly degraded ones, because their cell walls are disintegrated and therefore permeable to water and conserving agents. For very degraded wood therefore, the species of the wood has little influence on the success of conservation.

The cell wall structure after treatment varies according to the method of conservation used. In air dried and freeze-dried samples the fibrils of the cell walls cohere to form a brittle, homogeneous mass (**Figures 7a,7b**), and the walls tear during shrinkage, so that cracks appear. If the wood is soaked in PEG before freeze-drying, the plastic material fills the cavities between the fibres in the cell walls and subsequent shrinkage does not result in tearing, so that no cracks are formed. The original subfossil form of the cell wall is preserved by treatment with PEG, freeze-drying or alcohol/ether/resin (**Figures 7b,7c**). Shrinkage and tension stresses are carried by the spaces in the cell walls and few cracks are produced.

Conclusion

It would be of great value if further ongoing investigations of this type could be carried out by other laboratories. Our experience shows that the following points are particularly important:

- collaboration between those laboratories working on conservation of wood and those working on wood anatomy
- comparison studies of material as similar as possible (e.g. of the same wood species, without knots, in the same state of preservation, having the same orientation of the annual rings to the form of the artifact etc.)
- examination of commonly occurring forms (e.g. branches)
- investigation of the economics of the various methods
- observation of long term changes in conserved samples under various museum conditions.

I should like to take this opportunity to make a strong plea for collaborative work between North America and Europe -we in Switzerland are very eager for such collaboration and there is much material available.

References

Braeker, O.V., & J. Brill, "Zum derzeitigen Stand der Nassholzkonservierung," Z. fur Schweizerische Archaologie and Kunstgeschichte, vol. 36, no. 2, pp. 97-145.

Haas, A., H. Muller-Beck & F. Schweingruber, "Erfahrungen bei der Konservierung von Feuchtholzen mit Arigal C (CIBA)," Jahrbuch des Bernischen Historischen Museums in Bern, vol. 41/42, 1961/1962, pp. 509-537.

Kramer, W. & B. Muhlethaler, "Uber die Erfahrungen mit der Alkoholathermethode fur die Konservierung von Nassholz am Schweizerischen Landesmuseum," Zeitschrift fur Schweizerischen Archaologie and Kunstgeschichte, vol. 25, 1967/1968, pp. 78-88.

Rathgen, R., Die Konservierung von Altertumsfunden, Teile II and III, 2 Auflage, (Berlin/Leipzig: 1924), bes., pp. 133-140.

Questions

(ed. note: Fritz Schweingruber was not able to be present at the meeting, his comments were added afterwards)

Colin Pearson: On the subject of the colour of treated wood; there is no point in comparing it to wet untreated wood, because archaeological wet wood is very different in appearance to dry wood. This is not a fair comparison. We have to decide whether we want the treated item to appear "as found" or "as original."

Ian Wainwright: Many people in conservation have a very primitive understanding of the meaning of colour; how it is measured and how the ambient light affects it etc. Many who use colour patches to assess changes, don't apparently understand the basic physics of the principles of colour.

Howard Murray: Looking at samples on display here and elsewhere, it seems to me that one of the basic problems of PEG (even with freeze-drying treatments employing PEG) is that heating causes considerable darkening. We avoid heating our wood; it is never allowed to go above 40°C. We have a set of standard samples of fresh seasoned wood with which to compare our treated wood. We try to make the treated wood look like these.

Allen Brownstein: PEG darkens considerably if heated strongly over long periods, particularly in the presence of oxygen. If air can be excluded, or water soluble anti-oxidants used, darkening may be prevented.

Ian Wainwright: I was interested that the aesthetics were not considered here. (Aesthetics is a word I don't like because I don't think most of us know what it means). Are there any comments about this?

Victoria Jenssen: Sometimes wood treated with Arigal C is bleached, did this take place in Schweingruber's work?

Fritz Schweingruber: Arigal C and lyofix bleach the wood. This phenomenon can be prevented if the surface of the wood is treated with (Methyl? ed.) cellulose before polymerisation.

Mary-Lou Florian: Schweingruber stated that he had no difficulty in making species identification microscopically in all cases; I have had a lot of trouble in doing this with wood soaked for long periods in PEG.

Fritz Schweingruber: Persons familiar with the anatomy of modern and subfossil wood have no difficulties in species identification of preserved wood.

Ian Wainwright: Do you find that some of the papers presented here are too "theoretical" and the approaches described not applicable in the real world?

Mary-Lou Florian: We need theoretical frameworks to be able to carry out practical analyses. Certainly a lot of the very involved interpretation may not be achieved practically.

POLYETHYLENE GLYCOL TREATMENTS FOR WATERLOGGED WOOD AT THE CELL LEVEL

Gregory S. Young
Ian N.M. Wainwright
Analytical Research Services Division,
Canadian Conservation Institute,
1030 Innes Road,
Ottawa, Ontario, K1A 0M8

Introduction

The success of conservation treatments in providing dimensional stability to waterlogged wood derives from the interaction between consolidating materials and wood, and on their distribution in the wood at the cellular and ultrastructural levels. The natural dimensional instability of wood is a function of its total water content, particularly the hygroscopically bound moisture in the wood ultrastructure, i.e. within the cell walls. Below the fiber saturation pont (FSP) of wood, swelling occurs to accommodate an increasing moisture content and shrinkage results with loss of this moisture. So the success of a treatment will depend on the extent the consolidating material replaces the water and where replacement occurs in the wood structure. Of particular importance, then, in studying treatments, is the determination of the location of the consolidating materials in the wood. Some materials are expected to bulk cell lumina with solid material and in so doing resist the shrinkage forces that develop as wood dries below the FSP. Other consolidants are expected to curtail these forces by substitution of the bound water in the cell walls. These remain in the wood, during evaporation of the excess water, to keep the cell wall capillary system bulked.

Polyethylene glycol (PEG) could be expected to either bulk the cell lumina or infiltrate the cell walls, or do both. The final location of the PEG depends on its physical properties and these depend on molecular weight (MW). The lower MW PEGs in the range 200 to 600 replace some of the bound water during soaking treatments. With drying it is assumed that they remain in the cell wall. These PEGs impart substantial dimensional stability even at concentrations in the wood, below amounts equivalent to the water content at the FSP. For example, the freeze-drying methods after Ambrose (1,2) employ PEG 400 at 10 and 15 percent in aqueous treatment solutions. (At these concentrations appreciable bulking of the cell lumina is not expected.)

Solid PEGs in the MW range 1000 to 3000 are again expected to infiltrate the ultrastructure but at high concentrations they are expected to bulk the cell lumina as well. PEGs with a MW above 3000 are expected only to resist the shrinkage forces by bulking cell lumina because their solvation sphere size is thought to be too large, preventing infiltration into the cell walls.

An important factor, therefore, in addressing the question of how PEG prevents shrinkage is to determine the degree to which PEGs of different molecular weight occupy the cell wall spaces and how much of the bound water is replaced, or alternatively how effectively the PEG occupies the cell lumina. Previously, locating PEG within the cell wall layers has not been possible and the determination of consolidant infiltration in general has been a complicated task involving wood extraction procedures (3,4), infiltration of mixtures consisting of PEG and silver nitrate for microscopy (5), or studies of wood dimensional change after PEG treatment (6,7).

A new, simple stain has now been developed for the light microscopical differentiation of PEG in wood. Based on cobalt thiocyanate it is apparently selective (8) under the experimental conditions described here. The present paper concerns its use to determine the location of PEG; specifically PEG 400, 3350, and 540 Blend; in transverse sections of treated wood.

Cobalt thiocyanate has been used for the detection of trace amounts of wetting and emulsifying agents in aqueous systems, and these agents include the non-ionic polyoxethylene derivatives (9). For PEG determinations in wood, cobalt thiocyanate is a microscopical stain when dissolved in non-aqueous histological grade cedarwood oil (10). The oil (cedrene and cedrol) is immiscible with both water and PEG. It dissolves the cobalt thiocyanate, but readily gives it up to the PEG in wood sections.

Experimental

(a) The Wood and the Conservation Treat-

ments:

The wood samples were cut from those prepared in a comparative study of popular conservation methods (11). Samples of Populus spp. (aspen, poplar, or cottonwood) were treated with PEG 400 at a 35% concentration in water before freeze-drying, PEG 3350 in either a methanol or t-butanol solvent system, or PEG 540 Blend in water.

For the PEG 400 treatment, the water saturated wood was immersed in a 35% solution and soaked for three months. The solution was kept at room temperature. Afterwards the wood was frozen and freeze-dried at -20°C. The average antishrinkage efficiency (ASE) was 116%; some swelling had occurred.

Treatment with PEG 3350 involved first the dehydration of the wood with either t-butanol or methanol. PEG was introduced initially at a low concentration, but then small increments over several months put the final concentration at 60% w/w. The temperatures of the impregnation baths were 55° to 60°C. The wood was allowed to air dry to constant weight and the resulting average ASE was 45% for the methanol system and 85% for t-butanol.

The wood treated with PEG 540 Blend was placed in water and the PEG concentration raised in small steps over 18 months to 64%. As with PEG 3350, the treated wood was allowed to air dry to constant weight. The averge ASE was 98%.

(b) The Microscopical Preparations:

(i) Sectioning:

Samples of the treated wood were cut as 14 mm diameter plugs obtained by drilling in the wood's longitudinal direction. A slow drilling speed was used to avoid heating, nevertheless some plugs were slightly warm afterwards. Small rectangular blocks approximately 3 mm in the tangential and radial directions were cut from the innermost portions of these plugs with the aid of double-edged razor blades. Transverse freehand sections, of thickness approximately 20 to 30 micrometers, were cut with a fiber microtome (12), and stainless steel surgical blades (13). The flat transverse surface of the wood was raised slightly above the flat steel surface of the microtome and the blade, at an angle of 45°, was drawn across with the cutting edge flush to the surface of the microtome. This produced a thick microscopical section. The PEG 3350 treated sections curled during this procedure but each was flattened with a fine bristle brush before complete excision. The sections were placed into microscope slides and coverslips were applied for protection against dust accumulation.

(ii) Staining and Production of Permanent Microscopical Slides:

Successful staining of PEG with cobalt thiocyanate took place only when the staining chemicals, and the wood sections, were free of excess moisture. Dryness was accomplished by using a glovebox at approximately 4% RH, to precondition the chemicals and wood sections.

The stain was produced by adding approximately 0.2 g of dry ammonium thiocyanate and 0.2 g cobalt nitrate hexahydrate in a test tube containing 10 ml of cedarwood oil. After 10 minutes of mixing (which gave a deep blue colour) and after all particulates had settled, a 1:4 dilution was made with more cedarwood oil. The colour of the cobalt thiocyanate in this working dilution proved to be unstable and faded over 18 hours. Refrigeration slowed the fading, but fresh solutions were needed for each staining operation.

In the dry glovebox dehydrated sections were placed in small screw capped vials containing the stain. These were left overnight at 3°C. At this point either fresh stain could have been substituted and staining continued if the results were deemed unsatisfctory, or the procedure continued towards the production of permanent microscopical slides. Stained sections were cleared with xylene by filling, then partially emptying the vials three or four times. Care was taken to keep the sections immersed at all times and therefore quite anhydrous. Permanent slides were made by quickly transferring the sections, along with a small excess of xylene to a drop of resinous mounting medium on a slide and applying a coverslip. A pair of microforceps and an abundance of Pasteur pipettes were essential in this procedure.

(iii) Microscopical Examinations

Observations and photomicrography were carried out with a Leitz Orthoplan light microscope with Leitz plano-oblectives. For fluorescence microscopy a Leitz vertical fluorescence illuminator (after J.S. Ploem) was used with a HBO 200 watt high pressure mercury vapour lamp as illumination source, a BG 38 red suppression filter, a KP 490 exciter filter, a PK 410 dichroic mirror and a K 515 barrier filter. This combination provided incident illumination with a maximum at 400 nm, (blue).

Results

The staining of the microscopical cross-sections from Populus spp. wood samples treated by each of the conservation methods is shown by pairs of photomicrographs under two light conditions; conventional transmitted light (brightfield) and fluorescence. With conventional illumination a red filter provided enhanced contrast of the blue stain for black-and-white photomicrography. The darker, high contrast photomicrographs are of the natural fluorescence of the wood as observed by fluorescence microscopy. Untreated and unstained wood is shown in **Figure 1.** With transmitted light, **Figure 1a,** the cell walls and lumina appear quite light, and whereas with fluorescence microscopy, **Figure 1b,** the natural fluorescence is exhibited and the full thickness of the cell wall is easily seen. Identical results are obtained for unstained PEG treated wood. PEG has no effect on the fluorescence. Control sections of untreated, air dried wood subjected to the staining procedure, **Figure 2,** retain no stain and appear very similar to the unstained wood shown in **Figure 1.**

The stained, PEG treated sections show different results. the PEG 3350 t-butanol treated sections reveal cell lumina filled with a dark, solid, crystalline complex; **Figure 3a.** In this example the larger pores along a growth increment are completely filled, as are many of the fibers and parenchyma tissue. In contrast with the lumina, the cell walls appear quite light. Furthermore the fluorescence image, **Figure 3b,** displays the full thickness of the cell walls. Changing the solvent in the conservation treatment from t-butanol to methanol had little effect on staining, **Figure 4.** (A difference in the distribution of the solid PEG was apparent. Microscopically, the methanol-PEG treated wood was more thoroughly bulked but areas of wood devoid of the PEG separated bulked portions of growth rings).

The microscopy of the stained PEG 400 treated sections showed contrasting results, **Figure 5.** Cell lumina contain no complex, but the cell walls include dark rings located closest to each cell lumen, **Figure 5a.** These areas were stained blue. The presence of the cobalt thiocyanate stain in the cell walls dramatically inhibits the fluorescence, **Figure 5b,** and the full thickness of the cell walls is no longer apparent. Much of the fluorescence of the secondary wall is lost, but in the middle layers, including the middle lamella and particularly at the junctions of 3 and 4 contacting cells, it is still present.

Since cobalt thiocyanate is selective for polyethylene glycol under the experimental conditons and does not stain the wood alone, these results indicate the presence of PEG 400 within the cell wall capillary matrix. The idea that an inhibition of fluorescence occurs only when the stain is actually incorporated is supported by observations of sections from PEG 3350 treated wood that were stained and then exposed to room atmospheric moisture, which liquified the PEG-stain complex, **Figure 6.** In these, beads of PEG 3350 -cobalt thiocyanate complex adhere to the outside surfaces of the cross sections, **Figure 6a,** but no inhibition of the natural fluorescence of the wood occurs, **Figure 6b.** Therefore, for cobalt thiocyanate to inhibit fluorescence, it must be incorporated into the cell wall; furthermore, for it to be incorporated it is apparent that polyethylene glycol must be present in the wood.

Sections treated with PEG 540 Blend show the presence of crystalline PEG in the cell lumina and of PEG infiltrated into the cell wall, **Figure 7. Figure 7a** also shows the presence of heavily stained partially delaminated G layers in gelatinous fibers. This is new evidence for the open porous nature of these structures. The photomicrograph, in **Figure 7b,** shows a substantial loss of cell wall fluorescence indicating the presence of PEG within, even into the middle lamella.

Discussion

The results indicate that PEG 3350 did not have access to the wood cell wall capillary system when dissolved in t-butanol or methanol, but showed substantial bulking of the cell lumina. PEG 400 did have access, at least into the secondary cell wall, but no bulking of cell lumina was evident. Since the

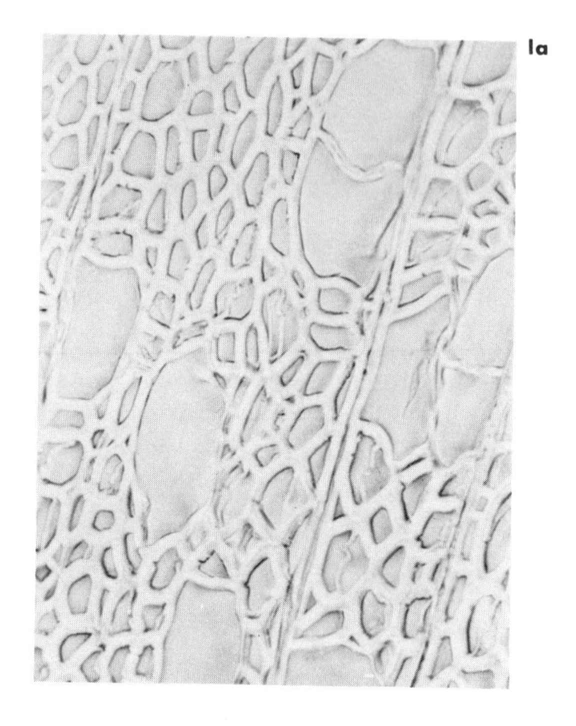

1a

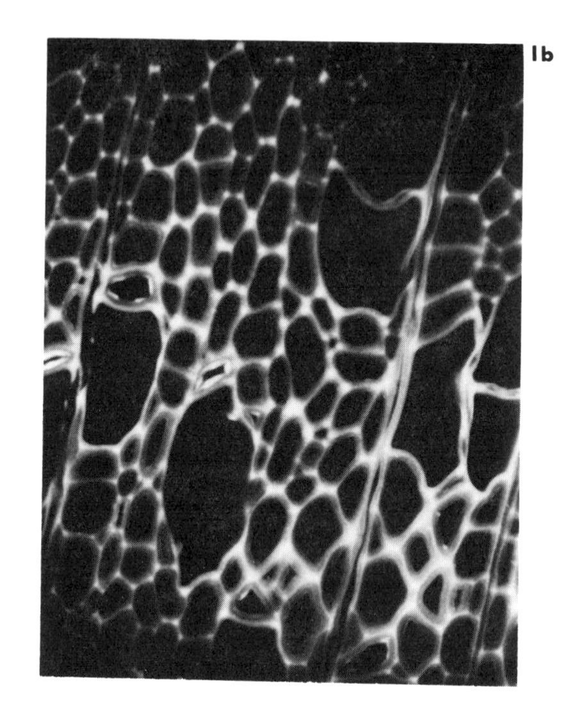

1b

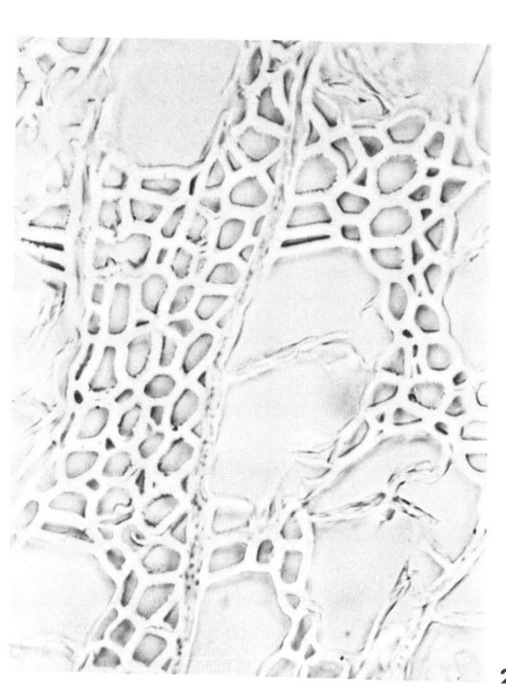

2a

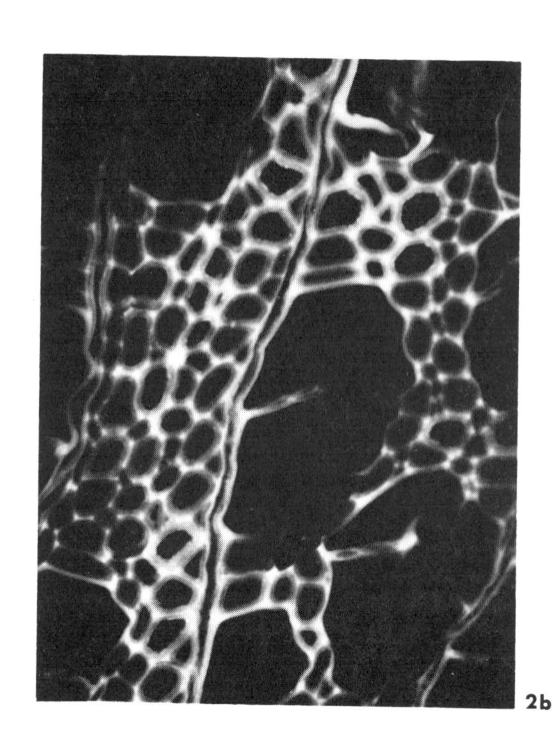

2b

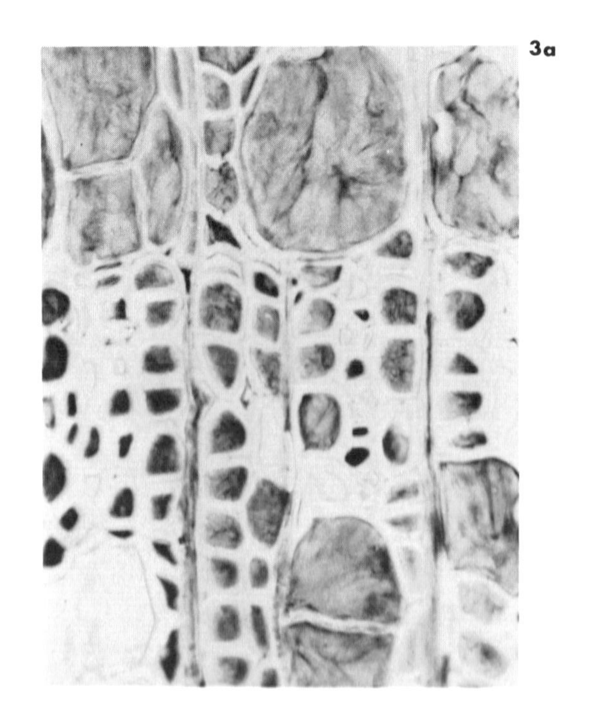
3a

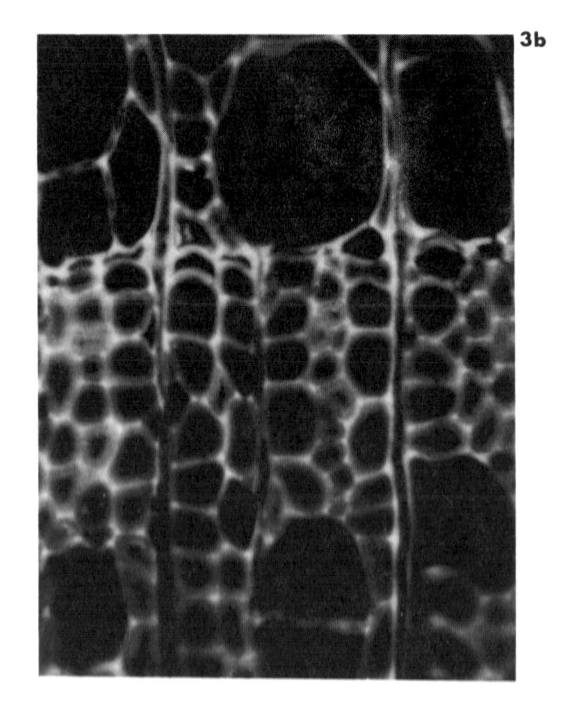
3b

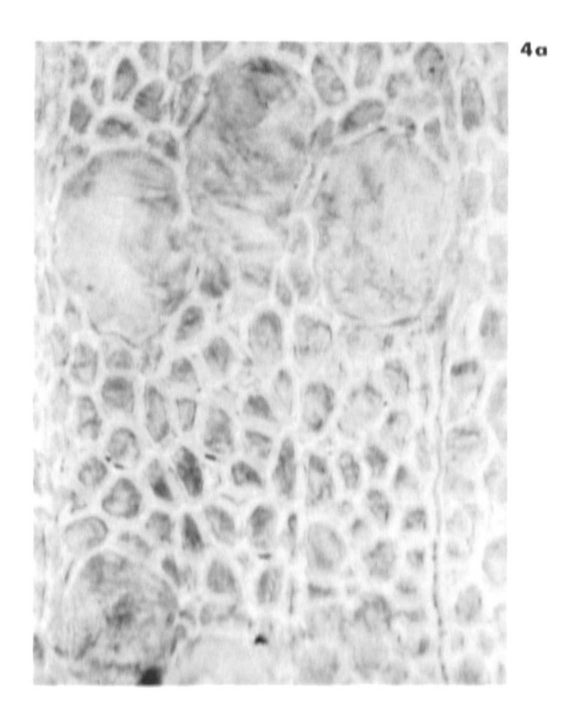
4a

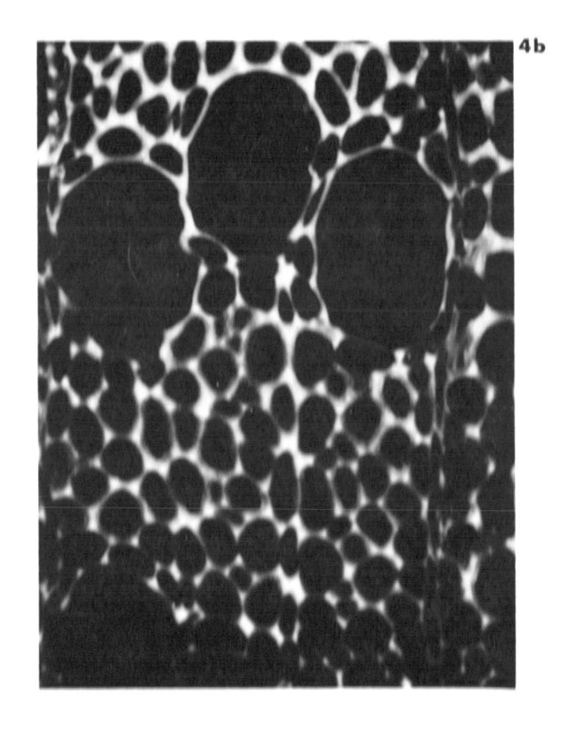
4b

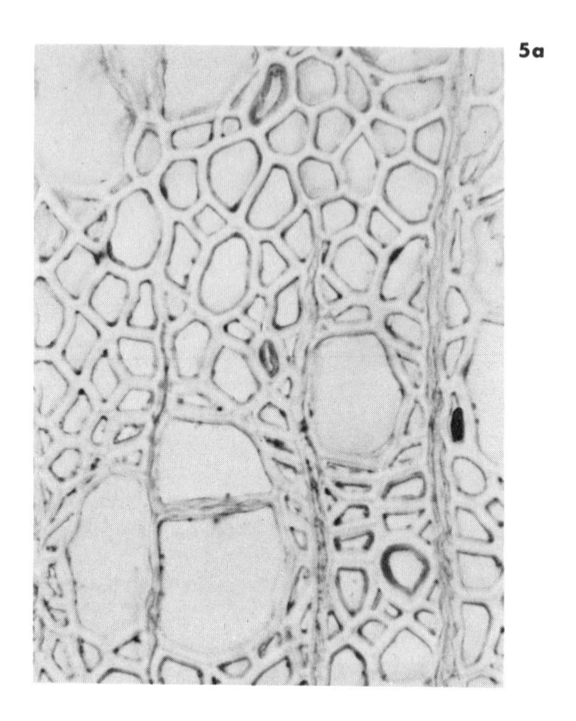
5a

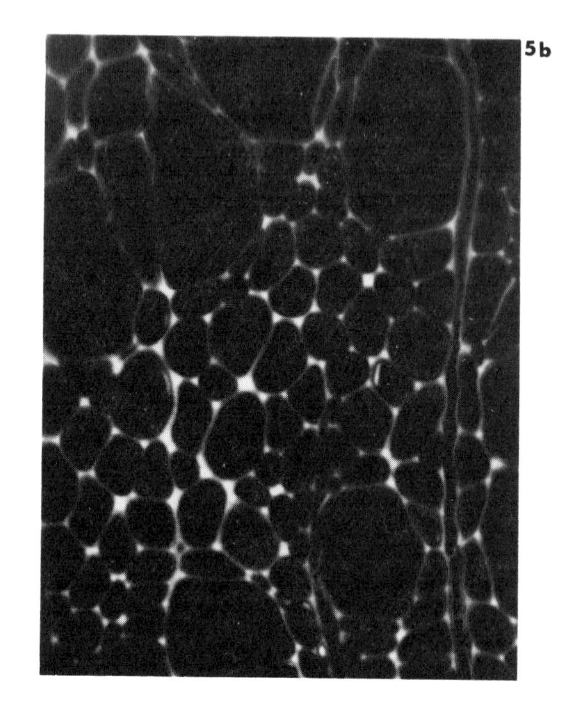
5b

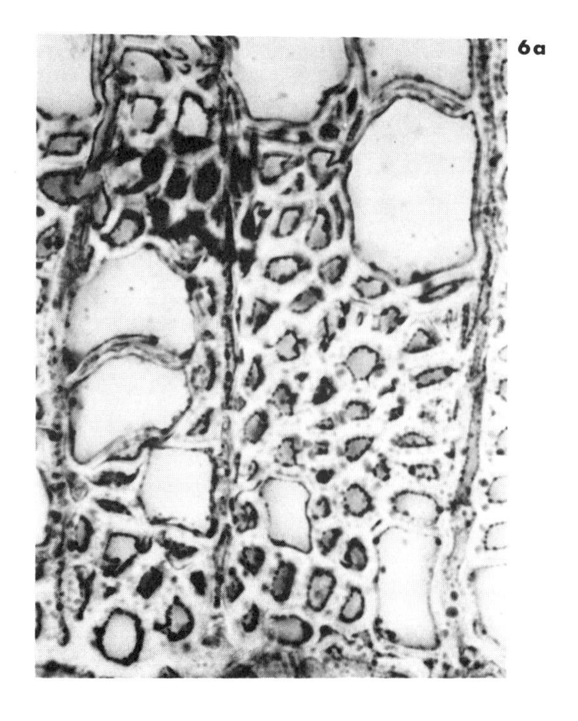
6a

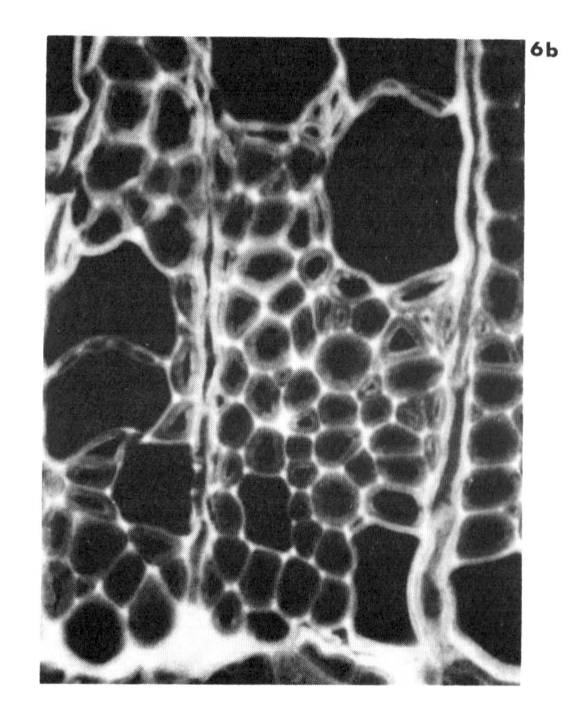
6b

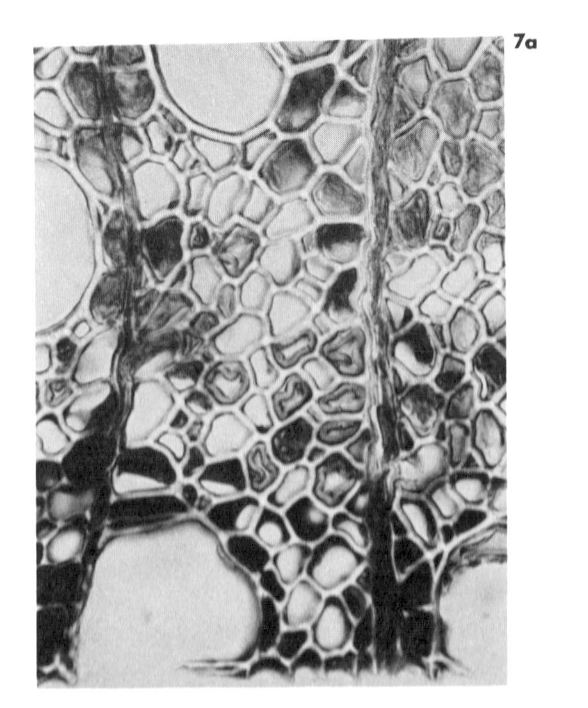
7a

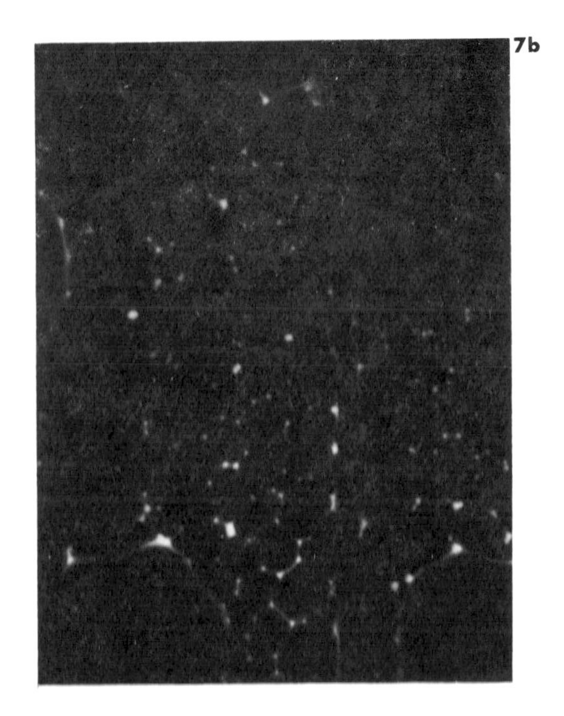
7b

microscopical preparation procedure made no attempt to retain any excess liquid PEG 400, no bulking of the cell spaces could have been expected. PEG 540 Blend treated wood showed both crystalline material filling the cell spaces (probably the 1450 molecular weight fraction), and infiltration into the cell walls. Since PEG 540 Blend is an equal mixture (by weight) of two molecular weight fractions, 1450 and 300 MW, it is not possible to attribute the infiltration to one or the other fraction alone. Both are expected to infiltrate the cell wall.

These observations of PEG distribution corroborate studies that used well fractionated polyethylene glycol to determine that above a molecular weight of 3000 infiltration of the cell wall does not occur into intact wet wood at 22°C (14). In addition, above weight 2000, apparently penetration is progressively more inhibited (4).

Other microscopical studies have used electrolytes (Ag^+, Au^{3+}, Fe^{2+}, $Cr_2O_2^{2-}$, Cu^{2+}) and precipitation reactions to reveal the capillary structure of the cell wall (15, 16, 17, 18, 19). Although research into the number and size of capillaries remains controversial, there is much agreement that the middle lamella of wood is more porous than the other layers of the wall, but that the average pore size is smaller in this area. From the present work, it appears that whereas PEG 300 has access to the middle lamella PEG 400 has inhibited access. There is a greater reduction in fluorescence of the outer wall layers, including the middle lamella, for PEG 540 Blend compared to PEG 400 treated wood, after the same staining time. At present, this infers more accessibility for PEG 540 Blend i.e. the PEG 300 component. However a differential diffusion rate of the stain might exist for the two PEGs, so that a longer staining time may show otherwise. How the thoroughness of bulking of the middle lamella by lower MW PEGs will affect the success of conservation treatments is presently unknown, but it seems that molecular weights below 400 will penetrate better. The bulking of the compound middle lamella is suggested to have little affect on improving dimensional stability (6,20) since PEG 400 shows stabilization very close to the possible maximum. Its effect must be measured in terms of some other variable.

Of particular interest is the fact that staining always initiated at the cell wall lumen surface. Work by Wardrop and Davies (19) supports the contention that the stain is actually staining PEG and not merely absorbing to the wood structure. They applied the Coppick and Fowler (21) stain reagents to the wood blocks which were sectioned afterwards and this produced results similar to those presented in this paper. Conversely, when the wood was stained after sectioning, substantial colouring was produced and was uniform across the cell wall thickness. In the present study, the cobalt thiocyanate stain was applied after cutting the microscopical thick sections, but the PEG was introduced beforehand and, of course, its only access to the cell wall was via the cell lumen.

Variables that govern the staining process include: contact time, temperature, concentration, and water content of both the staining solution and the wood sections. Generally, the longer the contact time the more staining that occurs. Staining is usually carried out over 18 hour periods. At this time no systematic evaluation beyond 24 hours has been performed, so an "end point" has not yet been rigorously determined. As mentioned above, the blue colour of the stain fades after 18 hours and this complicates procedures for extended staining time periods.

Refrigeration during staining helped to slow the fading phenomenon so that a useful solution was maintained throughout the contact time. Unfortunately, the reduced temperature also slowed the staining of the PEG treated wood sections, however a staining rate advantage was definitely provided at the lower temperature.

Greater quantities of cobalt thiocyanate in the stain accelerated staining, but no optimal concentration has been established. Observations of wood sections subjected to concentrations higher than that recommended revealed a dissolution phenomenon of the higher molecular weight PEG's (1450 and 3350). In the limits of this study this effect was undesirable and so was avoided.

Water appears to play an important role in the staining procedure. During developmental work, scrupulous dryness was not maintained. Cobalt nitrate hexahydrate and ammonium thiocyanate, both of which deliquesce, were used directly from their storage bottles. The

PEG treated wood sections were equilibrated to ambient humidity (circa 55%). Under these conditions, a rapid dissolution of PEG 3350 was likely to occur and dissolution always occurred with the 1450 molecular weight in the PEG 540 Blend; even when an appropriate staining concentration was used. Moreover, the staining of the PEG 400 treated wood sections was uniform through the double wall thickness between adjacent cells. Air dried control sections also showed some staining, but not as much as the treated samples. Microscopical observations of the dissolution of stained PEG 3350 implicated atmospheric moisture as the cause of this effect. As a consequence, efforts to isolate the staining sections from sources of moisture were incorporated into the procedure, and these were described herein.

Close observation of PEG 400 treated sections at intervals during staining revealed what appeared to be a loss of stained PEG from the sections. As previously suggested this appeared to be mostly excess liquid PEG from the lumina. It is probably a phenomenon equivalent to the dissolution of PEG 3350 in stain containing moisture. PEG 400 continues to be sensitive to some dissolution in the staining medium.

Other preliminary results suggest that the water of crystallisation in cobalt nitrate hexahydrate, plays no part in this problem. When cobalt nitrate trihydrate was used for staining (produced by subjecting the hexahydrate to 170°C for several hours) no difference was observed in the resulting stain distribution. (Small amounts of water in the cedarwood oil were probably of importance.)

Collaborative work with Grattan (11) has shown the relationship between cell wall infiltration and dimensional stability. The combined results show that polyethylene glycol must substitute the water in the cell walls to provide the best stability. With PEG 400, the highest antishrinkage efficiency (ASE) was achieved when the amount taken up by the wood samples approached the amount of water present at the fiber saturation point. With PEG 3350, where the staining results showed no wall infiltration, this beneficial relationship was not as evident and even high levels of bulking beyond 50% did not provide the ASE achieved with PEG 400 used in soaking baths at 25 and 35 percent. Since freeze-drying alone, with no polymer pretreatment, provides low levels of stabilization, the difference in ASE between high and low molecular weight PEGs is more comparable between PEG 3350 and 540 Blend; where air drying brought the wood to the dry state for both methods. The double bulking, of the cell walls and perhaps 30% of the cell lumina compared with over 50% for PEG 3350, provided the best ASE at 98%. Although a minority of the cell lumina were bulked all of the cell walls showed substantial infiltration. Again the latter was apparently of greater importance in providing dimensional stability.

Acknowledgements

The authors are indebted to David Grattan of Conservation Processes Reseach and to Marilyn Laver, Scott Williams, and Jeremy Powell of Analytical Research Services at the Canadian Conservation Institute and to the CCI for its support. This work was carried out as partial fulfillment of a research contract with CCI.

References

1. Ambrose, W.R., "Stabilizing Degraded Swamp Wood by Freeze-Drying," in: ICOM Committee for Conservation, 4th triennial Meeting, preprints of the conference, (Venice: ICOM, 1975), session 8 paper 4.
2. Ambrose, W.R., "Sublimation Drying of Degraded Wet Wood," in: The Pacific Northwest Wet Site Wood Conservation Conference, proc. of the conference, 19-20 Sept., 1976, (Neah Bay, Washington), pp 7-14.
3. Tarkow, H., W.C. Feist and C.F. Southerland, "Interaction of Wood and Polymeric Materials, Penetration Versus Molecular Size," Forest Products Journal, vol. 16, no. 10, 1966, pp. 61-65.
4. Ishimaru, Y., "Absorption of Polyethylene Glycol on Swollen Wood. I. Molecular Weight Dependence," Journal of the Japan Wood Research Society, vol. 22, no. 1, 1976, pp. 22-28.
5. Furano, T. and T. Gota, "Structure of the Interface Between Wood and Synthetic Polymers V., The Penetration of Polyethylene Glycol into Woody Cell Wall,"

Journal of the Japan Wood Research Society, vol. 20, no. 9, 1974, pp. 446-452.
6. Stamm, A.J., "Dimensional Stabilization of Wood with Carbowaxes," Forest Products Journal, vol. 6, no. 5, 1956, pp. 201-204.
7. Stamm, A.J., "Effects of Polyethylene Glycol on the Dimensional Stability of Wood," Forest Products Journal, vol. 9, no. 10, 1959, pp. 375-381.
8. Brown, E.G. and T.J. Hayes, "The Absorptiometric Determination of Polyethylene Glycol Mono-oleate," Analyst, vol. 80, 1955, pp. 755-767.
9. Rosen, M.J. and H.A. Goldsmith, Systematic Analysis of Surface-Active Agents, (New York: Interscience Publishers, Inc., 1960), pp. 30-32, and 187-193.
10. Walker, G.T., "Cedarwood Oil," Perfumery and Essential Oil Record, May 1968, pp. 347-350.
11. Grattan, D.W., Studies in Conservation, in press.
12. Fiber Microtome, cat. no. 201-A, Mico Instrument Co., Cambridge, Mass., U.S.A.
13. Surgical Blades, no. 11, Feather Industries Ltd., Japan; available from Fisher Scientific Co. Ltd.
14. Aggebrandt, I.G. and O. Samuelson, "Penetration of Water-Soluble Polymers into Cellulose Fibers," Journal of Applied Polymer Science, vol. 8, 1964, pp. 2801-2812.
15. Murmanis, L. and M. Chudnoff, "Lateral Flow in Beech and Birch as Revealed by the Electron Microscope," Wood Science and Technology, vol. 13, 1979, pp. 79-87.
16. Yata, S., J. Mukudai and H. Kajita, "Morphological Studies on the Movement of Substances into the Cell Wall of Wood. 1. Methods for microscopical detection of diffusing substances in the cell wall," Mokuzai Gakkaish, vol. 29, no. 9, 1978, pp. 591-597.
17. Timmons, T.K., J.A. Meyer and W.A. Côté, Jr., "Polymer Location in the Wood-Polymer Composite," Wood Science, vol. 4, no. 1, 1971, pp 13-24.
18. Rudman, P., "Studies in Wood Preservation, Pt. III, The Penetration of the Fine Structure of Wood by Inorganic Solutions, including Wood Preservatives," Holzforschung, vol. 20. no. 2, 1966, pp. 60-67.
19. Wardrop, A.B. and G.W. Davies, "Morphological Factors Relating to the Penetration of Liquids into Wood," Holzforschung, vol. 15, no. 5, 1961, pp. 129-141.
20. Sadoh, T. and H. Urakami, "Rheological Properties of Wood Treated with Polyethylene Glycol. II, Effect of Moisture Content and Moisture Content Change," Journal of the Japan Wood Research Society, vol. 13, no. 8, 1967, pp. 327-330.
21. Coppick, S. and W.F. Fowler, Paper Trades Journal, vol. 109, T.S.: 135, 1939.

GENERAL DISCUSSION PERIOD
SESSION II
ANALYSIS & CLASSIFICATION OF WOOD

Questions concerning Per Hoffmann's Talk

Per Hoffmann to Allen Brownstein: Can you verify the diameters of the various PEG molecules that I calculated? (i.e. PEG 400, 2 x 0.25 nm; PEG 1000, 4.5 x 0.35 nm; PEG 4000, 18 x 0.35 nm). This was in aqueous solution and I considered the lower molecular weight PEG to be in simple chain form, and above about 600 to 1000 that the higher molecular weight molecules were in a zig-zag arrangement.

Allen Brownstein: Such calculations can be misleading. The PEG chain is flexible and is capable of numerous twists and folds, particularly as the molecular weight increases. Our work has verified this, however, most of the work was performed on aqueous solutions of extremely high molecular weight material.

Richard Jagels: One of the reasons why the hemicellulose/cellulose ratio might be unaffected despite the hydrolysis of cellulose during degradation is the possibility of formation of complexes between the hemicellulose and lignin which protect the lignin. Is this a likely explanation?

Per Hoffmann: You would expect the cellulose to be less quickly degraded than the hemicellulose, but that was not so , both were degraded at the same velocity. No, I don't think your explanation is likely.

Richard Jagels: I wonder if the explanation for the equivalent rates of degradation of cellulose and hemicellulose is a physical rather than a chemical effect? In a study that I have recently carried out on the bleaching of fibres with SEM combined with X-ray elemental analysis, I looked at the movement of chloride ion into the cell wall. We obtained X-ray distribution maps of chloride ion in thin section and looked at the change in distribution with bleaching time. Physical effects control the movement of the ion from the surface of the cell wall inwards, and I suspect that the same is true for the factors controlling the degradation of wood, i.e., cellulose is degraded as much as the hemicellulose because the hemicellulose is physically protected from the initial hydrolysis whether chemical or bacterial in origin.

Ian Wainwright: Were the concentrations of chloride ion, sufficiently high as to give an unambiguous interpretation of distribution, and what was the embedding technique?

Richard Jagels: Sections were embedded in a chlorine-free water-soluble methacrylate resin. The pulp was bleached at a 20% chlorine level which allowed us to achieve unambiguous peaks. Lower chlorine levels were difficult to discern from background noise.

Questions to Greg Young

Mary-Lou Florian: Does the fluorescence of cellulose derive from the crystalline structure?

Greg Young: No, it is unrelated to it, I think it is associated with the lignin.

Herman Heikkenen: Have you applied for a process patent, and has the paper presented this morning been copyrighted? Because what you have here is a very brilliant discovery that certainly should be protected, and I really urge you to get matters moving. You are to be congratulated.

Ian Wainwright: Thank you very much. We have not yet addressed these problems, but we intend to publish an article in Nature. We are more interested in distributing the information than pursuing a patent.
(ed. Heikkenen was referring to patent and copyright law as it exists in the U.S.A., Canadian law is very different: a process can be patented up to two years after publication: copyright is also quite different)

Colin Pearson: I want to ask a question about the penetration of different blends of PEG. What would happen is you impregnated with a mixture of PEG 400 and PEG 4000? Would the 400 penetrate the cell walls, leaving the 4000 in the lumina?

Greg Young: I did not try this, and only tested 540 Blend.

Colin Pearson: It seems to be the next logical step.

David Grattan: We have been discussing this and plan to test it.

Suzanne Keene: Suppose you used PEG 1500 instead of PEG 4000 could you get round the crystallization problems?

Greg Young: There is one problem with the staining technique. When cobalt thiocyanate is combined with PEG, the melting point is reduced. This probably results from the complexation of the cobalt salt by the PEG resulting in a perturbation of the properties of PEG, as described here by Allen Brownstein. With molecular weights around 1000, they are barely solid at room temperature, thus I am not really clear about the results with PEG 540 Blend; (which of course contains PEG 1500). I have not done any experiments with PEG 1500 by itself and don't yet know whether this penetrates the cell wall.

Howard Murray: We have experimented with the approach of putting PEG 400 in the centre of an object, and then introducing higher grades until we reached PEG 6000 on the surface. We tried to do this in two ways (i) by adding all grades together contemporaneously and (ii) exposing the object to the different grades in separate solutions one after the other; i.e. going from a bath of PEG 400 to 1540 to 4000 and finally to 6000. Neither worked and in the latter approach, the exposure of the object to a new solution, leached out the previous PEG of a lower grade!

Allen Brownstein: Is there any way in which the staining technique can be quantified? Can you put a photon counter on the microscope and get some information about how much PEG is actually in the wood?

Grey Young: I have been thinking about methods to - in histological terminology -"fix" the material before I continue staining.

Allen Brownstein: As you know, ammonium cobalt thiocyanate forms a blue complex with PEG. The intensity of the colour can be used to determine the PEG concentration.

Katherine Singley: It would not be much use in our brown solutions!

Ian Wainwright: Coming back to the problem of quantitation for thin sections. A quantitative interpretation would depend on section thickness, and thus elaborate sectioning would be required. I doubt this would be cost effective in conservation, but industrially it might be quite useful.

To continue this discussion I would like to raise a number of non-technical points.

I mentioned in the introduction the problem of a common language and was really quite delighted to hear this theme repeated by Dr's Hoffman and Schweingruber, both of whom are requesting more collaboration between laboratories in different parts of the world; particularly transatlantic collaboration. What we have heard this morning is a litany of hundreds of agents of deterioration, complicated by the number of species that can be encountered. I found it somewhat overwhelming and I wonder what you feel about the problem of using and distributing this kind of information at the level of the practising conservators, who in the long run have to do a lot of these examinations themselves; there are few scientific laboratories involved in this sort of work. I think therefore there are needs for literature reviews on deterioration studies and on analytical techniques for waterlogged wood. (We hope to carry out the latter and it should be published within a year)

Colin Pearson: To some extent the newsletter can assist collaboration, for instance it was used to distribute Per Hoffmann's chemical analysis scheme. However I'm not sure I share your view on the distribution of information, because I think that it is available to everyone. It is up to the individual to seek it out.

Ian Wainwright: Not everyone does a very conscientious literature survey, and in our experience it is all too easy to miss material published in other countries.

I'm not saying that we do a bad job: I think we do an excellent job! But is is very easy to miss work in say Scandinavia and which is only published locally. Could I suggest that someone should write reviews for the conservation community with recommendations for wet chemical techniques, or EM, TEM or SEM methods.

The development of techniques for preparing samples of waterlogged wood for microscopy is a very difficult problem, but Greg Young has satisfactorily solved it. It would be a complete waste of time if another laboratory say in Switzerland was to repeat all of that development work. We must make sure that this information is communicated.

Should we be considering the adoption of standards like TAPPI, or certain preparation methods; or for colour change measurement?

Howard Murray: How many people who actually work in conservation labs can actually afford the time to do all these tests? We must produce conserved objects, that is our main function. Most of this analytical work is simply done for academic publications.

Ian Wainwright: That is fair enough, that is the sort of reply I'm looking for. But what about techniques such as free-hand sectioning with conventional biological microscopy using quite conventional stains such as those used by Greg Young or Mary-Lou Florian.

Mary-Lou Florian: Many things can be done, but I think you are optimistic if you hope that you can generalize methods of analysis for such objects. The environment of the wood can vary so much, that it is really wrong to be thinking in terms of general techniques. Even in the case of chemical analysis, one cannot simply take recipes from TAPPI or ASTM and use them on wood that is full of iron or calcium and expect to get lignin and cellulose analyses. Data so produced will not be comparable with anything else. Before we can approach the problem two things must occur:

(i) Conservators must look to becoming materials scientists. The priceless cultural heritage materials with which they are concerned must be stabilized. They must therefore know what the result of treatment is going to be and whether treatment is worth doing at all.

(ii) Many analyses must be carried out on waterlogged wood before treatment. I don't know what are the implications of my analyses. Before I can do this, much hard core research on the effects of treatments must be done. 90 or even 99% of the analytical work that I did may be of no consequence, but until it can be properly interpreted it has to be done.

Ian Wainwright: I was thinking more of simple rather than general methods.

Richard Jagels: I agree partly with Mary-Lou Florian's comments, but I disagree with her basic starting point. If we were to take such a pessimistic view, we wouldn't be building houses out of wood. Despite the variable nature of wood, it hasn't prevented people setting standards for strength and kiln drying schedules. Obviously, we can't account for every single possible variation but I think that one can, by starting some kind of analysis, eventually come up with standards. We must begin with something that is workable. Eventually you will get an improvement in the qualitity of result. Even 20% would be satisfactory. Whatever increase you get is an improvement over a haphazard approach. In the pulp and paper industry when they increase the efficiency of a process by 1/2% they are as happy as can be. Here we are talking about being happy with a perfect solution or we are not happy at all, and I don't agree with this approach. What we have to do is to work towards getting some sort of common analytical methods which are used in all laboratories, so that at least we can begin talking.

Ian Wainwright: What proportion of your results are based on free-hand-section?

Richard Jagels: Everything that I talked about can be done with free-hand section.

Ian Wainwright: There is the barrier of the sample that Mary-Lou Florian referred to.

Richard Jagels: Yes, but if you have enough material for an identification you can do a substantial amount of interpretation.

Herman Heikkenen: The problem you face really reduces to the fact that for the conservation of wood many different areas of study, expertise and skills are required. In practice, the only way to approach these problems is to get a blend of people working together on them, as you have here at the CCI.

Charles Hett: I think perhaps we are being unduly pessimistic. For the sake of the archaeological interpretation a large proportion of the objects have sections taken, to help identification and this may as well be done before treatment. The information that has arisen has been pretty useful. At CCI we have continued to use the approach of Mary-Lou Florian. From the microscopy of the section we always get information on the degradation and other features in addition to the wood identifiction. I am sure that Mary-Lou Florian's estimate of 90% of the analyses being useless is not correct. In practice it is much less than that.

Victoria Jenssen: The approach we are considering is going to be a lot of work for somebody. In the ICOM Metals Working Group, they spent three years filling in forms and came up with much statistical information. Perhaps we should consider doing this now, to be ready in three years time?

Colin Pearson: Its a pity that people don't take the time to do analyses, and spend all their time producing objects.

Per Hoffmann: Nobody asks how much time the analysis takes!

If you are able to remove 1 cc of wood you can measure the following: (1) max. water content (2) dry the sample and get shrinkage (3) one or two of the chemical analyses (4) ash content.

I don't propose that all of my scheme is necessary, my hope is that some people will take it up so that in several years we will get the answer of whether it is worthwhile. Of course, it would be better to use non-destructive methods but I haven't found any. So as Richard Jagels said, we must start with many "entrances" to the problem and then one day perhaps we will be lucky enough to reduce this to one or two tests.

Colin Pearson: Might I suggest that Hoffmann's and the Swiss scheme be taken up by people here and that people get together and decide to cooperate by exchanging wood and trying each other's techniques. Large labs such as CCI should definitely be doing this since they act as service departments for the smaller institutions.

Richard Clarke: Obtaining wood samples underwater is very difficult. We have adopted an apparatus called the pilodyn, devised originally for testing utility poles (Manufactured by Proceq SA, Riesbachstrasse 57, CH-8034 Zurich: ed.). It is just an impact density tester; a blunt pin is driven into the wood under a constant energy spring. Its not very good for objects, but for large ship timbers it seems to be quite good. It is speedy, large timbers can be rapidly surveyed and untrained staff can use it.

Kirsten Jespersen: I have tested the Pilodyn on Dutch and Danish wood, without seeing any significant variation. I'm told it should be calibrated for each wood species, but I don't feel it is very useful for waterlogged wood, at present. To use it there must be some resistance in the wood, if you have a deteriorated piece you will go right through it!

Richard Jagels: I wouldn't give up trying indirect methods such as ultrasonic techniques, or the "shigometer™" (a resistance measuring probe which gives an indirect assessment of extent of decay).

SESSION III
CONSERVATION FOR FROZEN OR WET LAND ARCHAEOLOGICAL SITES

John Coles
Session Chairman

Introduction

John Coles: In the last two days we have heard the views of wood specialists and conservators. We've been talking about qualities of wood, degrees of degradation and the problems of waterlogged wood conservation. And there have been a few murmurings here and there, on the general ignorance on the part of archaeologists in setting up their field or underwater projects and the lack of consultation and understanding about the problems of conservation. I think this is reasonably fair criticism. Today we have a combination of papers by archaeologists and conservators; archaeologist setting out their problems and the possibilities of their wet sites, conservators outlining their solutions to their own particular range of material.

Now we should today try to see a bridge, if not being built, at least being planned between archaeologists and conservation agencies, to allow these two to come together, and I hope today in our discussions that this will be one of our major subjects.

Our common ground is the recognition that in many areas of the world there are, or at least were, in some cases many wet sites; wet in flowing water or in the moisture content of the soil or the peat where organic materials of all kinds can survive. Archaeologists have for many years called for a greater data base on which to build their models about the workings of ancient societies. This greater data base cannot be interpreted as a higher structure, i.e. more and more of the same old evidence. What is required by archaeologists, when they talk about a greater data base, is a broader base which has many more different types of evidence introducing new ranges of evidence that allow archaeologists to develop from them complex or simple models about the workings of ancient societies. And it is the wet sites that are going to provide this broader data base for archaeolgists. Now everywhere in northern latitudes of the world, archaeologists are calling for wet sites, where organic material has survived - about environment, economy and other human activities. The danger is, as I am sure we will be told today, that such sites may not survive long enough for us to retrieve the evidence that we want. The fact remains that in some parts of this northern world, steps are being actively taken now to ensure that funding will be available for the retrieval of this new type of evidence. And this is when archaeologists and conservation experts must come together and plan.

The first paper will touch on this and also talk about the problems which are being posed today about set sites.

THE MANAGEMENT OF WET SITE ARCHAEOLOGICAL RESOURCES

George F. MacDonald
National Museum of Man,
Ottawa, Ontario,
K1A 0M8

Much attention in recent years has been devoted to the management of archaeological resources in both North America and Europe. Management concepts include the compilation of resource inventories; impact assessment of proposed construction on these resources; research design strategies for inter and intra site sampling; conservation and restoration procedures and legislation requirements to protect sites. All of these questions have been thoroughly explored as they relate to traditional archaeological site resources, that is, dryland sites and underwater sites.

Water saturated or wet sites, it appears, are rarely included in any of these resource management considerations. It seems that there is a conceptual problem associated with all of the terms "wet-site, wetlands, waterlogged or water-saturated site." Each one conjures up images of diving suits and aqua-lungs, the accoutrements of underwater archaeology, even in the minds of fellow archaeologists, who have not had direct contact with our area of special concern. It hangs in limbo between the traditional concepts of dry land and underwater archaeology. The reason that most archaeologists hesitate to support or engage in wet-site archaeology is possibly rooted in the differences in techniques that do exist between dry land and wet-site archaeology. North American archaeologists have been trained for the variety of problems encountered in dry-sites, but are not trained, in most cases, to work, or even recognize, wet-site potential. The equipment difference between wet and dry-site archaeology is intimidating to them. Dry-site archaeologists are used to very simple tools, like trowels, paint brushes and shovels. Rescue archaeology has brought general familiarity with the grader and back hoe, but these machines come with their own operators and maintenance personnel through local contractors. Wet-site archaeology requires considerably more mechanical abilities than most archaeologists possess, to run and service small engines, build settling tanks and cope with numerous problems of hydraulic engineering.

Furthermore, there is a widespread fear I detect among archaeologists in North America, to open up a wet-site and face the moral responsibility for the extraction and post excavation care of the artifacts and cultural features. There has even been a moratorium on excavating wet-sites in some areas of Canada in the belief that we are not yet equiped to handle them properly. What appears to be an outstanding wet-site near Port Coquitlam on the Fraser River, B.C., was encountered during salvage excavations for a highway but was resealed to protect it. The problem remains, however, that highway and other construction threaten to alter the delicate balance of the water table that might result in the disappearance of the artifacts if the deposit becomes aerobic. Another problem is that no archaeologist, to my knowledge, ever puts enough into their budget to cover the high cost of preserving the contents of a wet-site. Finding a rich wet-site condemns the excavator to an endless search for money, first to equip a laboratory, since no services are available for handling whole site collections, then for the payroll for staff, and finally to cover the high cost of permanent display or storage of the resultant finds.

Much pioneering has to be done by the archaeologist who chooses to venture into this field. He or she will be obliged to work out new sampling and survey strategies for their region. The search for crew with some prior experience in wet-site work will begin; new sources of project funding have to be explored, new equipment and procedures for their use have to be adapated to that region. Field laboratories have to be set up and staffed on a much more elaborate scale than is necessary for dry-sites.

Post excavation requirements are another story - when the fledgling wet-site archaeologist has finished coping with all of these problems, he finds that he has in his charge a collection of exotic items for which he has no comparative material from his region. The uniqueness of the material recovered is, of course, part of the attractiveness of wet-site archaeology, but it can also provide some serious problems to the archaeologist when it comes to cross-linking with existing collections for the area. It takes a special type of researcher to want to open this Pandora's box. In my view, however, the rewards of opening the box can be commensurate with the problems encountered. Much of world prehistory has been at least sketched out in terms of lithic and/or ceramic technologies. The new frontier of lithic technology studies are now glimpsed through the lens of an electron microscope. While traditional studies of prehistoric material culture are becoming ever more myopic, or at least microscopic, the potential of organic material culture from wet-sites is as yet still unplumbed. What, for example, is the prehistory of masking complexes in prehistoric North America? What kind of baskets or other containers were used by the Paleo-Indians who made fluted points? Was the woodworking of Adena and Hopewell of comparable style and quality as their work in mica, copper and stone? What great carved monuments of the northwest Coast Indians have yet to be found in wet-sites on the Queen Charlotte Islands? The list could be expanded infinitely.

At the present time the attitude of the public to wet-site archaoelogy is completely the reverse of that of the professional archaeologist. In my experience it is the public who is anxious to proceed with this kind of archaeology. Wet-site collections and structures can make very interesting displays and on-site interpretations. There has been a noticeable trend to return the artifacts, after analysis and conservation, to an interpretive centre or museum near the site. The public is particularly interested in seeing history as close as possible to the spot where it happened. Popular books, museum exhibits and on-site interpretations must link together in this regard, and the archaeologist interested in this field should pay more than lip service to these public responsibilities, since wet-site archaeology is very expensive and it is the public that ultimately decides if the cost has been worth it. It may be the wet and frozen site archaeologists who save the neck, in terms of public support, of those archaeologists in North America who have retreated to the computer and the microscope, and whose results are totally incomprehensible to the public.

By far the major portion of money spent on archaeology in the U.S. and Canada come from funds that are legislated by Federal or Regional jurisdictions for mitigation work associated with construction and development projects. The totals of these funds worldwide is staggering. In terms of Canadian dollars it is tens of millions, for many European countries; more than 50 million for Britain and France, over 100 million for West Germany and Japan. Canada is not in this league at all, but the U.S. is in the high stakes range, at more than $100 million/year for rescue archaeology. A truly disturbing fact is that only a miniscule portion of this money, in the U.S. for example, is going into wet-site archaeology, despite its potential yield in terms of information and interpretive materials of great public interest. It is not that wet-sites are rare, it is due mainly to the fact that wet-sites are missed in the surveys, both in the past, and even at present.

A case in point is the mammoth archaeological mitigation project on the Tombigbee River in the Southeastern U.S. The U.S. Corps of Army Engineers is constructing a large canal from Memphis, Tennessee to the Gulf of Mexico, at an estimated cost of more than four billion dollars. To date some sixteen million dollars have been spent on archaeological surveys and excavations. None of this money has been dedicated to wet-site archaeology, despite the great potential interest of wet-site finds from the Mississippi culture, which was the highest cultural phase attained in the prehistoric United States, and about which, little is known of the expressive culture beyond cemetaries and post moulds. A single water-logged canoe find was made by a dragline operator and reported to one of the project teams which took protective action.

The most significant point in regard to the Tombigbee project involving researchers from a number of different universities in the south, is that the detailed strategies for site sampling throughout the impacted zone, the lagoons and abandoned meander channels of the Tombigbee River where wet-sites are certain to be located, are considered as having no archaeological potential. In this particular example, the money the preservation environment encountered in the wetlands, and the cultural potential of the area, are all at a very high level, and yet we are likely to learn absolutely nothing about perishable material culture from this major transect of the southeast U.S. As far as I am aware, there are no plans to revise the strategy of the project despite the fact that wet-site archaeology symposia and conferences, like this one today, have been a feature of the North American archaeology scene for over a decade.

The problem of site survey strategies ignoring wet-site possibilities is endemic throughout North America. To my knowledge, there has never been a survey designed specifically to locate wet-sites anywhere in northeastern North America. Neither has there been any major excavation of a prehistoric wet-site in this vast area which includes six Canadian provinces and 18 States of the Union. The wet-site survey which Barbara Purdy and I conducted in Florida last year, was the first one in the southeastern U.S. Barbara will present the results of that survey later in this session. Dale Croes of Washington State University organized the first wet-site survey in the northwest, also last year, in Washington State. Attention must be focused on two priorities if wet-site archaeology is to flourish. The first is for archaeologists who are already convinced of the potential information value of wet-sites to conduct regional, state or province wide surveys specifically to locate such sites. This is the most efficient type of survey since all the equipment and expertise of wet-sites can be focused exclusively on this kind of site. There is some evidence that state and provincial authorities are willing to fund this kind of survey.

Secondly, there is the need to sensitize colleagues who have no experience in wet-sites at least to look for and record them during their own regular site surveys. Offering them advice at all stages of their survey is an important service that we can provide to them.

The major objective of any resource management plan is to identify and to inventory the resource in question. Throughout North America and Europe, our wet-site finds come to us as the result of someone else's activity. Peat cutters or dredge operators are not paid to find archaeological sites, and for most of them, stopping their equipment or notifying proper authorities of chance finds is only a costly nuisance. I have noted in my own wet-site work in the northwest and southeast areas, that heavy equipment can crush organic materials into unrecognizeable fragments even before they are tipped out onto a spoil pile. Movement of water-saturated soils by machines generally creates a mud slurry at the working face that often coats and obscures artifacts or features. I had an experienced wet-site crew at the Lachane site in Prince Rupert, B.C. that followed the bulldozers and backhoes as they destroyed a known wet-site. This site had produced over 500 artifacts in a six week salvage excavation before the construction contract commenced, but not a single organic artifact was found during the construction phase that followed. I returned to the site a few weeks after the bull-dozer departed, when the rains had exposed fragments of a half dozen wooden artifacts from site areas that we had not had time to excavate. I feel it is next to useless for us to sit back and hope for wet-site reports from machine operators or even from archaeological monitoring crews during construction on potential wet-sites. They simply do not notice anything to report.

There is a need to prove that wet-sites are not nearly as rare as we might suspect from the small number of wet-sites in the inventories at present. This imbalance between wet and dry land sites has now been recognized and should be quickly filled. This can only be done by legislation, and effective legislation proposals can only be made when we have characterized more accurately the resource we propose to manage through legislation. It will also be necessary, once wet-sites have been recorded through specialized survey, to establish priorities for successive surveys to cover the temporal range as well as the regional one. This becomes an important factor in management when decisions are to be made of which wet-sites most warrant excavation (and expenditure of funds).

Along with the need for developing survey strategies, there is also the need to develop site testing procedues. This includes both remote sensing as well as direct testing. Direct tests are the simplest, with coring and augering devices. An earth auger is appropriate for extensive testing on shallow sites but, as we saw yesterday at the Roebuck site, it is hard to pull out over 1 1/2-2 metres. Mechaniclly aided coring devices are available for deeper sites. In the summer of 1981 I tried out such a device on a frozen house site at

Point Barrow on a project of a colleague, Ed. Hall. When working well, a completely intact frozen core could be removed, but when gravel layers were encountered the drill bits, which cost $150. each, deteriorated rapidly. Probes that can be sunk into the site to measure humidity and decay factors are also useful in a survey.

Another point in terms of resource management of wet-sites that warrants further comment, concerns a current reaction from resource management archaeologists with state/provincial agencies, that wet-sites must be somehow isolated and mothballed due to the lack of local expertise or funds. Here some recent observations are relevant. Wet-sites found in urban areas may be impossible to mothball for the future. If development in the general area progresses, the water table may be reduced by drainage systems and wells to the point that the site in question is ultimately affected. Once the water table has been lowered the site is doomed to destruction in a relatively short period. Wet-sites are destroyed underground by silent agents like bacteria, without anyone's knowledge. Dr. Purdy and I found evidence of this at the sites Gordon Willey reported from the Works and Progress Administration days, at the time of the depression, around Lake Okechobee. Willey reported layers of organic muck with preserved wooden artifacts at the Belle Glade site, but felt at the time he was not prepared to excavate into that layer. When he relocated the layer in auger cores at the site last year, it had been reduced to a structureless black lens. This was due to the draining of surrounding soils for agricultural purposes that lowered the water table by twenty feet and more and caused soil collapse of up to eight feet around basements of houses. Confirming evidence from John Coles in the Somerset Levels project indicates that drainage of wet deposits can introduce aerobic bacteria that will destroy organic structures in as little as 2-5 years. Soil scientists in Florida maintain that destruction of organic structures in peat can occur in zones up to several miles on either side of drainage canals. One only has to look at the maps of southern Florida to realize that wet-sites with organic preservation have been obliterated over vast areas of the central and southern part of the States. Only the Everglades remain in Florida as a major area of water saturated soils that has not been drained for agricultural purposes.

In addition to the threat on a massive scale to wet-site preservation by agricultural drainage programs, other equally extensive threats exist. With the current energy crisis throughout the world, peat is becoming an increasingly desireable fuel. Much of Northern Europe contains peat bogs that have yielded outstanding individual finds and sites with excellent preservation in the past, but as I am sure John Coles will tell us, in the vast tract of the Somerset Levels, the peat cutters are posing a major long term threat, despite their co-operation in the rescue work.

Canada has vast peat resources in the Atlantic Region, on the shield and on the Northwest Coast that are being eyed as future fuel sources. There is some legislation protection for the waterfowl and fauna of the wetlands, but none against the destruction of archaeological wet-site resources threatened by drainage plans. Sites in Florida peat bogs are doubly threatened by mining peat for agricultural and horticultural purposes, as well as the fuel threat. Over thirty dug-out canoes were recorded from bogs and lakes in Florida in a single survey by Barbara Purdy a few years ago.

The dredging of waterways is also posing an increasing threat to wet sites. Dredging for navigation purposes has been around for a long time. Dredging of a ferry crossing on the St. John's River, Florida, brought up house posts that led to the discovery of a sizeable wet-site in a nearby lagoon. A relatively new factor, however, is the dredging of shallow waterways for pleasure craft. This includes deepening and opening up new channels and short cuts, straightening waterways, building marinas and installing various navigation aids and amenities. Mitigation archaeology is usually considered unnecessary in these cirucmstances due to the general ignorance concerning wetlands site potential.

On a larger scale, recent port development has been extensive, at least in Canada, to accommodate ever larger bulk carriers. Several super ports have been built in B.C. and others are planned in many coastal areas of North America. They can affect underwater sites, dryland sites and very frequently water saturated sites adjacent to the original shoreline. A simple coring program on proposed port sites would reveal buried wet zones of cultural material in advance of construction, but this is rarely done. If material is discovered, it is in the jaws of the earthmovers, but most often it is destroyed and never recognized. In Canada pieces of existing legislation to which regulations to protect wet-site resources should be appended, include the Navigable Waterways Act, the National Energy Act, and various pieces of environmental legislation protecting marshlands and scenic waterways. The need for comprehensive legislation providing funds for archaeological mitigation tied to federal development programs is a particular need in Canada. The U.S. already has such legislation on the federal level, although with widespread budget cutbacks of the present administration there is much concern among U.S. archaeologists for the future.

It was surprising to me in the opening remarks of our conference to hear someone complain that archaeologists had become over zealous and were taking on too many wet-sites for conservators to handle. This is certainly not the case in Canada. There has not been a prehistoric wet-site excavation anywhere in the country for close to a decade, although several frozen sites in the Arctic were excavated during the period that kept conservators occupied. Since the level of construction and drainage projects was high during that time, I am positive that many wet-sites were destroyed without our knowledge.

On a world-wide level there have been a considerable number of wet-sites excavated in the past ten years, and new areas have been found to contain high potential for wet-sites. Many areas that have not yet been tested that are of high potential include the Amazon, Niger, Yucatan, Amur River Basin, to name a few.

The next decade, I would predict, will see another surge in wet-site excavations. Much of this will be mitigation or rescue excavations since large scale projects, particularly in the energy resource field, are already scheduled to happen. The challenge to the archaeologist is to find these sites before they are destroyed, and to improve the record of the proportion of the site that is salvaged before the inevitable construction begins on-site. The challenge to the conservators is to be able to cope with the amount of material that comes in. Speaking for my colleagues, I think there is general satisfaction with the treatment procedures, although improvements can be made, but it would be the wish of most archaeologist that bulk processing facilities be developed to handle the multitude of small funds from wet-sites. At present however, the archaeologist would prefer improvements in the quantitative rather than qualitative conservation techniques.

Conclusion

To recap the main points I have raised about the management of wet-sites, I see the first task to be the improvement of site detection skills, particularly through the use of remote sensing techniques. With improved skills thorough regional surveys will be facilitated. Knowing the extent of the resource is the first step of a good management plan for a particular area. Setting priorities on where the staff and financial resources should go when mitigation work is necessary is then an easy step. Anticipating increased need in the near future for rescue archaeology on wet-sites because of energy and other development schemes; legislation is required that will provide the substantial funds that are needed for the recovery and preservation of material from wet-sites. Tax reforms are needed in most countries to encourage donation of wet lands for long term preservation purposes. I would suggest that the newsletter of the WWWG devote some space to the international needs for special legislation to protect wet-site resources.

Finally, I would stress the responsibility of both excavators and conservators in providing maximum satisfaction to the public who provide the funds for their work. The public is interested in what wet-sites can reveal to them, particularly about prehistory, but will not support work indefinitely without feedback on levels they can appreciate. As Graham Clarke so frequently said, "There is no free lunch in archaeology," it is all paid for by the public. I think the route we must follow is clear, and that efforts of the archaoelogists and conservators must be well co-ordinated in regard to wet-site archaeology over the next decade. Much of our knowledge of prehistory around the world will probably be substantially rewritten during that time, and in much finer detail than before. It is our obligation to put some flesh on the dry bones of archaeology through careful management of wet-sites and resources.

Comments and Introduction to John Cole's Paper

John Coles: I think that you have set out very clearly the potential of and the threat to wetland sites, and the potential for conservators in the future. I would only take issue with your optimism about site detection. Perhaps we can pursue this in discussion later. I doubt very much whether any machinery or expertise that we possess at the moment will allow us to detect those important sites in sufficient time to save them, but this is obviously a field for research. This seems to me to be a good follow-on from Dr. MacDonald's talk. He talked in general terms of the world-wide requirement on the part of archaeologists, for wet-land archaeology: some of the problems, costs, detection, ignorance on the part of archaeologists, priorities for funding and conservation. What I can give you this morning is a short talk about one specific example where many of the things that Dr. MacDonald was talking about can be seen to have happened and be happening. So the paper is called "The Somerset Levels, a Waterlogged Landscape," but it would be better entitled now "The Somerset Levels, the Death of a Landscape." You can judge for yourself.

THE SOMERSET LEVELS: A WATERLOGGED LANDSCAPE

John M. Coles
Department of Archaeology,
University of Cambridge
Cambridge, England

The Somerset Levels are a unique landscape in southern England and archaeological, environmental and conservation studies have all been developed in response to the unusual opportunities offered to scientists. The area measures 20 km x 40 km overall and today is dissected by slow-moving rivers and streams which flow around and between the many islands of calcareous rock and Late Pleistocene sands. Formerly an inlet of the sea, bounded by the Mendip and the Quantock Hills, its basal deposits consist of flooding marine clays and silts, interspersed with peat-beds formed during episodes of low sea-level or localized ponding. By 4500 BC the pattern of drainage had altered and the Levels became an estuarine swamp colonized by reeds. Freshwater from the hills and rivers gradually replaced the brackish water, and the flooded landscape began to accumulate mattresses of peat formed by the undecayed plant communities which grew and fell annually into the swamp and marshlands. In the natural succession of such an environment, the peats built up thick deposits which in places approached the highest seasonal water levels, and here a few birch and alder trees could colonize and form small areas of fenwood. By 3500 BC, the bulk of the reed swamps had been replaced by fenwood which created a tangled and still wet landscape between the hills and islands. Soon afterwards, climatic change led to increased rainfall and the consequent flooding drowned the fenwood which fell into the rapidly-developing peatbog which now provided a base for a bog flora of Sphagnum moss, heather and cotton grass. The rapid formation of this type of raised bog built up great thicknesses of peat, as much as 10 m in places, until AD 400 when conditions altered sufficiently to halt any further peat formation. During the period of raised bog formation, c. 3000 years, the bleak moors and marshlands had been virtually treeless, consisting of pools and hummocks with slow-moving streams ineffectually attempting to drain excess water. Seasonally flooded, from early autumn to early summer, the landscape would have presented an unlikely area for human habitation, yet the archaeological record of occupation in and around the Levels is extensive.

Archaeologists have worked in the Levels for many years and have attempted to explain the presence of man in terms of the opportunities offered by the environment and its natural evolution. A battery of scientific disciplines has allowed the details of this evolution to be revealed, and historic documentation of early forms of land-use permits a model of early exploitation to be presented. This model demonstrates that the 'concave landscape' of hills, downslopes, and marshlands offered a variety of resources based on differing environments and seasonal alterations (Coles 1978). The hills presented untouched dry woodland, and animals for the hunt; the upper slopes offered woodland management opportunities and rough forage for pigs; the downslopes provided fertile areas for cultivation; the base-slopes yielded a marshland-pasture interface for summer grazing and winter regeneration; and the water-logged peat moors supported a wide range of wild plants and animals for seasonal gathering. The key area was the water-meadows at the junction of slope and moor where herds and flocks could be fed and fattened in the summer; in medieval and recent times such common grazing attracted communities from all over south-west England, and winter hill-grazing and stubble-grazing were mere holding operations until the great meadows became dry enough to accept vast groups of animals. Such a model can help explain the presence of prehistoric communities in the area from c. 1000 BC when conditions began to assume those of medieval times. At earlier periods, conditions were different and the attraction of woodland, fertile sand islands, and wild resources were probably the major reasons for the presence of man in the centuries from c. 4000 BC.

In medieval times, the Levels were first attacked by their most serious threat to existence, drainage (Williams 1970). Monastic attempts to drain areas of the moor were rather feeble and inconclusive, but since then more and more effort has been put into the problems of water removal, and today vast stretches of channels, canals, and reservoirs are creating conditions where no wetland life can long survive. Peat deposits in part of the Levels have been cut, for burning and for fertilizer, since AD 1100 and this activity has also altered significant parts of the areas. The removal of peat exposes older and deeper peats and in these there often occur traces of human activity, a broken pottery vessel, an unretrieved arrow, an abandoned dugout, a forgotten hoard of bronze or flint tools, a stone axe dropped into the marsh, a hunting platform, a wooden trackway overwhelmed by water and then by peat, or a settlement abandoned by renewed floods or other pressures **(Figure 1)**. Each of these artifacts represents a human action held intact for us to uncover and interpret. Because the overwhelming peat was soft, these artifacts remain in their original form, in three dimensions, always in situ and in association with their contemporary local environment, and they therefore can cast unusual light upon ancient societies whose activities are normally only known from two-dimensional dryland sites where most organic material has been lost by decay. The acid nature of the peat has in some cases destroyed bone, but not always, and plant remains in particular are exceptionally well-preserved. Here of course is the point where conservation and recording assumes particular importance.

It is not necessary, nor is it appropriate, to describe the prehistory of the Somerset Levels here. To do so would require many volumes because the discoveries have been so abundant and because their explanation involves a full discussion of the essential and related studies of pollen, plant macrofossils, tree-rings, coleoptera, woodland management, and wood technology, as well as historic documentations and present land exploitations (Coles & Orme 1980). Instead, two sites will be briefly described and the problems of conservation assessed for each It must be remembered that these sites provide only a sample, and a small one, of the problems faced by conservationists in the Somerset Levels. Another site, of c. 1300 BC, has yielded well over 100,000 wooden artifacts, mostly of slender roundwood with axe-marks; of these only c. 500 have been saved for future study, because of the problems of conservation facilities.

All of the work described here has been carried out by the Somerset Levels Project, an archaeological research unit funded by the Department of the Environment, and the Universities of Cambridge and Exeter. The Project has a small staff of a field archaeologist, an environmentalist (pollen), and a research assistant. Contract work for specialists exists in several other universities and institutes, and student volunteers play an important role in seasonal fieldwork and excavations. The Project is directed by university officers without cost to the unit.

The earliest known structure in the peat of the Levels is an Early Neolithic wooden trackway constructed c. 4000 BC. At this time, the great swamps supported Phragmites, and the slopes and hills held unbroken stands of dryland trees such as ash, oak, lime, and hazel. Into this varied landscape came some of the first farmers, recently arrived in the British Isles, and they soon began to clear the woodland to make fields for their crops, to use the seasonal meadows for pasture and to exploit the wild resources of the swamps. In an effort to link settlements and fields on the hills and on one of the major islands, they constructed an ingenious trackway through the swamp for c. 2 km. This structure was first discovered in 1970 by a peat-cutter called Ray Sweet, hence its name the Sweet Track, and it has undergone excavation in several sectors (most recent publications: Coles & Orme 1979, 1981). It offers archaeologists a unique chance to see how and why Neolithic communities exploited their immediate landscape; nowhere else in Britain has such an abundance of evidence survived for 6000 years.

The track was built in a swamp where floodwater was deep, where the drowned vegetation base was soft and soggy, and where tall reeds and sedges provided thick protection for the route. Upon the marsh the builders laid long poles made from ash, oak and hazel tree trunks felled on the hills and slopes of the Poldens; these poles or rails were held in place by hundreds of pegs, of hazel, holly, alder, ash and elm, driven in obliquely on either side of the rail. Around and upon the rails were packed blocks of peat, and pads of vegetation, to form a raised base or platform about 20 cm or more above the soft marsh surface. Heavy planks of oak, ash and lime were then positioned on this platform, and held in place by the criss-crossed pegs and by new vertical pegs driven through holes or notches in the planks. The result was a narrow but firm structure, a

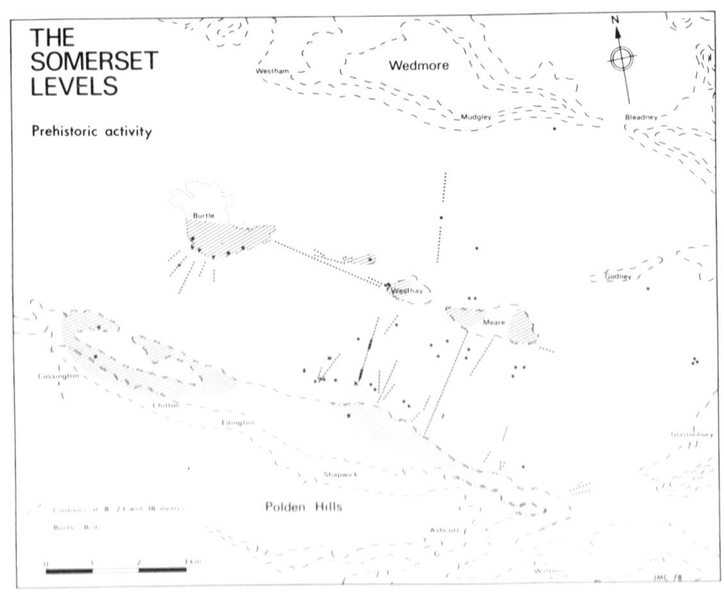

Figure 1. The northern part of the Somerset Levels, England, with some prehistoric sites and finds marked by dots, wooden trackways by dotted lines, and areas of activity (clearances, arable cultivation and pasture land grazing) hatched. The time-range represented here is c 4000-200 BC. The major area of discoveries falls within current peat-cutting extraction permits, and it is likely that equivalent amounts of information would emerge if extraction extended into other sectors.

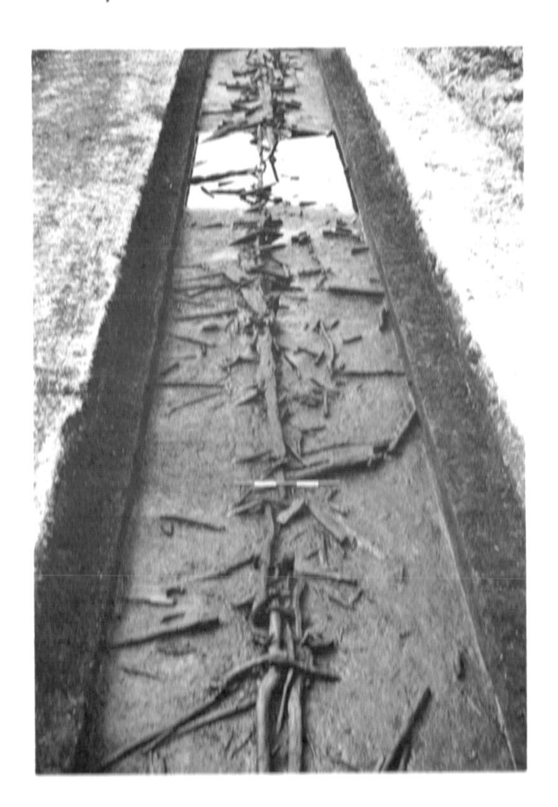

Figure 2: The Sweet Track, an Early Neolithic structure c 4000 BC. The components of the track consist of roundwood rails, held in place by criss-crossed pegs, supporting long narrow planks (the plank walk) anchored by thin pegs driven through notches and holes. In the foreground, the planks have been washed off the substructure; in the middle distance, the planks are still in situ; beyond, a shallow pool had formed and the track timbers have sunk along with much Neolithic debris. Besides the track lies other debris, including notched boards and unused pegs. At intervals, heavy posts were driven into the marsh during the setting-out of the structure alignment: one post can be seen in front of the scale, another is between the scale and the pool. The scale is one metre. All photos JMC.

plank walk, raised above the marsh surface, marked by numerous pegs which projected 10 cm above the walking surface (**Figure 2**). The amount of woodwork involved in this track is very considerable, and the workmanship is impressive. Oak trees were felled, by fire and axe, and split with wooden wedges and mallets to make uniform planks. Hazel and holly were selected for evenness of growth and long pegs were chopped out and sharpened by stone axes; perforations and notches in planks and rails were similarly cut. All of the timber and roundwood was collected from hills, slopes and swamp edges, and brought to the site, perhaps by raft.

This work was clearly well-organized, and archaeologists have been fortunate in finding some traces of the preliminary stages of the task. Neolithic surveyors determined the line of the track, cleared away the reeds, and laid down a preliminary track to give easier access and working conditions; this temporary track was made of huge planks of ash and lime held only roughly in position, and large posts of hazel were driven into the marsh at regular intervals, possibly to secure the rafts of building materials and perhaps also to support a handline or guideline of rope for this first penetration into the swamp.

By analysis of the annual growth rings of the oak trees used to make the planks of the Sweet Track, and the temporary track, we know that the whole enterprise was conceived and executed over one short period of time, and not over a period measured in decades (Morgan 1979). We also know that only a few best-quality oaks were selected for the planks, and that several different stands of timber were being attacked by the builders, perhaps one on the Polden Hills and one on the Westhay island. This structure is not the product of a group just arrived in the Levels; it must represent a permanently-based community with manpower and time and organization sufficient to plan and carry out a major undertaking.

Movement along the track is shown by the abundance of Neolithic artifacts dropped or abandoned along its length. Broken pottery vessels, one containing hazelnuts, another a wooden spurtle or stirring spoon, hint at everyday activities, as do broken wooden tools, flint knives, and finely-flaked flint arrowheads, several with parts of their attachment to the arrowshaft. More unusual are two axes, one of flint and one of jadeite, each completely unused and lacking any trace of handle or haft or box; it is tempting to think that such unusual and valuable artifacts would not have been placed beside the track except for some strong and symbolic reason at which we can barely guess. Equally unusual are a set of polished pins of yew, the only yew known from the entire track; these may have been decorative for hair or nose, or functional for clothes. All in all, this structure opens new insights into the workings of Neolithic society, as well as posing new questions about the value systems and motivation of such communities. It is salutary to remember that without the waterlogging nature of the peats which eventually submerged this track, after perhaps 20 years of use, we as archaeologist would have no trace of its existence except for a few potsherds, flints, and two unassociated stone axes.

The wooden components of the Sweet Track are varied and abundant (**Figure 3**). They range from planks 5 m long to small chips detached during carpentry. The woods include a range of species and complete stems as well as split pieces. Bark on roundwood is well-preserved, and axe, adze and wedge-marks are clear and distinct (**Figure 4**). It is possible to reproduce the characteristic facets with stone axes, but experiments thus conducted have demonstrated above all else the need for long practice and experience, the care in securely hafting the axes, and the extreme ease in splitting oak with wooden wedges and wooden mallets. The wood from the track is heavily waterlogged and the technique of recovery is fraught with difficulties. All excavators work from planks suspended above the structure, or from boards placed on the peat surface (**Figure 5**). No metal tools are used, but only plastic or wooden spatulae and human fingers. The site is watered continuously or very regularly, depending on the weather; fortunately the Levels tend to receive normal English weather, but exceptionally all work ceases in hot sunny conditions. Sheets of polythene are used to cover the site sections, and individual pieces are bagged or wrapped in situ while awaiting futher work.

Almost every piece of wood in the structure is now sampled for one reason or another, and potential conflicts with the conservation ele-

Figure 3: Part of the Sweet Track in an area where planks of oak, ash and lime lie in apparent disarray beside the track. The criss-cross pegs of hazel, alder, ash, holly and elm still hold the heavy rails of hazel, alder and ash in place. This site was flooded daily to aid the preservation of the timber and roundwood during its excavation and recording. The degraded condition of the timber in this sector discouraged attempts at full conservation.

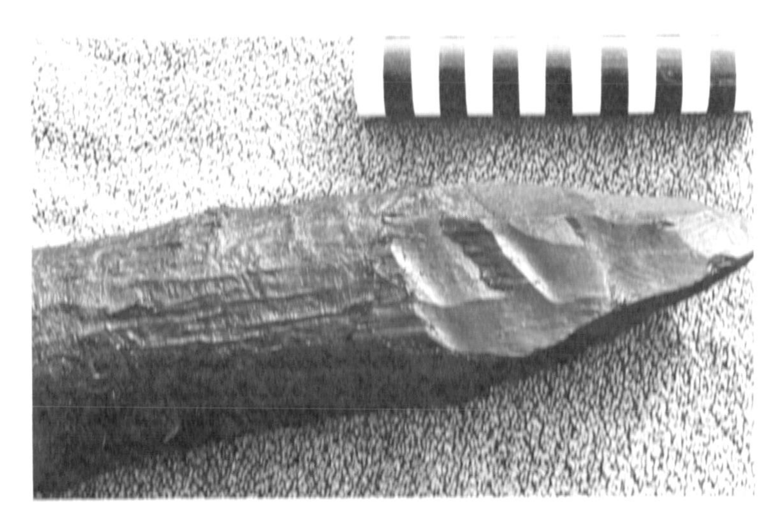

Figure 4: Sharpened end of a hazel peg from the Sweet Track, c 4000 BC. Note the sharpness of the facets, their clear indication of the manner of working, and the excellent condition of the bark. Unconserved. The scale totals 15 cm.

Figure 5: Excavation of the Sweet Track near its southern terminal. Most of the work is carried out from planks over the structure, as both the wood and its enclosing peat is soft. No metal tools are used. The upright pieces of wood are already beginning to be protected by the polyethylene bags, and the whole site will be watered continuously once more wood emerges. Environmental sampling is in progress from a cross-baulk.

Figure 6: Worked end of a peg of ash from the Sweet Track, after conservation by PEG 4000. The structure of the roundwood is still clearly visible and the facets are well-preserved although excess surface wax has gently smoothed the ridges. After cleaning with warm air, the piece should be judged as perfectly conserved with no measurable variation from the original.

ment in the programme are resolved by careful manipulation of the artifacts. Every significant piece is labelled with metal or plastic tape held by coloured pins, each colour signifying a particular sample and treatment. A roundwood peg or split plank, for example, would certainly carry pins of white (species identification), and (tree-ring), and black (conservation), and possibly also blue (woodland) and yellow (beetle and fungus). A tiny sliver will suffice for wood identifiction, and a small segment for fungus or beetle study; woodland study requires the entire piece but is non-destructive and carried out on-site or immediately off-site. Tree-ring samples must be taken in all cases from the best-preserved part of the artifact, and from the widest part where maximum rings are expected. The study is not destructive, contrary to popular thought, and in the Levels a sawn segment is sent wet to the specialist who freezes and slices a clean surface for measurement, then returns the piece virtually intact; it is pinned back in place before conservation, and the cuts are generally imperceptible (Morgan et al. 1981). The conservation of all of these pieces is therefore not hampered by other studies.

The woods recovered from the track are heavily waterlogged and severely degraded. Their porosity and permeability is very considerable, but their strength is correspondingly reduced. Finger pressure will mark a piece of roundwood, a finger nail will scratch and expose a light coloured streak of inner unoxidised wood, and some planks 50 mm thick can be perforated by a finger; other pieces are much more solid and can be lifted easily, but the weak artifacts will break unless bandaged and protected by foam or rubber cushioning when their removal from the structure is attempted. Every piece to be conserved is transported by van to the conservation laboratory in the Levels where it is further cleaned by water and paintbrush, recorded and photographed, labelled again and submerged in water in fibreglass tanks. No fungicide is used in these, and the water is changed regularly. Following the removal of all of the artifacts to this holding position, decisions are made about capacity, funding and potential future use of the artifacts, and at this point further samples may be taken for analysis, and certain artifacts discarded from the conservation programme. The remainder, and bulk, are then prepared for the process.

Each piece for conservation has a metal or plastic label pinned away from its worked end or ends, and is then bandaged firmly. The gauze bandage has two functions; one is to help hold the bark of roundwood in place during the circulation of the liquid in the tank, and the other is to help avoid any accidental damage to sides and ends during handling. The bandaged pieces are arranged on grid-frames 140 cm long and 50 cm wide and are loosely tied in place. Each frame when loaded is then ready to be lowered into a tank, and is held at a specific level by wire hangers or strips of bandage looped onto metal clips on the top edge of the tank. From two to four frames can be lowered into each tank. The tanks, of fibreglass, measure 150 x 52 cm, height 74 cm, and at the moment the Project has six of these, four with heaters and two used for excess liquid or wood. The heating wires in glass-reinforced plastic panels are placed to cover each tank bottom with a "thermowell" or sensor protruding from the tank end; the sensor is connected to a thermometer and thermostat on the wall beside the tank. A simple cylindrical dial and lights present information about the temperature in the tank and the point at which the heater switches itself on or off; the required temperature can be dialed. Each heating panel yields 575 watts, ample enough for a tank which holds up to 600 litres of liquid.

The regime adopted by the Project involves a nine-month process. The wood, on frames suspended in the tank liquid, remains unheated for two months. Thereafter, the heaters are turned on and temperatures raised by monthly steps of 7°C up to 55°C. The initial mixture of liquid is 25-35% PEG 4000 and 65-75% water. As the heat is increased, the plastic sheet covers which protect the tank and liquid surfaces are removed, and evaporation of the water begins. The water in the body of the wooden artifacts is gradually drawn out and replaced by the mix. The level of liquid drops with evaporation, and is made up by fresh neat PEG; the Project uses previously-circulated PEG to start the process off, but then uses new PEG for the topping-up. Careful note is kept of the amount of evaporation so that calculations can be made of the gradual loss of water; ideally this should be 100% but it is never so. When sufficient water has been evaporated, the wood is removed from the tank, unbandaged and wiped clean. It should be acknowledged that no conservation officer exists for the Project, and the programme is administered from Cambridge, 300 km away from the Levels. The recordings and adjustments are made by the resident Field Archaeologist as a part of his or her ordinary duties, but loading of tanks, essential repairs and modifications, and unloading, involve travel to and from Cambridge.

The results of this process are carefully assessed following each programme, and the details need not concern us here (Coles, 1981). Conservation by PEG is a well-tested process; for such heavily degraded wood its use is not common, however. The current procedure seems to work well and relatively little alteration in the artifacts occurs other than darkening of the surface, slight shrinkage transversely in a few pieces, and splitting of a very small number where conditions in the peat have affected the wood prior to excavation. In these cases, a period of alternating submergence by water and exposure to drying peat seems to have set up areas of resistance within the wood so that subsequent conservation by PEG seriously distorts the artifact. In other cases where the initial result appears unsatisfactory, resubmergence in PEG is practiced and can close small cracks or allow a twist to be manipulated out. Transverse shrinkage is a more major problem and is irreversible. Nonetheless, most of the Sweet Track artifacts, including those of oak, can be conserved successfully with this regime (**Figure 6**); the inadequate capacity for conservation has prevented such treatment for a majority of the artifacts from this unique structure, and less than 100 pieces have so far been treated alongside the hundreds of artifacts from other sites.

The Sweet Track has become the focal point of efforts in the Levels to preserve part of the prehistoric heritage. Working with the County Council, the Nature Conservancy and the Department of the Environment, archaeologists have identified all of the locations of the structure over 2 km, and have negotiated sectors for protection and monitoring. Pumping in of water to maintain waterlogging conditions has commenced in one area, but the future of this unique structure is still uncertain.

The Sweet Track is only one of many structures recently discovered in the peats of the Somerset Levels, and archaeologists have been able to recover a wide variety of prehistoric artifacts dating from c. 3000 BC to c. 200 BC. Among these are the earliest surviving hurdles from Britain, and their recovery has yielded unique information as well as posing particular conservation problems. By 3000 BC the Levels consisted of treacherous raised bog, with hummocks of vegetation and pools of acidic water dominating the otherwise featureless landscape. In parts of this gloomy moor, conditions for traffic were poor and Neolithic communities constructed various platforms and trackways, in effect bridging deep pools of water and soggy patches of vegetation. One of these structures, on Walton Heath, was first uncovered by a peat-cutting machine in 1975, although it had been heavily damaged in places by hand-cutting many years previously (Coles & Orme 1977). The trackway, only 40 m long, consisted of a series of woven hurdles or panels pegged down on the bog surface to provide a wide and immensely srrong passage across a particularly wet area of the moor (**Figure 7**). Many hurdles were brought down from a settlement for this purpose, some of them already old and damaged, others perhaps made specifically for this task.

The hurdles were made in the traditional Somerset way, by weaving long straight shoots of hazel around upright heavier shoots to form panels 2-3 m long and over one metre wide. The longitudinal pieces, called rods or ethers, were woven as singles across the upright sails, with a change of weave at intervals to prevent the hurdle bending (**Figure 8**); the process has been tested and reproduced experimentally (Coles & Darrah 1977).

From studies of the rods and sails of the Walton Heath hurdles we have learned much about the way in which Neolithic man was controlling the natural woodlands on the Polden Hills (Rackham 1977). The woodland consisted of 'coppice-with-standards', where oak and other large trees still stood, but cleared away sufficiently to allow quick-growing hazel to establish itself in the shade-free areas. The hazel was cut back regularly to encourage vigorous growth of shoots which could be harvested when their lengths and diameters were suitable. The shoots used to make the Walton

Figure 7: Part of the Walton Heath hurdle structure c 3000 BC. The platform and trackway was made by dumping broken hurdles into the marsh, and placing newly-made hurdles on top to provide a strong flexible surface. This site was heavily damaged by peat-cutting in 1930-1940, the effects of which can be seen along the left-hand edge in this photo. A modern peat-cutting in the background has totally destroyed the structure.

Figure 8: Junction of two hurdles on the Walton Heath Track. The hurdles were made almost entirely of hazel coppiced wood, growing and managed on the hills around the marsh. The sails were erected in the coppice, and the rods woven evenly around them. The coppice had been 'draw-felled' to yield shoots of uniform diameters and unequal ages. The panel on the right has been conserved by PEG, that on the left was dismantled and destroyed.

Heath hurdles were of uniform size (18-26 mm diameter) but were 3-9 years old; the craftsmen had 'draw-felled' the coppiced hazel, selecting from each stool only those shoots of correct size, leaving the others for future use. These Neolithic woodland practices are unknown from other areas of Britain, and only the water-logged peats allowed the hurdles to survive for archeologists to study. From the peats of the Levels have now come woven hurdles dated from *c.* 3000 BC at intervals to *c.* 300 BC, and the tradition persisted through medieval times up to the present day, a 5000 year-old industry.

Because the Walton Heath hurdles were in excellent conditon visually, it was decided to attempt to conserve one of the hurdles for museum display and future study. The hurdle consisted of 62 rods, 6 sails and several fragments of willow withies which had been used to bind the sail and rod ends together. The hurdle was undercut in the peat, leaving a peat base of 5-10 cm, and a wooden platform was built beneath it held together by steel bars **(Figure 9).** The whole structure was then dragged and lifted out of the soft peat onto a lorry, then a trailer, for transportation to the Conservation Laboratory of the National Museum of Antiquities of Scotland in Edinburgh, where a special contract had been arranged for its treatment. After over two years resting in water in a steel tank, the conservation began, but it had now to contend with the gradual break-up of the peat base due to its prolonged immersion (Bryce 1980).

The initial stage of conservation was a cold impregnating solution of 18% $\frac{v}{v}$ PEG 540 Blend, to penetrate both wood and peat, each now in severely weakened condition. This soak was only 5 weeks in duration to prevent further peat loss. After draining off, the hurdle was sprayed more or less continuously for two weeks with a solution of 20% $\frac{v}{v}$ PEG, and a humidity cover over the tank maintained appropriate conditions and prevented evaporation to excess. The spray was gradually increased over the next several months, by 5% every two weeks, while maintaining humidity above 80%. At 50% $\frac{v}{v}$ PEG, the spraying was replaced by sprinkling and at 70% and over, the wax was applied by brush. When the treatment with PEG 540 Blend was complete, PEG 4000 was brushed warm over the hurdle for two further months, and gradually this penetrated the surface of the wood and peat. The humidity cover was loosened to reduce the relative humidity to 60% and after a further three months the process was complete. The hurdle was eventually transported back to Somerset and installed in the County Museum, Taunton. The Walton Heath structure, with its 11 complete hurdles and 30 recognizable and recorded incomplete hurdles, was totally destroyed by peat-cutting in late 1975 and early 1976. Hurdles structures of the second milennium BC, with hundreds of woven panels, have similarly been destroyed in the past five years; archaeologists have been able to record segments of these structures for comparative studies, but not to attempt any further conservation of these unique artifacts.

These two structures, the Sweet Track and the Walton Heath hurdles, have revealed entirely new evidence, and new forms of evidence, about prehistoric woodworking and woodland management. The sites are, in fact, the two earliest examples of woodland control and manipulation known up to the present time anywhere in the world (Rackham 1980). Both also have posed new problems for convervation; their wood is as heavily degraded as that from any other archaeological site, and yet its superficial appearance is remarkably undamaged, and the finest working is preserved. That some measure of successs has been achieved with conservation of individual pieces from the Sweet Track and one hurdle from Walton Heath is the result of experimentation and persistence, some would say stubbornness. Yet as archaeologists we regret the need to continue to excavate, remove and conserve structures such as these; within the immediate future not only will most of the Sweet Track be destroyed by peat-cutting or progressive desiccation, but also every known, and unknown, structure from the entire area, and with this will disappear part of Britain's ancient heritage. A small part, we agree, but nonetheless a unique part; conservation is but a feeble attempt to save what we can from the onslaught of 'development and improvement' of the land.

References

Bryce, T., "Conservation of the Walton Heath hurdle from the Somerset Levels," *Somerset Levels Papers*, vol. 6, 1980, pp. 71-75.

Figure 9: Lifting one of the hurdles on Walton Heath. The panel has been isolated and undercut, with marine-ply boards pushed through to make a platform; the first of these boards is in process of positioning. The platform will then be bolted to a steel girder framework and made ready for hauling and lifting from the peat. The structure weighed c 700 kg and measured c 3 m x 1.3 m.

Coles, J.M., "The Somerset Levels: a concave landscape," in: Early Land Allotment, H.C. Bowen and P.J. Fowler (ed.) British Archaeological Reports, no. 48, 1978, pp. 147-148.

Coles, J.M., "Conservation of wooden artifacts from the Somerset Levels: 3," Somerset Levels Papers, vol. 7, 1981, pp. 7-78.

Coles, J.M. and Darrah, R.J., "Experimental investigations in hurdle-making," Somerset Levels Papers, vol. 3, 1977, pp. 32-38.

Coles, J.M. and Orme, B.J., "Neolithic hurdles from Walton Heath, Somerset," Somerset Levels Papers, vol. 3, 1977, pp. 6-29.

Coles, J.M. and Orme, B.J., "The Sweet Track: Drove site," Somerset Levels Papers, vol. 5, 1979, pp. 43-64.

Coles, J.M. and Orme, B.J., Prehistory of the Somerset Levels, (Cambridge: Somerset Levels Project, 1980)

Coles, J.M. and Orme, B.J., "The Sweet Track 1980," Somerset Levels Papers, vol. 7, 1981, pp. 6-12.

Morgan, R.A., "Tree-ring studies in the Somerset Levels: the Drove site of the Sweet Track," Somerset Levels Papers, vol. 5, 1979, pp. 65-75.

Morgan, R.A., Hillam, J., Coles, J.M. and McGrail, S., "Reconciling tree-ring sampling with conservation," Antiquity, vol. 55, 1981, pp. 90-95.

Rackham, O., "Neolithic Woodland management in the Somerset Levels: Garvin's, Walton Heath and Rowland's tracks," Somerset Levels Papers, vol. 3, 1977, pp. 67-71.

Rackham, O., Ancient Woodland. Its history, vegetation and uses in England, (London: Arnold, 1980)

Williams, M., The draining of the Somerset Levels, (Cambridge: University Press, 1970).

Questions

Barbara Purdy: You said that the dates ranged from about 4000 BC to the Roman conquest. Are the dates continuous? Is the 4000 BC material stratigraphically below the 2000 BC material or is it just in different areas? Do you think that the structures were built during extended dry periods when it was feasible to build them across a marsh like that, or during periods when they were flooded? Was the condition of the older wood different from the more recent, that was recovered?

John Coles: We have about 200 radiocarbon dates from our structures and we have evidence that something was going on throughout this period; there was no hiatus. We have only one place where there are structures which are grossly different in age, near enough so that you can actually see them stratigraphically separated. In most other examples, they are separated by several hundred metres, so normally it depends on their position in the peat and the radiocarbon dating, especially the former; so we don't have any juxtaposition of things like that. As for the building of the structures during dry or wet episodes, the area would have been flooded in the winter from October until March, rendering it easy to get about by boat. (Some dugouts have been found.) During the dry summer season of 1 1/2 to 2 months it is possible that much of the bog, in the later period from 2000 BC onwards, would be dry enough to walk on. So it may be, that it was only during the intermediate seasons that they actually had to construct things. That is the annual cycle. There are episodes in the third millennium BC, when increased rainfall created conditions for the formation of raised bogs, (these depend on rainfall as opposed to drainage waters), and in those cases, conditions are likely to have been wet all the time, and that is when we get our major episodes of platform and track re-building. As for the condition of the older wood versus the newer wood, the older wood is not found to be more degraded; within a site such as the 4000 year old one which we are digging at the moment, some of the oak planks are hard enough to stand on and are absolutely rock hard. Other planks two metres away, are soft enough to put a finger through. We find the same circumstances for sites of the 1st and 2nd millennia. Wood is extremely variable within a site. It doesn't depend on wood species. We haven't been able to work out why.

Colin Pearson: What are the pH levels in the bog?

John Coles: It is extremely variable, in some cases in the middle of the bog, conditions are very acid (I don't know the specific value). On the edges where you have mineral coming down from the lias slopes (calcareous material) conditions are much better for the preservation of things like animal bones; in the middle of the bog, animal bones do not survive. In experiments with a cow, a sheep, a pig and a dog, (Some deliberate and others involuntary!) by knowing when and how they were put into the marsh and by subsequent excavation little trace could be found. In two cases, the animals had been put in eight years ago, and no trace could be found -even the teeth had disappeared. On the edges, where we were digging a 200 BC Iron-age site, conditions for the preservation of bones were extremely good, and vast quantities were found.

John Dawson: How much pollen stratigraphy has been done in the area?

John Coles: A vast amount, Harry Godwin began and worked up to 15 years ago. The project has always had a full time, salaried, pollen analyst. At the moment, of the staff of five, two are pollen analysts, each working in different parts of the levels. We've had a succession of people; Alan Hibbert developed pollen transects, and then Steve Beckett came in and developed "episodes of clearance;" he used a lot of radiocarbon dates; episodes of major activity in the levels and then re-generation of woodland. Our present pollen analysts are not only questioning the early results but also developing new techniques, and new refinements. They work with us on the site, and tell us what we are allowed to do with the peat. This results in a sort of friendly conflict on the site, because the pollen analysts always want large sections.

Allen Brownstein: To what depths have the excavations gone?

John Coles: Perhaps 7 metres. In the area where we are working the peat was originally 10 metres thick, but it is now about 3 metres. Beneath it is a marine clay, and beneath that more peat, then gravels and then more peat and these have been cored.

Barbara Purdy: Were the pins you showed us bone, or wood?

John Coles: Wood; yew.

THE CONSERVATION OF TIMBER STRUCTURES AT YORK - A PROGRESS REPORT

James A. Spriggs
York Archaeological Trust,
47, Aldward,
York, England.

Introduction

In spring, 1984, one of the most unusual, and we hope, exciting displays of archaeological material in Britain will be opened to the public in York. This permanent exhibition, called the Jorvik Viking Centre, will contain the large-scale physical remains of 10th century Anglo-Scandinavian timber buildings, and other artefacts, actually on the site upon which they were discovered. The site is in a street named Coppergate, where government-sponsored rescue excavations have taken place continuously over the past five years prior to the redevelopment of the site as shops and offices, with the Viking Centre being situated in the basement. The name Coppergate translates from the Old Norse as 'street of the wood-workers', and as far as the excavations are concerned it has certainly lived up to its name.

The Viking city of Jorvik lies under the modern city centre, at the junction of the two rivers, the Foss and the Ouse. The Roman fortress and colonia, facing each other across the Ouse, were both built on higher ground, but the Viking and mediaeval Cities also covered the lower lying ground close to the junction of the two rivers. It is in these wetter, lower lying areas, always prone to flooding, that the greatest depth of anoxic waterlogged deposits are found. An unusually high water table in the area also contributes to the preservation of organic materials in these deposits, which can be up to 6 metres deep from modern street level to natural.

Since its formation in 1972, the York Archaeological Trust has always had a small conservation department, but the conservation of organic materials has only ever been possible on a very limited scale. The richness and importance of the finds from Coppergate have, however, encouraged the Trust to great efforts to provide larger scale facilities.

We now have a new timber conservation laboratory provided entirely by a fund raising appeal.

The Finds from Coppergate

Much of the interest in the Coppergate excavation lies in the discovery of a row of three mid 10th century Viking-period tenements, each comprising two post-and-plank buildings, which were originally built terraced into the natural slope of the ground; **(Figure 1)**. Each building, which was either used as a workshop, or as a combination of house and shop, measures approximately 7.5 metres long by 4.5 metres wide, and in some places buildings survive to a height of 1.5 metres. They are dug into 'clayey' silt, and were filled up to the surrounding group level with organic-based refuse on abandonment. The oak planks from the walls were anything up to 5 m in length and 8 cm thick, and were both radially and tangentially split from the original tree trunks. The planks were supported from the outside by the ground into which the buildings were dug, and from the inside by squared oak posts, mostly quarter cut, and measuring approximately 20 cm to 30 cm square in cross section. The bottoms of the posts were either supported by sill beams embedded in the ground, or simply slotted into post holes, and resting on wooden pads (Hall, 1982). The tenements were separated by wattle fences running between the workshops. Earlier structures lay beneath the post-and-plank workshops and were constructed entirely of wattle and daub, but the surviving remains of these were comparatively slight.

Why the Structures were Saved for Preservation

Prior to the work of the York Archaeological Trust, excavations and watching briefs in the city record the presence of timber and wattle structures. In the majority of cases, the standards of recording were poor, the remains were often misinterpreted, and there was little or no attempt made to save the remains. Thus the Coppergate excavation provided us with the best, and perhaps the only, opportunity for recording, researching, and preserving some of the last surviving remains of domestic architecture from this important, but hitherto elusive period of York's history. Despite the lack of funds or facilities

Figure 1: 10th Century post-and-plank workshop during excavation.

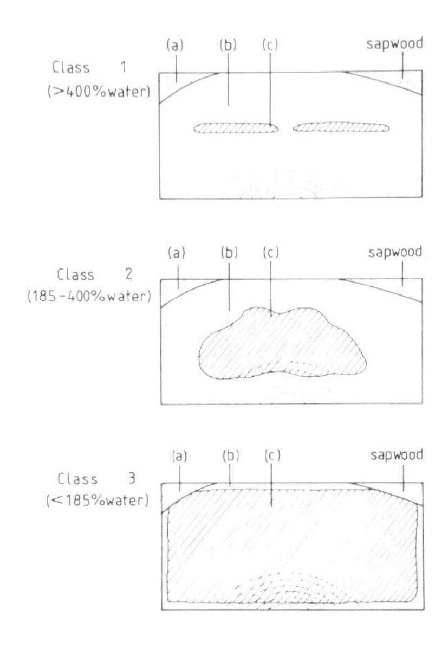

Figure 3: The three 'class' system for describing the condition of waterlogged oak (after de Jong).

Figure 2: 10th Century window shutter about to be transferred onto awaiting palette, with foam rubber padding.

to attempt the difficult and costly business of preserving these large timber structures, at a very early stage in the excavations the decision was made to save for conservation all the timbers from intact structures, as well as any unassociated structural timbers displaying any technological features of possible interest. The reasons for saving this large volume of material were threefold:

(a) The structures and many other individual timbers are of major archaeological and technological significance.
(b) The sheer rarity value of the finds, many being unique.
(c) The thought that the timbers could form the basis of a first-rate archaeological display, with considerable fund-raising potential.

Naturally, all the smaller objects made of wood which could be classed as 'small-finds,' are saved automatically, and over the period that the excavation was running (it ended just last week) we saved about 200 structural timbers, and over 15,000 small-finds, many of which are made of organic materials. We also have in store many timbers from other excavations in York, for example, several fine examples of Roman and mediaeval timber well linings, all of which it is hoped may be used as museum exhibits one day.

On-Site Techniques and Methods

The timber buildings took anything up to six months to excavate and record, often during the warm summer months. Keeping the timbers from damage by drying was a continual problem. At an early stage it had been decided that continual spraying (often the best technique) was neither conducive to good archaeology nor to the comfort of the archaeologists. A compromise was reached which permitted the timbers to be kept covered with 6 mm plastic foam sheet saturated with water, when the timbers were not being worked on. Both the timbers and the foam sheet were sprayed at least twice daily to keep the timbers as damp as possible. However, it was unfortunately necessary to expose the timbers to the atmosphere, and occasionally in sunlight. Many of the timbers did suffer some slight damage from partial drying, and the fact that drying cracks do already exist on some of the timbers has a bearing on our approach to the choice of a stabilization technique.

All the timbers to be saved for conservation were lifted onto supporting palettes, made-to-measure out of marine plywood and thin battening, the materials being first proofed against decay by the 'tannalizing' process. Foam plastic sheeting was used to surround each timber, to act both as padding, and as a sponge to keep the timber wet; **(Figure 2).** Attempts were made to lift areas of fallen wattlework, but due to the difficulties of undercutting such fragile material, all attempts failed. It was, however, possible to lift two lengths of standing wattlework by laying up a resin and fibreglass support against one side, and tipping the fence over. We have also taken several moulds of wattlework fencing, both vertical and horizontal, using a polysuphide rubber compound, trade-name 'Smootheon' (Brown & Peacock, 1981). This material will set perfectly in the wettest situations, and so long as there is a film of water between the object and the rubber to act as a separating agent, it was found easy to peel away. As supplied, it is a viscous liquid suitable for using on a horizontal surface. For vertical surfaces, however, it was found necessary to thicken the rubber by the addition of aerogel silica. These moulds will be used to recreate wattlework for display in the Jorvik Viking Centre.

Recording On-Site

All worked wood from the excavations, except stakes and piles) was allocated a timber record number whether it was to be saved for conservation or not. This number refers to a record sheet kept for each timber (Spriggs, 1980), on which is recorded all the relevant information about the timber. Instruction boxes at the foot of the sheet are filled out by the archaeologist in consultation with the finds-researcher, botanist and conservator, so that there is no ambiguity about how the timber is to be recorded, sampled, etc. A record photograph is taken of all structural timbers, whether they are to be saved or not, and many timbers are also drawn at a scale of 1:1 by tracing onto thin polyethylene film using spirit based felt-tip pens of various colours to denote different features.

Storage

Timbers to be saved for conservation were heat sealed into polyethylene tubing still on their palettes. Each package contained a dilute fungicide solution (normally, dichlorophenol: trade-name Panacide). We found that timbers could be stored like this for several weeks until enough timbers were ready to constitute a truck load. The timbers are stored in a large fire-reserve tank on an old war-time aerodrome outside York. The water is changed annually, and fungicide added at three-monthly intervals. A floating polyethylene cover keeps out some of the dirt that is blown in, but more importantly, keeps down the level of organic activity in the water by excluding light.

The Degree of Degradation

Before making a choice of conservation process, it was essential to have a clear idea of the condition and degree of degradation of our timber. We wanted to adopt a system and nomenclature which was not only reasonably simple to use, but one which could be directly compared with systems in use elsewhere. This would allow us to make use of other people's experiences, and obviate the need for a lot of experimental work which would probably only repeat that already done by others. After a literature search, we found that the system used by de Jong (de Jong, 1977) seemed to suit our needs. This system, based on the work of several other workers in the field, is designed specifically for the description of degraded oak, and is based on two physical properties - water content, and relative hardness. By establishing these two simple factors, individual timbers may be ascribed to one of the three 'classes' shown diagrammatically in **Figure 3**. Core samples were taken from representative parts of our Viking timbers from which water content was calculated. Thus, we were able to ascribe almost all our timbers to 'class 1' (400-669% water) except for some of the lowest planks, and the sill-beams, which were in 'class' 2 (233-400% water). The relative degradation of different parts of the wood, and the extent of these different 'conditions' (a,b & c in **Figure 3**) are ascertained by probing with a mounted needle. Condition (a) is extremely soft, allowing a needle to be pushed in with very little pressure, and is normally found in sapwood. A little more force is required to pierce wood in condition (b), this comprises the majority of waterlogged oak heartwood. Condition (c) is heartwood which has been virtually unaffected by burial, is very hard, and cannot be pierced with a needle. Through his experimental work, de Jong has shown that timber in the three different 'classes' are best treated by varying the standard PEG treatment to suit each state. We are hoping that through adopting this system, de Jong's experimental treatment results using PEG are now relevant, and can be applied (with care) to our own material.

The Choice of Conservation Treatment

The stabilization of the structural timbers required for the Jorvik Viking Centre has not yet quite started, although we have almost finished fitting out a Wet Wood Laboratory to receive them. Our choice of treatment method for the timbers had to fulfil several criteria:

(a) The treatment must efficiently stabilize the timber.

(b) It must be simple to organize and maintain.

(c) The equipment should not be costly, nor should the process be expensive to run or involve the use of large quantities of hazardous materials.

(d) That the timbers be ready to display by spring, 1984.

It seemed that there was really very little choice of process, bearing these requirements in mind, and that a variation of the standard PEG impregnation process would seem appropriate. We now propose to apply the following two-stage process:

1) A two year immersion in heated PEG 4000, the concentration of wax to water being gradually raised from 10% to 80%.

2) After bulking, the wood will be kept at an artifically elevated relative humidity of 70%. The humidity will then be dropped gradually over a two to three year period to allow the wood to dry out gradually, and to stabilize to the ambient R.H. This second stage will take place while the wood is actually on display.

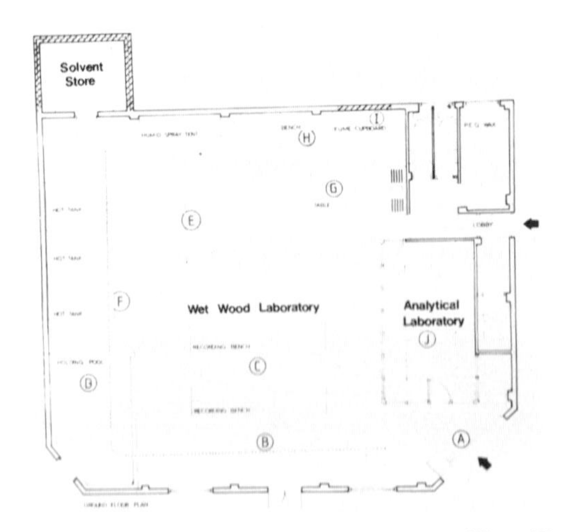

Figure 4: Ground plan of the new Wet Wood Laboratory at York.

Figure 5: Concrete tanks for hot PEG treatment at York.

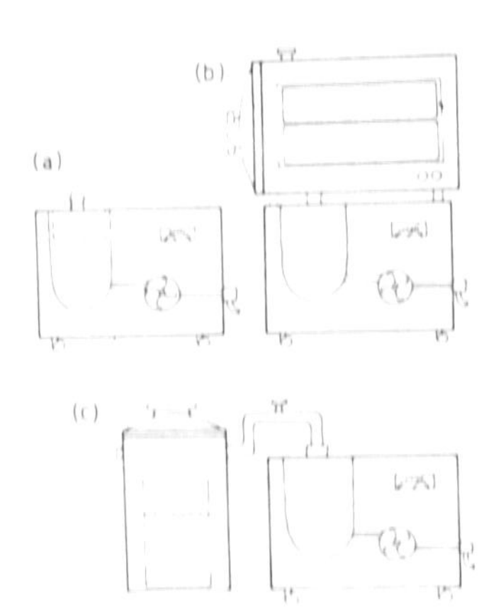

Figure 6: Heto CD12 freeze-drier set up for three different uses.

The Jorvik Viking Centre

The main attraction of the Viking Jorvik Centre will be two 'streets,' one of them, a modern full-scale simulation, and the other a reconstruction of the remains as they were found during the excavations. It is this second 'street' that will contain the post-and-plank buildings that we are about to conserve. We had at one time thought of isolating the whole of this 'street' by means of pressure barriers, so that a high humidity could be maintained in the whole area. But this idea soon became impractical for several reasons:

1) Not all the structures (e.g. wattle fences, and other small wooden exhibits in the area) require to be maintained at a high humidity. Indeed, some damage would be caused by this.
2) It is most likely that damage would be caused to graphics and other display materials, if exposed to a high humidity over a long period of time.
3) Any cold spots in the area would be likely to cause condensation, with the associated damaging effects, such as mould growth.
4) It became obvious that the expense of plant and running costs to maintain an elevated humidity in such a large area would be very great.

A more practical solution will be to encapsulate the timber buildings inside temporary glasshouses. These have yet to be designed, but will certainly enable a high humidity micro-environment to be maintained around the buildings much more easily and cheaply, and will allow the three timber buildings to be de-humidified at different rates, should this become necessary. The exhibition designers have been convinced that far from impairing the quality of the display, if the reasons for needing to keep the timbers at an artificially elevated humidity are explained to the visitor, then the presence of the glasshouses might positively enhance the exhibition.

The Wet Wood Laboratory

These new facilities are housed in a single story, brick building, providing about 200 m^2 (600 sq. ft.), **Figure 4.** Large sliding doors at one corner (a) give easy access to the main area, and unloading heavy timbers from transport is simplified by the provision of an overhead rail with sliding pulley system (b). This rail extends around the whole of the main area, and serves the other main timber treatment facilities. Two long tables (c) with suspended overhead plumbing fixtures, are for washing down, and recording purposes. A shallow concrete-lined pool (d), is used for temporary storage, and also serves as a location in which to wash timbers. It may also serve to pump molten wax into at the end of a treatment cycle, where it can cool and set prior to being broken up and stored again for re-use. Our two main treatment tanks (e), **Figure 5,** are constructed from hollow concrete blocks filled with iron reinforcing rods and liquid concrete, built onto a cast concrete base. They measure 4.5 m by 1.2 m by 1.2 m deep, and assuming that only two thirds of their volume is taken up with ancient timber, their working volume together will be 9m^3. The inner surface of the tank has a fine cement rendering, coated with 'Peridite' epoxy waterproof paint. The outer sides of the tanks are insulated against heat loss with a 150 mm thickness of polyurethane foam. The light timber lids are similarly insulated, are close-fitting all the way round, but can be adjusted to leave gaps between lid sections for evaporation. The water/PEG mixture will be heated and circulated around the tanks by means of a Conair-Churchill 'Mark 2' unit, with heating elements rated at 20 kw, and a maximum pump capacity of 100 l/min. The timbers will be supported inside the tanks on frames made of light battening covered with pierced 'Masonite' board.

Smaller wooden objects requiring treatment by a PEG process are impregnated in a smaller tank (f), of which the internal dimensions are 2.4 m by 1.2 m by 0.6 m deep. It was built from timber, with a fibreglass lining, and is heated by six 'over-the-side' mineral insulated submersible elements. Together, the elements are rated at 20 kw, and are controlled by a thyristor stack power control unit. Liquid circulation is by means of a multi-purpose 50 l/min oil pump. This tank is also well insulated, with close-fitting lids with adjustable louvres for ventilation.

Treatments for Small Wooden Objects

The remaining areas of the new laboratory

are mainly for the treatment of smaller wooden objects, and for waterlogged leatherwork. A square washing-down table (g), **Figure 4**, has suspended plumbing fixtures over it, and is used for washing and soaking small objects prior to stabilization. Objects soaking in acids, or being dehydrated through organic solvents are stored under the fume-hood (h). Adjacent to this are two fume-cupboards, one of which houses the 'acetone-rosin' tank (i). The 'acetone-rosin' process (McKerrell et al., 1972) is our preferred treatment for fragile wooden items, such as thin-sided bowls, and also for certain other waterlogged materials such as wickerwork, and ropework. The surface finish, strength, and ease of re-fixing broken fragments of wood afforded by this process accounts for our preference for this method over PEG or freeze-drying methods for these types of materials. Composite objects, such as iron-bound buckets and wooden-handled iron knives are also treated with the 'acetone-rosin' method as it is the only treatment for wood that we feel is even remotely conducive to the preservation of iron as well.

A small office area (j) is used for simple analytical work, and for storage of records and drawings, and it also houses the freeze-drying unit. We are using freeze-drying for the treatment of the vast majority of our small wooden items, as well as nearly all the leatherwork from York. Our freeze-dryer is a Heto unit, custom-designed for use in conservation and archaeological work (At the time of writing Heto are the only manufacturers to supply Freeze-Dryers specifically for archaeological purposes). Our freeze-drier was designed so that it could be used in three different ways, (see **Figure 6**):

(a) Using the condenser alone, with the chamber removed. Fresh biological specimens are often very slow drying, and need to be kept artifically frozen. They are placed in a tray fitted into the top of the condenser.

(b) With the chamber mounted on top. This arrangement is used for all archaeological material.

(c) With the chamber dismounted, and up-ended. In this position the system can be used for the vacuum impregnation of dry, fragile objects for example, in consolidants.

The freeze-drier unit consists of a standard Heto CD12 base unit, which contains the vacuum pump, and a freeze condenser with a capacity of 6 kg of ice. On top of the base unit is mounted the chamber, which was specially designed and built to our specifications. The chamber is cylindrical, measures 75 cm long and 50 cm in diameter, and is fitted with three sliding shelves. A square outer casing contains a system of copper tubing to enable heating or cooling fluids to be circulated around the chambers. Ports were fitted into the top of the chamber to enable temperature and pressure sensors etc. to be added to the system as required. We did add temperature sensors at an early stage, and we also found it necessary occasionally to supply extra energy to the load by pumping hot water around the water jacket.

Conclusion

In this paper, I have tried to summarize the most recent developments in York in the conservation of wooden artifacts, both big and small. In particular, I have described the development of the new Wet Wood Laboratory, now one of only four places in Britain capable of treating large timbers on a routine basis. I would like to thank my colleagues, both past and present, in the Conservation Unit who have helped to plan, and to construct the new facilities, and we are indebted to our fund raising team. Cultural Resource Management Ltd., and to the many donors and sponsors who have made the project financially viable. Finally, I would like to thank the York Archaeological Trust, The British Council, and Heto Ltd. for making it possible for me to attend this conference.

References

Brown, C., and E. Peacock, "Smooth-On Flexible Mould Compound for Wet Site Consitions," Conservation News, no. 15, 1981.

Hall, R.A., "10th Century Woodworking at Coppergate, York," in: Woodworking Techniques Before A.D. 1500, ed. by McGrail, S., (Greenwich: National Maritime Museum, 1982), British Archaeological Reports, International Series, Archaeological Series, no. 7, 1982, pp. 231-244.

de Jong, J., "Conservation Techniques for Old Waterlogged Wood from Shipwrecks found in the Netherlands," in: Biodeterioration Investigation Techniques, ed. A.H. Walters (London: Applied Science Publishers, 1977), pp. 295-338.

McKerrell, H., Roger, E., and A. Varsanyi, "The Acetone/Rosin Method for the Conservation of Waterlogged Wood," Studies in Conservation, vol. 17, 1972, pp. 111-125.

Spriggs, J.A., "The Recovery and Storage of Materials from Waterlogged Deposits at York," The Conservator, no. 4, 1980, pp. 19-24.

List of Suppliers

Peridite (epoxy sealant) -
Ault and Wiborg Paints Ltd.
28 Wadsworth Road
Greenford, Middlesex, U.K.

Masonite (perforated board) -
Masonite Wikstrom
70 South Street
Lancing, West Sussex, U.K.

Panacide (biocide) -
MacFarlane Robson Ltd.
Hedgefield House
Blaydon-On-Tyne,
Tyne and Wear, U.K.

Smooth-On (moulding rubber) -
Walter P. Notcutt Ltd.
44 Church Road
Teddington, Middlesex, U.K.

Conair-Churchill (heater-circulators) -
Conair Churchill Ltd.
Riverside Way,
Uxbridge, Middlesex, U.K.

Jabsco (general purpose pumps) -
ITT Fluid Handling Ltd.
Belcon Industrial Estate
Bingley Road
Hoddeston, Hertfordshire, U.K.

Heto (freeze-driers) -
Uniscience Ltd.
8 Jesus Lane
Cambridge, Cambridgeshire, U.K.

Helinex (Tracing Film) -
Bexford Ltd.
Brantham
Nanningtree, Essex, U.K.

Equipment

Heto freeze-dryers are available from: Bach-Simpson Ltd., 1255 Brydges St., London, Ontario, N6A 4L6.

Questions

David Grattan: Concerning your problem with the drying out of the wooden structures during excavation. Did you consider the possibility of spraying them with a PEG solution during the excavation? It occurs to me that PEG might have had three beneficial properties; (i) it would act as a humectant and keep the wood moist, (ii) it would keep the surface of the wood in expanded condition and stop cracking and (iii) it would begin the conservation treatment. I wonder if it would have given the wood some defence during that very tricky stage of the excavation?

Jim Spriggs: Yes, we did consider doing that, but at the time I had no idea what technique we were going to use for conserving the wood, and therefore we decided not to in case it was necessary to remove the PEG later on in order to commence a different treatment. I think it would have helped, and I wish that we had done it.

Barbara Purdy: How would you go about this?

David Grattan: I would use a process similar to the brushing/spraying method of Victoria Jenssen, Lorne Murdock, and Robson Senior. I would spray with 10-15% solutions of PEG 540 Blend, rather than PEG 4000, because of the danger of dehydration of the interior of wood with the latter; unless I had some information about the condition of the wood. Perhaps somebody with direct experience of this method would be better able to answer this question.

Lorne Murdock: The danger of continually spraying wood to keep it wet is that the mechanical action tends to disintegrate fragile surfaces. The wetting-drying cycles during the course of this process will also tend to cause cracking and fragmentation. If as David Grattan suggests you could get some PEG on the wood surface, this should reduce the amount of spraying required and possibly help to stabilize the wood.

John Coles: We observe these kinds of effects when we expose waterlogged wood from the peat deposits in the Somerset Levels. On a "drying" day if we spray, then leave the wood for ten minutes and then spray again and continue this throughout the day, by the end of the afternoon we do see deterioration of the wood. The only solution we can suggest is to use a constant spray so that the wood is kept constantly wet and not exposed to drying mechanisms, which occur when the surface becomes drained and the sun shines on it.

Suzanne Keene: Another approach might be to keep structures covered with foam rubber, which does absorb a lot of water, dries extremely slowly, and which can easily be maintained in a wet condition by trickling water over it. Although it is very difficult for vertical structures!

John Coles: We do use this method, but there are problems as you suggest. Some of the vertical objects have very degraded top surfaces which are very easily damaged, such as pegs with tops hammered 3000 years ago, or the rubber may stick to the surface of a rotten plank and there's much difficulty in removing it and we find you may lose some edges.

Victoria Jenssen: You could use some cotton flocking on the surface or make up a viscous solution of PEG to cover the delicate surfaces to act as a buffer.

Richard Clarke: Another point in favour of holding objects by PEG spraying is that it allows you the opportunity to choose between three of the most effective conservation methods.

(i) For air drying, you already have some impregnant in the wood that will give it a bit of mechanical support in the surface layer.

(2) For freeze-drying, the PEG impregnation has for the pre-treatment been commenced and can continue.

(3) A standard PEG treatment can be used.

And certainly, if you have a large structure and are worried about cost, you could commence by PEG spraying and should sufficient money be obtained, switch to a freeze-drying process. The most important thing, is to choose the correct grade of PEG.

David Grattan: If you use PEG 540 Blend, there's a bit of everything in it!

Judy Wight: You will run into two problems with using PEG spray solutions for holding large structures in the field:

(i) If there is any iron in the soil it will be damaged.

(ii) For the people excavating, they are going to be very annoyed if they are going to be coated with PEG.

Lorne Murdock: What we mean is not to use a continuous spray of PEG solution, but to spray occasionally. What really upsets archaeologists is when we wrap up their objects or on-site features in plastic sheeting; because they do not like to see the site covered up, they prefer to be able to observe it at all times, am I right?

John Coles: No!

(Laughter)

Jim Spriggs: There is one objection I can think of for using a PEG spray apart form the fact that it makes everything rather slimy, and that is that the archaeologist wants the surface to look good for photography before listing. If you have a black slimy deposit on the surface they are not going to be too pleased.

David Grattan: This is not necessarily so, if you use a dilute solution of PEG, waterlogged wood will be almost unaffected in appearance.

John Coles: But do you really think that photographers will object to losing the "glorious site photograph."

Jim Spriggs: Our archaeologists are particularly keen on this. Has anyone had any experience of using it in the field?

Jamie Barbour: We have employed it on the Hoko river and Ozette sites. There's one other side effect, it consolidates the sand and silt around the wood. It makes the object rather like a brick although it can easily be washed out later. However it does create more problems for the excavators since it hardens the soil around the structures.

Jacqui Watson: I was wondering how you would contain it. It might affect the carbon 14 dates of the objects it contacts.

(ed. A question arose at this point on the practical difficulties of containing PEG solutions on-site, and the difficulties they would pose if the site became contaminated with them)

Victoria Jenssen: To me, it is simply a matter of technique and it can be made to work.

Jamie Barbour: I agree, if you have to save an

object sometimes it's necessary to take actions which may have other side effects. I can't see why anybody working on a wet site would worry about getting messy. I think it is more a matter of which is more important, the archaeologist wanting to take a nice picture of the site, or having the object afterwards. If it's a question of losing the object or taking the pictures the archaeologist has to decide.

CONSERVATION OF WOODEN ARTEFACTS

Larry Titus
Department of Archaeology,
Simon Fraser University,
Burnaby, British Columbia,
V5A 1S6

Abstract

During the summer of 1978 approximately 500 kg of midden material was removed en bloc from a Western Thule house on Herschel Island, Yukon Territory. The midden matrix is a degraded fatty material containing well preserved wooden, bone and antler artefacts and large quantities of other organic materials.

A conservation procedure involving organic solvents, ethyl hydroxy cellulose (Ethulose - 100), low pressure impregnation, and treatment with an anti-bacterial surfactant has been successfully employed. After prolonged open storage the wooden materials show no superficial degradation or warping.

The ease and practicality of the method for conserving wooden artifacts begs for implementation by the archaeologist, yet a recent survey has shown that conservation is not considered a high priority by "practising archaeologists." This position is obviously untenable for archaeologist and conservator alike and demands correction.

Conservation of Wooden Artefacts

One of the main reasons that amateurs, like myself, become involved in conservation work is that there is to-day an abrogation of responsibility on the part of conservators and archaeologists alike. To be blunt, when it comes to the protection of cultural heritage we often act in a negligent manner. Having cast the first stone, I shall expand on this position drawing from my own experiences.

During the summer of 1978 approximately 500 kg of midden material was removed en bloc from a Western Thule house on Herschel Island, Yukon Territory (Yorga, 1980). The midden matrix is a degraded fatty material containing well preserved wooden, bone, and antler artefacts and large quantities of other organics.

The midden material was excavated later under more comfortable conditions in the laboratory at Simon Fraser University. Among the preserved remains were many wooden artefacts ranging from ulu knife handles to wick trimmers to the ubiquitous wooden peg. From past experience it was realized that some conservation work would have to be done on these objects before they became enveloped in too thick a layer of oleophilous mould.

It should be noted here that the bulk of the artefacts recovered from the site were sent to the Canadian Conservation Institute for conservation and are currently undergoing treatment. The material on which this paper is based is restricted solely to those wooden objects recovered from the midden that were shipped out of the field situation.

Structurally the artefacts appeared to be quite sound, having been oil soaked and frozen in an essentially anaerobic environment for over a 1,000 years. Furthermore, it was probably the oil soaking which was responsible for their continued excellent condition. So, the problem arose of how to treat these objects.

No one, at that time, in the Museum of Archaeology and Ethnology at Simon Fraser University could give me any advice on conservation procedures. Indeed, even now there is no conservation work being done on extant collections in that museum.

Other researchers in the Department of Archaeology had never treated such materials but were nonetheless very generous with their time and advice. Unfortunately, from an examination of the materials that they had recovered from waterlogged sites in coastal British Columbia it was apparent that conservation work had been superficial and largely unsuccessful. Cordage and basketry was rapidly degrading in their storage trays and large wooden objects which had been painted and/or soaked in Polyethylene Glycol were flaked and checking badly, the PEG having crystallized on the surface of the artefacts.

The next step was a literature search. Numerous monographs and journals were read, recipes copied out, and advice duly noted. However, there never was anything specifically related to the types of problems I was encountering.

In desperation the following procedure was initiated. Wooden artefacts that were apparently sound were soaked in acetone for at least 24 hours. This removed the hardened fatty midden matrix from the surface. The objects were then air dried for 5 to 10 minutes. The surfaces were then thoroughly painted with neatsfoot oil in which was dissolved a small amount of the biocide Dowicide. The artefacts were then placed on absorbent paper to air dry for about a week.

Objects which were not obviously sound, required one additional step. After cleaning with acetone the artefact was immersed in a solution of ethyl hydroxy cellulose (Ethulose - 100) in water (about 5 grams to 1 litre). A partial vacuum was created in the container by use of an aspirator. When the object ceased to give off bubbles it was removed and laid out to dry on a large mesh screen. Thus excess solution could drain off. After about a week of air drying the artefact was lightly cleaned with acetone, air dried, and then painted with the oil and biocide mixture.

The treatment appears to have been effective up to the present. After 10 months of open storge at approximately 15°C the artefacts show no signs of superficial degradation or warping. However, the initial structural integrity of the wooden artefacts is undoubtedly the reason behind any success so far.

The question immediately arises, why did I not contact a conservator? The answer is twofold; firstly I did not know how to even start contacting someone interested in the problems I was facing and secondly, there was a hesitancy on my part because I felt that I did not know enough about the material to even ask the right questions.

This situation is, I came to find, widespread among archaeologists today. In speaking with various archaeologists in Canada and the United States almost all invariably said that they had never actually done any conservation work and really would not know where to go to have it done.

Legally and hence financially there do not appear to be any prohibitive factors for conservation work. Antiquity laws and various provincial, state, and federal regulations often include clauses requiring the preservation of cultural property. One example is the United States Department of the Interior's "Recovery of Scientific, Prehistoric, Historic, and Archaeological Data: Methods, Standards, and Reporting Requirements," (36 CFR Part 66).

Sections of this document state,

> "If portions or elements of the property under investigation can be preserved, the program should employ methods that make economical use of these portions or elements."

and later,

> "...it is important that the data and material resulting from data recovery programs be maintained and cared for in the public trust. ... They shall be maintained by the Government or on behalf of the Government by qualified institutions through mutual agreement. ... in general, it is necessary for a qualified institution to maintain a laboratory where specimens can be cleaned, labelled, and preserved or restored if necessary ..."

Put into practice, these regulations would provide a firm foundation for conservation work in the United States but at the same time provide for major practical problems in curation, since according to Robert Dunnell there is no "qualified institution" as defined by this document in the United States today (February 26, 1981).

In Canada no such codified regulations exist at the federal or provincial levels, however by the very existence of facilities such as the Canadian Conserviton Institute it is apparent that no legal or financial hindrances exist, ideally at least, to conservation work being done.

The problem, and I would argue that it is the crucial one here, is communication between conservators and archaeologist. The one side says that the other is too wrapped up in an Arts History way of thinking, only really concerned with "museum pieces" (this latter used amazingly enough in a derogatory sense). The other side responds that the former lack any conception of the practicalities and requirements of modern conservation work.

Such criticisms are not only useless but in the long run are damaging for both parties and ultimately for our non-renewable archaeological resources. This latter point is perhaps the most damning if we, as professionals devoted to the understanding and preservation of cultural heritage, cannot or will not reach some common ground.

The role of the archaeologist in conservation has been summarized as follows.

"In addition to the responsibility for discovering and interpreting archaeological data and for insisting upon accuracy in preservation projects, the archaeologist must often also be a scientist-conservator. While in the field, he may have to face the same conservation-preservation problems regarding archaeologically recovered artifacts as does the conservtor working in the laboratory. And, when the archaeological program does not include the services of a staff conservator, the field archaeologist is required to perform necessary treatment or to stabilize the object so that it can be examined and treated later" (South, 1976: 40).

However, when faced with thousands of organic artefacts in the field, a six to eight week field season, and a contractual obligation to produce a final report within a short time, advice to stabilize and bring the artefacts back for treatment begins to pale. So, the archaeologist is often forced into the situation of taking a "cookbook approach" to field conservation.

The classics of conservation are pulled out of the library and recipes extracted, museums are contacted, and anyone who has ever been connected with conservation work is recruited. The pattern should by now be familiar. Given a good deal of planning and forethought this approach can be effective in the field situation.

During June, July, and August of this year I was fortunate enough to have been able to work on the Utkiavik Archaeology Project at Barrow, Alaska. The project was run by the State University of New York at Binghamton with support form the Bureau of Indian Affairs. One purpose of the excavations was to mitigate the impact of an underground gas pipeline to be constructed this winter. To this end, selected house mounds dating to the 19th Century were excavated.

Over 26,000 objects were catalogued in the field lab including several hundred wooden pieces. These included foundation timbers for the houses, floor and wall planks, buckets, arrow shafts, knife handles, ice probes, drill bows, and scraper handles to name only a few types.

Approximately 150 wooden artefacts were treated in the field with the following procedure. Objects were air dried and then brushed to remove surface dirt. The piece was then soaked first in water and then in a 50% w/v solution of water and PEG 4000, although PEG 1000 was used initially due to the non-availability of the former. The period of immersion varied from 48 to 165 hours depending upon the initial state of the piece.

The description of the artefact, it's find locale, specific provenience, PEG solution percentage, beginning and end points of treatment, and the duration of treatment were all noted.

The object was then removed from the solution, any excess PEG wiped off, and allowed to air dry on a large mesh screen. Finally, the artefact would be lightly sprayed with Lysol (ortho phenylphenol) and packaged for shipping.

This procedure was usually successful, at least up to the point when the artefacts left the field. Notable exceptions were composite pieces such as wooden and ivory float and pieces which were at the time of excavation in a badly degraded state.

The conservation work done at the Utkiavik site is, then, the cookbook approach at its best. On the other hand all of us here are quite familiar with horror stories concerning the misapplication of conservation techniques or the all too familiar situation of no conservation. As many other participants at this conference have pointed out, there is no neat and pat answer in conservation, there is no cookbook. Yet among archaeologists it is a common belief that "armed with the right lab manual, archaeologists are competent to care for what they have unearthed" (Bourque, et al., 1980: 795).

Clearly, there must be an increase of communication between archaeologists and conservators. Since excavation is by its nature destructive archaeologists must be "committed to the responsible preservation of archaeological data and conservation of archaeological resources" (Smith, 1974: 377). To this end conservation advice must be an integral part of the overall research design from beginning to finish.

Despite the formal and fictive ties between various archaeological and conservation establishments there is little or no communication now. Indeed, the former Provincial Archaeologist for British Columbia in discussing the role of the Provincial government in Canadian archaeology categorized conservtion as not within the realm of academic responsibility (Simonsen, 1980: 5-6).

What is required at this time is a concerted effort to re-educate conservators and archaeologists alike to the needs, capabilities, and constraints of the other party. This is, of course, nothing new as several conservators and archaeologists have recognized this need for years. The need now is to build upon this realization and attain a full collaboration in order to preserve the cultural heritage.

Clearly, we are approaching a crisis in the management of cultural resources. These non-renewable resources are being depleted at an alarming rate and yet in a survey of faculty and graduate students at Simon Fraser University's Department of Archaeology conservation was not considered a high priority in post-fieldwork responsibilities. This position is obviously untenable for archaeologists and conservators alike and demands correction.

We do have, I believe, a duty to the Past, to the Present, and to the Future. If we are to live up to such an onerous responsibility then we must work together towards common goals for the common good.

References

Bourque, Bruce J., et al., "Conservation in Archaeology: Moving Toward Closer Cooperation," American Antiquity, vol. 45, no. 4, 1980, pp. 794-799.

36 CFR Part 66, "Recovery of Scientific, Prehistoric, Historic, and Archaeological Data: Methods, Standards, and Reporting Requirements Proposed Guidelines," United States Department of the Interior, Federal Register vol. 42, no. 19, January 1977, pp. 5374-5383.

Dunnell, Robert C., "Cultural Resource Management," Seminar given in the Department of Archaeology at Simon Fraser University, February 6, 1981.

Simonsen, Bjorn O., "The Role of Government in Archaeology," Datum, vol. 5, no. 4, 1980 pp. 5-6.

Smith, Robert H., "Ethics in Field Archaeology," Journal of Field Archaeology, vol. 1, no. 3-4, 1974, pp. 375-383.

South, Stanley, "The Role of the Archaeologist in the Conservation-Preservation Process," in: Preservation and Conservation: Principles and Practices (Washington: The Preservation Press, 1976), pp. 35-43

Yorga, Brian, "Washout: A Thule Site on Herschel Island, Yukon Territory," National Museums of Canada, Mercury Series no. 98. 1980.

Questions

Colin Pearson: Concerning the point about education, that is extremely important. If you look at the courses for teaching archaeology; how many of them include any materials science? The archaeologist must understand the materials he or she encounters. Then he or she will appreciate the problems more. Conservation of any sort, even simply preventive, should be taught in an archaeology course, furthermore the training techniques themselves need re-examination. I know this is a lot to ask, because there is a vast amount of other material that the archaeologist has to learn, but I think it should be included.

Larry Titus: During my entire undergraduate career at the University of Toronto, I never heard conservation mentioned once.

Barbara Purdy: I would just like to say that I have the dubious honour of being the assistant secretary treasurer and head of the membership committee of the Society of Professional Archaeologists in the United States. I have been asked to prepare a list of requirements for people who wish to apply for underwater archaeology emphasis. I would think that anybody who is going into the field of excavating wet sites and doing a good job at it would probably qualify as an underwater archaeologist, even if their sites have to be pumped dry in order to retrieve the material. It doesn't have to be a shipwreck. One of the things that we discussed at our last membership meeting, was that in order to fulfill these requirements, that we would absolutely insist that applicants be aware of conservation techniques and trained in them, so maybe that will be something that will reverse the situation that you talked about.

Larry Titus: I hope so. One of the archaeologists in our department went into historical archaeology recently, and had no idea that he would find materials that needed to be preserved. There was no basis for conservation work at all through the entire project. It is only recently that he has become interested.

Herman Heikkenen: I would like a make a comment to Barbara Purdy. If your organization has sufficient strength to form an accreditation committee for the various departments then you "fight city hall" with universities, but from my world in forestry, the Society of American Foresters has a well established accreditation committee and reviews are held every five years for every forestry department, and the curriculum is approved by the accreditation committee. If you do not have this, I would suggest that your society "tackle that bear" and bring your program up to standard.

Barbara Purdy: We are also going to try and apply pressure to some of the State agencies that hire archaeologists to do mitigation kinds of excavations and insist (or try to insist or apply pressure) that these people have passed these sort of requirements, in order to go out. This doesn't just mean underwater archaeology. Right now, the States are pretty weak in terms of requirements for the people they get to do the work.

Bob Barclay: I would like to say to Larry Titus, that I find it very difficult to understand why you couldn't get better conservation advice in 1978 than that you apparently got. Aside from the existence of CCI which included an underwater laboratory, there was also the Conservation Division of Parks Canada; furthermore at the British Columbia Provincial Museum they have an experienced conservator and excellent facilities and also at the Centennial Museum in Vancouver. I think that any one of those bodies could have given you some sound practical advice on the treatment you could have done. I think that the use of neatsfoot oil, to say the least, is somewhat unorthodox.

Larry Titus: I'm sure that's true, that other opportunities were there. The situation in British Columbia, is that there is very little cooperation between institutions, which is sad.

Cliff McCawley: About four years ago the Canadian Archaeological Association held its meeting in Vancouver. During which there was a session devoted entirely to convervation. It was notable for one thing, mainly conservators were there, there were very few archaeologists. Those archaeologists present were those who have been interested in conservation for a long time and who make the effort to attend. Really there was just no interest whatsoever from the bulk of the archaeologists.

Charles Hett: I was at the same conference, and I attended a session on the training and education of archaeologists, and spoke on the need for further knowledge of material culture, and conservation as part of the basic training of archaeologists. The comments of a senior historic archaeologist hardly bear repeating, that we treat bottle caps like Rembrandts, never get anything done etc. Since then, I know of only one university in Canada which proceeded to have conservation as part of their basic field techniques (Memorial University in Newfoundland). There may be others that I am unaware of, but certainly no other department in Canada has approached me and asked for that kind of assistance.

Mary-Lou Florian: I'm really quite surprised. I feel very guilty personally. Victoria Jenssen had told me about Larry Titus before I went out west to Victoria, British Columbia, but I didn't hear from him until after the fact, when I received a box of wood from him. Is that not correct?

Larry Titus: Yes. I did send you some material.

Mary-Lou Florian: Larry Titus mentioned that there are no communications between institutions in British Columbia. What he is saying is basically true, but only in a sense; there are reasons for this lack of communication. The British Columbia Provincial Museum has a mandate for its own collection. It also does a lot of extra work for the Museums Assistance Programme educational services in which we have workshops throughout British Columbia. Our conservators go out to these workshops, so we do have this type of communication. Because our museum has its own collections it's not possible for us to get involved in university or other such institutions and look after their artifact material. But, we do give assistance, e.g. verbal apprenticeships and training we often have volunteers come in. We certainly could have assisted you. But our institution cannot take upon itself the care of a collection belonging to another institution.

Suzanne Keene: In the United Kingdom we have made a small attempt to address this problem and two years ago the United Kingdom IIC held a seminar to which we invited representative archaeologists and museum curators. We wrote up the proceedings of this: "Conservation Archaeology and Museums," cost $4. including postage available from the Tate Gallery, London, SW1P 4RG. It is, we think, a small step in the right direction.

Brian Yorga: I was the senior archaeologist at Herschel Island in 1978. I think that Larry, in an attempt to make a very important point missed two essential minor details. One is that we did have very good conservation expertise at Herschel Island, in the person of Charles Hett of the Canadian Conservation Institute. Who was there for over a week and aided us considerably. If not for that, we may never have removed that material en bloc and Larry Titus may never have had the opportunity to examine it.

Charles Hett: I would like to comment on those specific treatments. When wood comes from the middle of a block of fat, little treatment is needed. This comment is based on the observation of other similar small objects impregnated in fat, which have survived remarkably well without treatment. It is important to realise, that for objects such as these, no treatment would probably have sufficed. When one considers the relative merits of treatment, "no treatment" should always be considered.

David Grattan: I would like to mention that some of the fat impregnated wood from Herschel Island has been tested for treatment in various ways by myself, some of it is on display here. I tested (i) Air Drying, (ii) Freeze-Drying it from 15% PEG 400 solution and (iii) Impregnation in aqueous PEG 4000 up to a concentration of 55% w/v without application of any heat. The wood (all spruce) could be divided into two categories, whole round portions which were very sound, and split planks which were very degraded. For the former, the freeze-drying technique was most successful, followed by air drying, and for the latter impregnation in aqueous PEG 4000 was best, followed by the freeze-drying method. It should also be noted, that use of snails to keep the wood clean (described by Dawson, Ravindra and Lafontaine - this publication) also resulted in much of the fat being removed. The snails loved it! No impregnation problems were encountered.

SURVEY, RECOVERY AND TREATMENT OF WOODEN ARTIFACTS IN FLORIDA

Barbara A Purdy
University of Florida,
Department of Anthropology,
Gainesville, Florida,
32611

Abstract

Florida has extensive organic soils in which prehistoric wooden artifacts have been preserved, including the largest number of prehistoric dugout canoes in the world. This maritime heritage is as plentiful, as ancient, and as elaborate as the famous Lake Dwellings in Switzerland. There is limited evidence that is 10,000 years old but, beginning 6500 years ago, vast quantities of human skeletons, artifacts, and flora and fauna have been preserved in areas where people were utilizing aquatic environments. In addition to 107 dugouts (**Figure 1**), the wooden objects include toy canoes, canoe paddles, totems, figurines, masks, bowls, pestles, adze handles, spearshafts, float pegs, fire starters, cordage, etc., often associated with burials as grave goods.

Florida is undergoing rapid growth. In addition to the use of land for agriculture and the attraction of Florida to tourists, industrial and residential building has been encouraged and is increasing at an exponential rate. There are also indications that Florida peat deposits will be used as energy sources. Development projects often occur in or near wetland areas such as mangrove swamps, lakes, rivers, necessitating draining, dredging, filling, and asphalting operations.

In this paper I (i) describe efforts to access the prehistoric cultural potential of Florida's wetlands, (ii) discuss the issues involved such as time, money, unawareness, and conflicts of interest, and (iii) outline procedures we have initiated to educate the public about a unique, non-renewable, invisible heritage that should be systematically recovered, preserved, and studied.

Introduction

The cultural material from wetlands in Florida, along with the flora, fauna, and climatic record, is endangered; it is an invisible heritage that is being destroyed often because developers are not aware of it. As a matter of fact, all of the wooden artifacts, with one exception, have been found during draining, dredging, and removal operations rather than as a result of systematic archaeological excavations. When removed from the matrix that has preserved them so long, these ancient, degraded, waterlogged woods deteriorate rapidly if preservation treatment is not started immediately.

The objectives of a project funded by the National Trust for Historic Preservation and Heritage Conservation and Recreation Service were to:

1. Survey, inventory, and evaluate those areas of Florida where prehistoric organic cultural materials were found previously in order to determine under what circumstances preservation had occurred and to develop predictions about wet site locations. A summary of the results of the field survey and responses to a request for information are given in this paper.
2. Establish correspondence with other researchers in the field of wooden artifact preservation and compile a bibliography of publications relating to wooden artifact preservation procedures.
3. Request information from Florida museums and institutions to ascertain the existence, location, form, and condition of wooden artifacts recovered from Florida wet sites, and establish a master file of all pertinent data.
4. Assess the preservation requirements of wooden artifacts housed at the Florida State Museum, and analyze and initiate preservation treatment of a large quantity of wooden materials received in May 1980 from a salvage site near Naples (the Bay West Site 8-Cr-200) and from a site excavated in December 1980 on the St. Johns River near Deland (the Hontoon Island Site 8-Vo-202).
5. Begin a comparative collection of thin sections of Florida wood species. This objective has not been realized. It has not been possible to implement a comparative collection of wood thin sections because published materials about noncommercial Florida wood species are not readily available and because no trained personnel are present at the University of Florida who can assist in this endea-

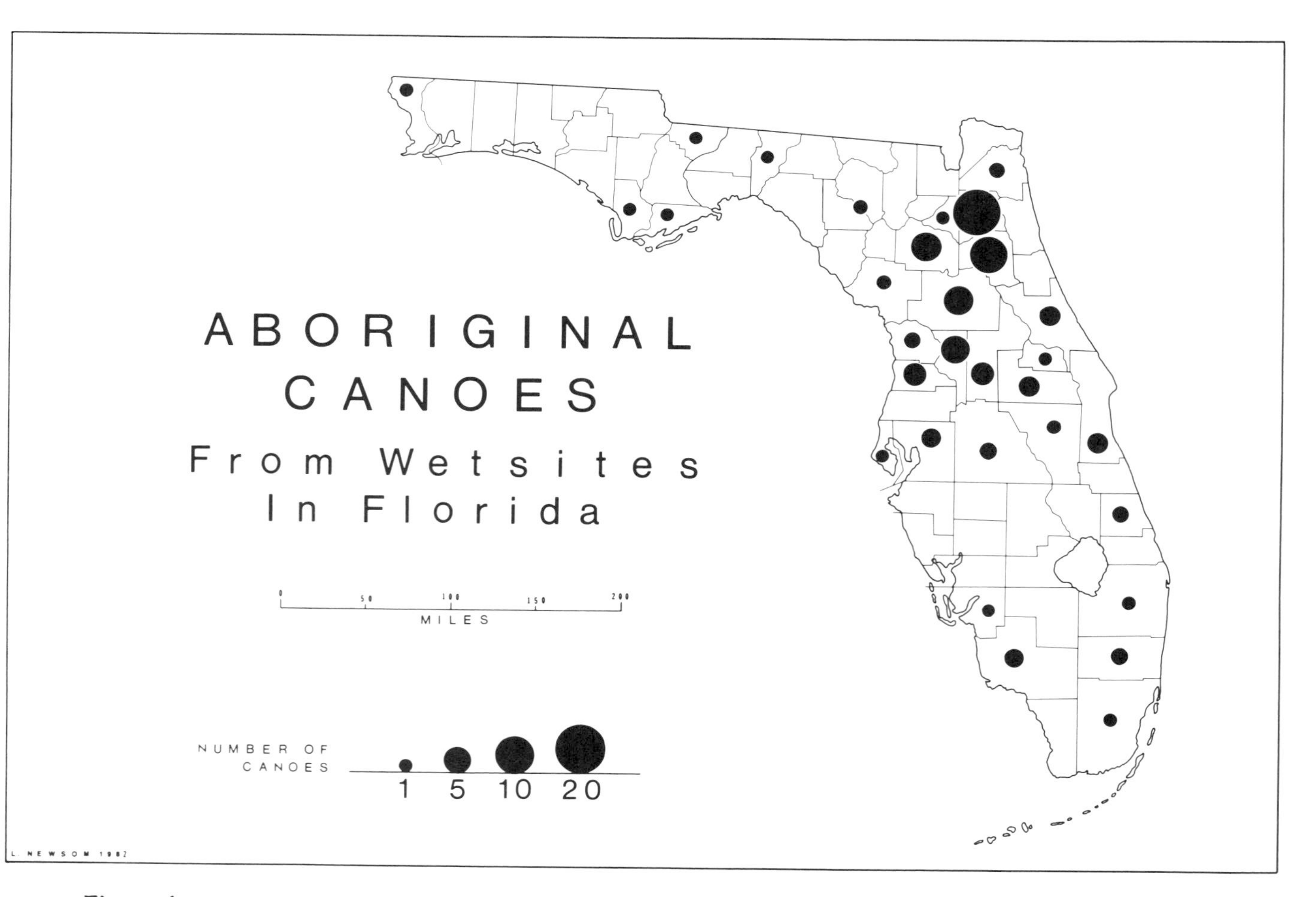
ABORIGINAL
CANOES
From Wetsites
In Florida
0 50 100 150 200
MILES
NUMBER OF
CANOES
1 5 10 20
L. NEWSOM 1982

Figure 1.

vor. Dr. William Stern and Dr. David Hall, Department of Botany have provided identification of individual specimens. Such a collection would be of great benefit to future wooden artifact investigations. (I am taking sabbatical leave from the University of Florida in the 1982-83 academic year, to compile a library of wood sections and to identify wood species.)

All other objectives have been met. A data retrieval system, INQUIRE Database Management system, has been initiated and will eventually contain data pertaining to all phases of Florida's maritime heritage. Treatment of wooden artifacts at the Florida State Museum has continued in spite of limited facilities and storage space. An immediate need is a facility large enough and equipped to house, preserve, analyze, interpret, and display all of these unique cultural items. Wooden materials from the Bay West site near Naples have been preserved and returned to the Collier County Museum. Approximately forty of these specimens are undisputedly artifactual and should contribute greatly to a comparative analysis of wooden materials dating to the same period (ca. 6500 BP). Organic materials from the Hontoon Island Site near Deland, Florida have been categorized, partially analyzed, and preserved. Additional comments about both of these sites occur later in this report. It is apparent that Florida holds a wealth of potentially invaluable cultural material which, if recovered and properly preserved, could provide data for the investigation of a variety of important anthropological and historical problems.

Preservation in the Natural Environment

The primary focus here will be on environmental conditions that favor preservation of organic materials, particularly those items utilized by human populations. Theoretically these very perishable materials need only be deposited in still water deep enough to discourage microbial destruction, but it is unlikely that organic remains in Florida, especially woods and other flora, could have survived if they had not been buried in water-saturated, oxygen-free conditions. The majority of specimens for which there is specific recovery information come from peat or muck deposits.

The peat deposits of Florida have been discussed in numerous publications since the 1800s. The most complete account was authored by David (1946). Recent excellent articles about Florida peats, their development, ages, uses, and quantity, appear in Gleason (1974).

Peat is formed under several environmental situations; its composition depends upon vegetation, soil, bedrock, moisture, topography, geographic location, etc. Muck may be derived from fibrous or woody peat that has disintegrated into smaller fragments brought about by changes in moisture content and for other reasons. In this case, the conditions that cause the change from peat to muck may also cause a disintegration of organic cultural items incorporated in it. There are, however, other kinds of mucks, e.g., soil muck and diatomaceous earths or mucks, that are not derived from peats. It is possible that organic artifacts may be found in these kinds of mucks if there was human habitation in the area. Sometimes conditions for preservation are created as a result of human activity. At the Fort Center Site near Lake Okeechobee on Fisheating Creek, for example, a pond dug by the prehistoric inhabitants for ceremonial purposes gradually filled with mud from slope erosion and debris from the midden and house-cleaning activities. This artificial "bog" contained well-preserved organic artifacts (Sears, 1971). At the present time, adequate information has not been compiled whereby correlations can be made between organic soil types and preserved organic artifacts. More than twenty varieties of organic soils are listed on soil survey maps for the state of Florida. Archaeologists are not qualified to assess the distinctive vegetational and depositional environments in which different peat types were formed, nor are they qualified to assess the subsequent changes in the deposits. I would recommend that soil scientists, paleobotanists or other specialists be consulted who can provide information about the history of these deposits. In the meantime, archaeologists would do well to refer to such deposits as organic soils rather than using a more specific connotation such as peat, muck, etc. As I discovered in my statewide request for information, I may have received more complete accounts if I had asked about organic soils rather than peat deposits.

Overwhelming evidence suggests that severe

aridity existed in Florida when glaciation occurred elsewhere during the Pleistocene. As sea levels gradually rose in the early Holocene and Florida became warm and moist, various effects on the land mass occurred. Of application to this study is the fact that drainage was slowed down as the water-table rose and favoured conditons for peat accumulation especially in areas where the Hawthorne Formation or other impermeable deposits formed an effective aquiclude. The inundation of the Everglades, for example, began about 6000 years ago (Fairbridge, 1974: 227), and Lake Okeechobee originated about 6000 years ago (Brooks, 1974). Except for older deep lake deposits, 6000-8000 years ago marks the onset of the development of widespread organic soils and our knowledge about modern flora, fauna, and climate in Florida (Watts, 1969, 1975, 1980; Clausen et al., 1979). There is unsettled controversy about transgressions and regressions of the sea in Recent time and about climatic changes during the past 10,000 years. Most authorities concur that Florida has been warm and wet with at least one period, about 5000-8000 years ago, of weather warmer than today. They also agree that there were probably minor regressions that caused lowered water tables for a long enough time for peat fires to leave a record of severe burning in various strata throughout the Everglades (Cohen, 1974: 213) and elsewhere (Brooks, in Gleason, 1974: 311).

The archaeological record adds supporting evidence. A great many areas remain to be sampled, but there appear to be hiatuses in the archaeological record at approximately the times suggested that Florida experienced dry periods during the Holocene; and organic artifacts have been recovered that fall within the postulated wet periods. Four sites that date around 6500 years ago contain human skeletal remains and organic artifacts recovered from an organic soil matrix. The Atlantic (Altithermal) period dates to this time. Other organic artifacts have been recovered dating between 2500-3000 years old, the time of the Sub-Atlantic period. There are many sites 1000 years old, the date of the Neo-Atlantaic period. The Atlantic, Sub-Atlantic, and Neo-Atlantic Periods were times of warm, wet weather in Florida. Few sites have been discovered with organic artifacts dating in between these times. Is it possible that cultural material of other time periods was destroyed when Florida experienced extended drought conditions and peat deposits burned or dried to below 40 percent moisture and could not reabsorb water (Davis, 1946, 46)?

Survey

Davis' (1946: Fig. 1) map shows that peat deposits in Florida are nearly statewide. In an effort to determine the areas where evidence of human occupation might be found in peat (organic) matrices, I wrote to all of the County Extension Directors (supplied by the Institute of Food and Agricultural Sciences, University of Florida, Gainesville), the Soil Conservation Offices (Soil Conservation Service, Florida Directory 1979), and to all of the peat operators in Florida reported in "Peat Producers in the United States" (Bureau of Mines, 1979). With Dr. George F. MacDonald, Dr. James Shaeffer, and various field assistants, I visited 35 sites to test deposits and talk to local people. At one site, Hontoon Island, a 3-meter square was excavated in December 1980.

In general, the response to my request for information was excellent from the state and federal agencies. Only seventeen counties did not reply. When necessary, soil survey or state planning maps were obtained from the Soil Conservation Service in Gainesville. None of the peat producers in the state replied to my inquiry. A county by county summary of the results of the survey was compiled and tabulated.

With the information gathered from the field survey, maps, and local cooperation, we should now be in a position to make recommendations about the need for additional investigations. Priority areas were selected for future testing because (i) limited information suggests that organic cultural remains may be present; individuals need to be contacted personally who have extensive knowledge of the terrain and site visits should be made to test the soils; (ii) locations where numerous wooden artifacts have been recovered in the past should be studied more systematically; for instance, in Clay, Putnam, and surrounding counties where so many canoes have been found, the history of drainage needs to be studied and at least one site excavated to determine if artifacts other than canoes might be present; (iii) rivers that may have been entry routes into Florida that contain organic soils should be tested for the added dimensions organic artifacts might supply about trade routes, etc.; (iv) organic soils are present, adjacent to well-drained land in locations undisturbed by modern development (or are endangered by development projects). Even disturbed areas may sometimes remain partially intact.

As a general rule, Florida has become wetter as the Holocene progresses. Areas now listed as swamps were probably not as wet earlier on. The well-developed channels, even though sluggish today, indicate to me that channelization occurred when greater slope permitted drainage patterns to form. The early sites (6500 years old) seem to be areas where the accumulation of organic sediments, or ponding, kept pace with a gradual sea level rise and land became inundated that contained very little moisture earlier.

In many cases, the key seems to be the slope and the nature of the soil on the slope adjacent to the wetland or organic soils. We can test the hypothesis, that areas listed as peat or muck soils adjacent to areas with sloping sandy soils in locations of extensive wetlands such as marshes or swamps, will contain evidence of human occupation, if people were ever there at all. The value of such a prediction will be especially high in areas with access to a river system. A number of factors, both natural and cultural, are interrelated: water supplied from seepage through the porous sand supplies the moisture to support vegetation adapted to an aquatic environment, which in turn supports fauna; and man at the end of this food chain, can live quite comfortably with the resources present in these areas. As an example, nearly all of the food remains and the materials utilized to manufacture tools at the Hontoon Island site were typical species of an aquatic environment. In addition, if an area is adjacent to a river, it provides a means of communication, trade, and energy conservation. I recently estimated that four people transporting goods in a canoe for fifty miles would save 25,000 Calories (ed. i.e. kilocalaries; equivalent to 39 Horse Power hours) over that needed to take the same trip overland.

Since people use their culture as an adaptive mechanism, in an area with abundant resources but not surrounded by well-drained living quarters, human habitation may begin to build up its own "uplands" and subsequent drainage from these "uplands" will create a ponded micro-environment in which organic artifacts are preserved. This situation may have occurred at the Key Marco Site and other of the large mound sites in Florida.

Tidal areas that are flooded daily might not be good places for preservation to occur because too much energy is involved. But one should not rule out the possibility that areas with organic remains might be found. Some maps distinguish between tidal marsh and tidal swamp. Tidal marsh has more possibility for preservation because it is not subject to wave action like tidal swamp.

Many of the survey maps are said to be outdated and are not available for distribution. It is interesting to contemplate why a soils map would become outdated. Under normal, natural processes, it is unlikely that this should occur in 50 years, 500 years, or longer. In some cases, of course, early surveyors may not have taken the care in mapping that is expected today or had access to technology such as aerophotos. The recent soils maps reflect drainage of about 80 years and it may not be apparent from these maps if organic soils were more abundant earlier. There has been a great deal of subsidence where peat lands have been used for agriculture or drained for other reasons. The water table in south Florida has been lowered 2 metres in the last 40 years, for example. There has probably been extensive development in some areas since publication of many of the maps, e.g., Dade County, where Miami is located, has not been mapped since 1947. It is imperative in areas that may contain valuable information about aboriginal cultures, that the history of drainage be studied and individuals with vast knowledge of local conditions be consulted. It is important also to discuss with soils specialists and paleobotanists the qualities of the various organic soils in Florida with regard to preservation. Many of the soils maps indicate that soils listed as swamp are unclassified; there is a good possibility that these contain potential for organic preservation.

Future studies of any area should include the history of drainage projects, a search for pre-drainage period maps, and the history of peat mining. Davis (1946: 7-13) discusses attempts to develop nonagricultural uses of peat. The first use of peat for fuel in Florida was in 1905. Most of the peat mined today is sold to nurseries for humus and fertilizer.

Organized efforts are presently underway to mine peat for energy production on a large scale in Florida and studies are being conducted to determine its feasibility (e.g. Williams Brothers Environmental Services, 1980). Peat is one of the most abundant and attractive of the "synfuels," which to date have not been developed because of the availability until recently of cheap oil. Florida is estimated to have approximately 1.2 million hectares (three million acres) of peat containing 5.4×10^9 metric tonnes (6.0 U.S. billion U.S. tons) of the mineral. This quantity could deliver the equivalent of nearly 13.7 thousand million (U.S. billion) barrels of oil. In other words, it would take a tonne (ca. 1 ton) of peat to produce little more than the equivalent of two barrels of oil. The total energy available from all U.S. peat resources is estimated to be equivalent to about 240 thousand million barrels of oil or enough to furnish all of the United States natural gas requirements for 35 years. Peat energy can provide an important domestic substitute for imported energy sources. According to Williams Brothers' report, there are 38 counties in Florida that contain organic soils with a minimum thickness of 41 centimetres. A 60-MW plant, to operate for 20 years, would require 1400 hectares (3500 acres) of peat, 2 metres deep.

Thousands of years of accumulated history of vegetation, climate and human utilization of areas will be lost if the peat is mined. It is a matter of fact that nearly one hundred percent of the wooden artifacts we have recorded during this study were found during dredging operations or other types of development projects. In areas where drainage has occurred, subsidence is estimated to average 2.5 cm per year. At this rate, it takes only about six years to erase the accumulated history of 100 years of "rapid" peat development (ca. 16 cm/century).

Preservation

Since Frank Hamilton Cushing retrieved the spectacular wooden masks, figurines, and utilitarian objects at Key Marco (Cushing, 1897), people have been aware that the prehistoric inhabitants of Florida had an elaborate woodworking technology. For a long time the Key Marco site was a unique site because of the nature of the artifacts found there. Human remains and occasional artifacts have been found sporadically in bog deposits in the Old World, and the Swiss Lake Dwellers are well known to any student who has taken a course in Old World Prehistory. But the Key Marco site had no equal until Daugherty excavated the Ozette Village Site on the Olympic Peninsula where a mud slide entombed and preserved a village, its occupants, tools, and ceremonial objects (Daugherty, 1974: 1980).

Individual wooden artifacts have been found throughout Florida; several specimens were taken from the Chosen Mound at Belle Glade (Sterling, 1934; Willey, 1949); and Sears, in the last decade, has retrieved a great number of wooden artifacts at the Fort Center Site which he has preserved with liquid polyvinyl acetate (Sears, 1971). Other examples are discussed elsewhere in this report. In addition, I believe Florida has the largest collection of prehistoric watercraft in the world. The broad geographic distribution, the diverse environmental situations, and the great time span of the artifacts recovered thus far, suggests that a great wealth of wooden artifacts may still be undiscovered. In fact, in comparing Florida's prehistoric wooden artifact inventory to that of all other regions, it is not premature to conclude that Florida is unsurpassed in regard to quantity, variety, and antiquity of this rarely preserved heritage.

With the recovery of this heritage comes a major responsibility: the application of preservation techniques. Nearly all of the wooden masks, figurines, and other objects from the Key Marco Site (Gilliland, 1975; Purdy, 1974) have disappeared or degraded so badly that one can scarcely handle them or recognize them as artifacts. The whereabouts of many wooden objects that were reported in the early literature are not known, e.g., those from the Chosen Mound.

We have a great deal yet to learn abut preserving prehistoric wooden artifacts. One major problem is a matter of semantics. We speak of waterlogged wood being preserved at a saturated site. Yet the wood is obviously not in pristine condition and one need only compare the changes that take place in old waterlogged wood if it is permitted to dry with a piece of recent wood exposed to the same conditions. The point is that, in most cases, the wood is not preserved at all. We can still see it because its macroscopic, physical dimensions have remained pretty much intact in an oxygen free environment. This only means that organisms needing free oxygen have not been able to degrade the wood. Other mechanisms are operating, however, because otherwise why is it that the lignin or cellulose are leached out and replaced with water giving the impression, when still waterlogged, that the piece is structurally sound?

All of the information we have obtained, through correspondence and publications suggests that polyethylene glycol (PEG) is the product of choice to preserve large, old, degraded, waterlogged wooden objects. It is safe to use and easy to handle. It penetrates the wood, replacing the water that has been holding the wood in shape since the leaching out of lignin and the loss of cellulose (supportive cell wall). Other products might be superior in some cases, perhaps with smaller specimens or a limited number of objects. Polyvinyl acetates, however, are usually not recommended because their molecules are too large and will enter into the cell structure except under very unusual circumstances, e.g., in the case of extremely degraded wood.

We have been using a "cookbook" type of procedure for preserving wooden artifacts with PEG but it should be kept in mind that the treatment of each specimen is dependent upon many variables, including the specific type of wood, its degree of degradation, depositional environment, and how the wood has been altered from its natural form. Different types of wood have differing resistance to decay which can significantly affect the rate of degradation.

The most imperative initial step is to make sure that the wood is kept saturated. We constructed a preservation vat in which a solution of PEG and water were kept warm and circulating to hold the PEG in solution. The molecular weight of the PEG depends upon the condition of the wood: the more degraded the specimen, the higher the molecular weight. The completely waterlogged object should first be placed in a 10% solution of PEG and water. As yet we have not done microscopic studies to determine how long it takes for the PEG to penetrate the mass or how long an object should remain at each concentration. The thicker the artifact, the longer the recommended duration. A canoe, for example, should probably remain in 10% solution for a month, 25% solution for six months, and 50% solution for one year. If one is anticipating that these unique specimens will last for thousands of years as a result of treatment, one should not be in a hurry to take them out of the preservation vat. When removed the object should be wiped gently with clean water to remove the excess wax, and then room dried. If bacterial or fungal growth is noted in the solution during preservation, it can be skimmed off or a very small amount of a biocide can be added.

We have preserved a number of individual wooden artifacts during the past year as well as two large collections from the Bay West Site (8-Cr-200) and the Hontoon Island Site (8-Vo-202).

Approximately 200 bags of unprovenienced wooden material (ca. 500 items) from the Bay West Site near Naples, Florida were received at the Florida State Museum. Approximately 40 of these specimens were culturally modified including 4 fire-starters, 1 ladle fragment, 2 atlatl or spearshaft section, and 2 atlatl spears (antler with wood centers).

Only two specimens showed any significant modification after treatment. Both were tapered objects (artifacts). These objects were a darker wood color originally than the other objects and it is believed that they are of a different species which may account for their unsuccessful treatment. A further study of this phenomenon may reveal something about the properties of the wood selected to perform specific tasks. One important specimen, the atlatl or spear shaft fragment with incised lines, was unaccountably left out of the water bath in which it had been soaking and, as a result, became greatly shrunken. Fortunately, it had been measured, weighed, and photographed before this incident occurred.

Along with the wooden materials from the Bay West site, a large quantity of human skeletal remains was also recovered and sent to the Florida State Museum for analysis by Dr. William R. Maples. A small amount of faunal material was found among the human bone. Some of this appeared to have been culturally modified. Forty-three species were identified in the Zooarchaeology Laboratory at the above museum by Steve Hale! See Beriault et al. (1981) for additional information about the Bay West Site. Because of the nature of the salvage operation at the site, it is difficult to determine exactly the function of the site. It is suggested that it was a burial area in which bodies were placed on wooden frames for decomposition before final burial and that some of the artifacts may have accompanied the deceased as burial goods. It is hoped that the accidental recovery of so much potentially informative material from the 6500-year-old Bay West Site might alert professionals and lay persons to the likely existence of similar materials elsewhere.

A 3-meter square was excavated at Hontoon Island in December 1980 from which thousands of pieces of detrital wood and wooden artifacts were recovered as well as pottery, stone, flora, and fauna. The wood is presently being preserved. Since it cannot be analyzed until it is preserved, little can now be concluded about its use at the site. A canoe paddle fragment that was found indicates that canoes were used in the lagoon, Lake Beresford, and the St. Johns River. The detrital wood chips, in fact, might be the by-product of canoe manufacture. In addition, many charred post pieces were recovered and might indicate the construction of structures over the water, either habitation structures or platforms used to more easily exploit the aquatic resources. Some of the wood may have resulted from housecleaning activities where objects were "swept" into the water from the large occupation midden adjacent to the lagoon. Other sources of wood in the excavated square probably resulted from inclusion of natural debris falling from trees or small animals dying, etc. The composition of the deposits in the square is very complex. Margaret Mead said that when she lived among the Manus of New Guinea in 1928, she stayed in a house with a high, thatched roof and a slatted floor through which small objects fell into the water (Mead, 1981). I can envision precisely this same kind of thing occurring at Hontoon. When preservation of the wood is completed, an attempt will be made to identify species, the qualities of different woods that caused them to be selected by people for different tasks, and the effect that the selection and alteration of the woods has on the laboratory preservatives that are applied, or, indeed, whether the wood was preserved at all. My impression is that artifactual wood does not degrade as quickly as naturally deposited debris.

It would be desirable if I could conclude this section of the report with a detailed comparison of the wood preserved at the Bay West Site and that preserved at the Hontoon Island Site. Unfortunately, that is not possible. The Bay West Site is much earlier by approximately 5500 years and was probably a mortuary site. Hontoon Island probably represents a habitation area and an area of intense activity. The "excavated" area (dredged) of the Bay West Site was much larger than the excavated area at Hontoon Island, but the Bay West materials have limited interpretive value because of the conditions of their retrieval; the Hontoon Island materials have limited intepretive value because only one square was excavated.

The results of a wood degradation study indicated that the wood from Hontoon Island is very fragile losing, in some cases, over 90% of its wet weight when allowed to dry which resulted in checking, cracking, splitting and warping. It was my impression in handling the Bay West material after it had been removed from PEG that it would be in even worse condition than the Hontoon wood if it had dried out. While it is true that the Bay West materials were stabilized with PEG, the exterior of many specimens were already flaking or shedding so badly before treatment that they probably should now have some surface preservative applied to "glue" the surface to the rest of the object.

Considering how long people have been using wood (hundreds of thousands of years) it is quite possible that, long before the Western Hemisphere was inhabited, woodworking techniques and wood properties were already known elsewhere. In that case, would we expect there to be a vast difference in woodworking skills between Archaic Period populations in Florida and people utilizing the resources at Hontoon Island about 800 years ago? It is true that no works of art or carvings were found at the Bay West site compared to Hontoon Island where an owl, frog, and pelican totem were found. This could be due to accidents of recovery or to socio-economic situations rather than to lack of ability.

We have not yet turned the first corner in

Florida with regard to problems of preservation, wood identification, and cultural preferences. Future emphases should be on training students in wood identification, thin section analysis, preservation methods, prehistoric wood manufacturing techniques, and finally, on a comparison of changes in styles and preferences through time. Only then will we begin to understand the requirements and the behaviour of the people who were producing items of wood in Florida.

Summary and Recommendations

The focus of this report has been on conditions in the natural environment that preserved organic cultural remains and the necessity to treat and study the materials if they are found. These unique specimens are found only in wetlands and they provide information about past technologies and art styles that is not available elsewhere. It is only in these areas that there is an opportunity to discover the relationships between resources available and resources utilized by prehistoric groups of people. Despite the excellent preservation in Florida wetlands, flesh and hair are not preserved because of the acidity of the soil and other compositional factors. At terrestrial sites, pottery, stone, and shell are all that survive. Their importance to the people who used them may have been minor but they are assigned great importance by archaeologists who analyze them.

Wetlands and wooden cultural material in wetlands are found nearly statewide in Florida. Organic artifacts in Florida are known to be more abundant than anywhere in the world at the present time, including the largest number of prehistoric watercraft. Wooden cultural material has been recovered that is 10,000 years old, with large quantities beginning at 6500 years. These objects include canoes, toy canoes, canoe paddles, totems, bowls, pestles, figurines, masks, spearshafts, adze handles, float pegs, fire starters, cordage, etc., often associated with skeletal material as grave goods. The cultural material from wetlands, along with the flora, fauna, and climatic record, is endangered; it is an invisible heritage that is being destroyed, often because developers are not aware of it. Florida is a rapidly growing state necssitating extreme modification of certain areas, particularly waterways, that are attractive to visitors, future residents, and industry. A conflict of interest often arises between groups who need to dredge and drain in order to utilize (or even create) acreage profitably and those who wish to maintain the land in its natural state. It is impossible to calculate how extensive the destruction of sites has been as a result of draining, dredging, and removal operations. Studies have demonstrated that drainage operations will affect soils (and artifacts in them) many miles away.

The occasional references to the Swiss Lake Dwellers that are made throughout the report are not coincidental. In June 1981, I travelled to Switzerland and visited many of the lake sites where wooden artifacts were found and museums where the specimens are displayed. I was amazed at the similarity of the terrain surrounding the lakes compared to the terrain in Florida adjacent to wetland areas. The Swiss Lakes probably developed from glacial scouring. Many of Florida's lakes were created when the underlying limestone dissolved and the Karst topography resulted in an undulating landscape. The outcome of these different origins was much alike: seepage from uplands created ponding and permitted the growth of vegetation which, in turn, preserved in it evidence of human utilization of the area. This observation needs more investigation to test this hypothesis.

Another interesting point is that the dates of the Swiss Lake sites and the artifacts from them are very similar to those in Florida. It appears that as the glacier receded in Switzerland, the ice-scoured areas became lakes which were fed constantly by the melting snow of nearby mountains. In Florida, the rise of sea level and the warm moist climate occurring at the end of the Pleistocene caused the creation of wetland areas that receive water from seepage of adjacent uplands. Water flows from these porous sand ridges even during periods of drought. In both cases, the age is nearly identical for the earliest remains found in any quantity, about 6500 years ago. In both cases moreover, the objects tend to be utilitarian, such as house posts, bowls, shafts, etc., rather than works of art. In fact, the earliest wood carving found in Switzerland so far is only 2000 years old. Again, this situation needs more investigation.

Areas for systematic study need to be selected very carefully, because it is unlikely that many opportunities will arise or be funded to conduct thorough investigations. It is impossible to overstress the importance of local knowledge about past and present activities relating to the land.

I attempted to concentrate only on the archaeological implications of wetland areas but it is impossible not to consider other parameters as well: climate, time, vegetation, hydrology, geology, soils, etc. Furthermore there is the possibility that dendrochronology studies of cypress which are now being conducted, may provide a means to date wetlands deposits. Because so many parameters are involved, this preliminary investigation seems to deal principally in generalities. Armed with this information however, I believe we are now in a position to conduct more thorough studies of specific locations and to make recommendtions about sites and about preservation of wooden artifacts.

Most projects are of a "start-stop" nature and this one is no exception. At the time of writing, all financial support has ceased although the investigations are far from completed. Additional funds, facilities, and personnel are needed to preserve, study, interpret, and display organic artifacts in Florida. They are a unique and non-renewable heritage; the public should know about them.

A thematic nomination of wetlands to the National Register of Historic Places would be a means of protecting these sites and their cultural history. These areas are part of the climatic and cultural history of our nation. By destroying them we lose an important link to the past and processes in operation that provide a preview of coming events.

References

Beriault, J., R. Carr, Jerry Stipp, Richard Johnson and Jack Meeder, "The Archaeological Salvage of the Bay West Site, Collier County, Florida," Florida Anthropologist, vol. 34, no. 21, 1981, pp. 39-58.

Brooks, H. Kelly, "Lake Okeechobee," in: Environments of South Florida: Present and Past, Edited and Compiled by Patrick J. Gleason. Miami Geological Society, memoir 2, 1974, pp. 256-286.

Cohen, D., "Evidence of Fires in the Ancient Everglades and Coastal Swamps of Southern Florida," in: Environments of South Florida: Present and Past, Edited and Compiled by Patrick J. Gleason. Miami Geological Society, memoir 2, 1974, pp. 213-218.

Clausen, C.J., A.D. Cohen, Cesare Emiliani, J.A. Holman and J.J. Stipp, "Little Salt Spring, Florida: A Unique Underwater Site," Science, vol. 203, no. 4381, 1979, pp. 609-614.

Cushing, Frank Hamilton, "A Preliminary Report on the Exploration of Ancient Key-Dweller Remains on the Gulf Coast of Florida," Proceedings of the American Philosophical Society, vol. 35, no. 153, 1897.

Daugherty, Richard D., (with Ruth Kirk) Hunters of the Whale, (New York: William Morrow & Co., 1974).

Daugherty, Richard D., "Wetlands Research at the Ozette Site," in: Florida Maritime Heritage, proc. of a conference, March 23-23, 1980, ed. Barbara A. Purdy. (Tampa, Florida: 1980).

Davis, John H., "The Peat Deposits of Florida: Their Occurrence, Development, and Uses," State of Florida, Department of Conservation, Geological Bulletin, no. 30, 1946.

Fairbridge, Rhodes W., "The Holocene Sea Level Record in South Florida," in: Environments of South Florida: Present and Past, Edited and Compiled by Patrick J. Gleason. Miami Geological Society memoir 2, 1974, pp. 223-232.

Gleason, Patrick J., (Editor and Compiler) Environments of South Florida: Present and Past, Miami Geological Society, memoir 2, 1974.

Gilliland, Marion Spjut, The Material Culture of Key Marco Florida, (Gainesville: The University Presses of Florida, 1975).

Mead, Margaret, "New Guinea Revisited," Odyssey, 1981, (reprinted by permission of Redbook Magazine, February 1965).

Purdy, Barbara A., "The Key Marco, Florida Collection: Experiment and Reflection," American Antiquity, vol. 26, no. 4, 1974, pp. 142-152.

Sears, W.H., "Food Production and Village Life in Prehistoric Southeastern United States," Archaeology, vol. 24, no. 4, 1971, pp. 93-102.

Stirling, Matthew W., "Smithsonian Archaeological Projects Conducted Under the Federal Emergency Relief Administration, 1933-34," Annual Report of the Smithsonian

Institution, 1934, pp. 371-400.
Watts, W.A., "A Pollen Diagram from Mud Lake, Marion County, North-Central Florida," Bulletin of the Geological Society of America, vol. 80, 1969, pp. 631-642.
Watts, W.A., "A Late Quaternary Record of Vegetation from Lake Annie, South-Central Florida," Geology, vol. 3, 1975, pp. 344-363.
Watts, W.A., "Late Wisconsin Climate of Northern Florida and the Origin of Species-Rich Deciduous Forest," Science, vol. 210, October 1978, pp. 325-327.
Willey, Gordon R., "Excavations in Southeast Florida," Yale University Publications in Anthropology, no. 42, 1949.
Williams Brothers Environmental Services, Energy Production from Central Florida Peat Deposits: Preliminary Study. (Prepared for Coastal Petroleum Copmpany, Tallahassee, Florida). (Tulsa, Oklahoma: Williams Brothers Environmental Services, 1980).

Questions

Anon: Did you measure the pH when you were working on these sites?

Barbara Purdy: At the Bay West site we measured it as 3.5, at Hontoon we didn't measure it. There is much published information which indicates that pH 3.5 is about the average value.

Victoria Jenssen: I think that when you were here previously John, (ed. John Coles), George MacDonald mentioned that coring was being done in Florida.

George MacDonald: I was talking about Alaska. By using jacketed coring devices, with liquid nitrogen cooling, you could pull frozen cores out of peat bogs in Florida. In this way deterioration of the cores in transit to the lab can be minimized.

Barbara Purdy: We did that on the survey in two places. We took long sections with the assistance of Dr. Deevey's wetlands laboratory at the Florida State Museum. With this technique whole column samples can be brought into the lab.

CONSERVATION OF WATERLOGGED WOOD AT A 16TH CENTURY WHALING STATION

James A. Tuck
Memorial University of Newfoundland,
Newfoundland, Canada

This short paper will describe briefly our excavations at a 16th century Spanish Basque whaling station in southern Labrador and our attempts to conserve the numerous organic specimens which have been recovered there. I should state at the outset that I am not a conservator, nor is there a trained archaeological conservator except for Parks Canada regional staff employed anywhere in Atlantic Canada, a problem to which I will return later in this paper. The Red Bay project has been fortunate, however, to have had the assistance of the Canadian Conservation Institute almost from its beginning in 1977. It is difficult to express our gratitude for we now know that without the cooperation of the Canadian Conservation Institute our work at Red Bay probably would have been impossible.

The Saddle Island Site

The chain of events which led to the discovery and excavation of the shore stations used to process whales during the latter half of the 16th century began with archival research in Spain by Mrs. Selma Barkham during the early 1970s. Mrs. Barkham was contracted by the Public Archives of Canada to search archives in Spain for documents pertaining to the early history of Canada. In this endeavour she has been very successful, and a large number of records pertaining to both the east and west coasts have been catalogued and photocopied. Many of the documents made mention of Basque whaling expeditions to a place known as the "Grand Bay" which, through the mention and description of several ports and sailing directions to them, Mrs. Barkham was able to identify with the Strait of Belle Isle, the narrow channel which separates northern Newfoundland from Labrador and mainland Canada. One of the ports most frequently mentioned was Butus, Buytes, Boytus or Buitres which is described as a good harbour sheltered by two islands from the often treacherous seas and weather of the Strait of Belle Isle.

Only one harbour in southern Labrador fits this description, a place now known as Red Bay, which still enjoys the reputation as the best harbour on the southern Labrador coast. Accordingly we planned a reconnaissance there for the summer of 1977. This survey was successful in locating traces of Basque occupation at a number of small harbours. For the most part these consisted of scattered fragments of red roofing tile but on Saddle Island, in Red Bay Harbour, thousands of tile fragments littered the beaches and traces of stone walls could be seen protruding through the surface vegetation. Cementing the rocks which comprised these walls was a fine clay infused with a black, carbonized substance later identified as mammal fat. It seemed a safe bet that this substance was whale oil and that we had discovered the sites of "ovens" where the actual rendering process had taken place. During the same brief survey we also noticed numerous small wood chips and fragments eroding from a small deposit near the present beach but they were so well preserved that we suspected they were the product of some recent log squaring or boat bulding operation.

The other evidence of 16th century activity was sufficient to encourage us to return to Red Bay in 1978 for a full months excavation during which time the presence of several ovens was confirmed, what appeared to be a dwelling was partially excavated, and further testing in the small waterlogged deposit noted in 1977 produced barrel hoops, fragments of oak and beech, shreds of baleen and other material which associated the deposit with the 16th century whalers who had worked there. At this point we contacted the Canadian Conservation Institute and within 48 hours a conservator was at Red Bay to direct the field treatment of waterlogged objects and, not incidentally, artifacts of a variety of other materials which probably would have disintegrated were it not for this timely assistance.

Since 1978 the Red Bay Project has grown in scale and now employs more than 20 staff during the months of July and August. Excavations are planned to extend through to the 1983 field season.

As a result of our work through the 1981 season we are now able to offer some suggestions about what one of the shore stations must have consisted. The heart of each station was an oven where the actual rendering took place. Those we have excavated thus far consist of a central wall up to 10 metres long running parallel to the shore. On the outside, or seaward side, of the wall were built circular fire pits with the central wall forming the back of up to five of these in a single oven complex. Atop these fireboxes were placed copper cauldrons in which minced blubber was placed to be "tried out" into train oil. Fragments of cauldrons have been found in the fire pits, many having burned fat which was used to fuel the fires still adhering to them.

Behind the stone wall there was a wooden platform from which the cauldrons were tended. This originally stood at the top of the wall, probably about one metre above the wall base. In one instance the platform, made of local softwoods and occasional reused oak boat planks, collapsed into a wet environment and has been preserved, although it has survived in an advanced state of decay. Tubs, made from barrels or barricas cut in half transversely, were set on this platform where their remains have been exposed. We suspect that these were half filled with water and used to cool and purify the oil since, when hot oil was ladled into the tubs, the dross would have sunk while the lighter oil, now cooled, would have floated. It could then have been ladled into casks for storage and shipment to Europe.

A second important activity at any shore station was the assembly, repair, and in some cases possibly construction of the casks in which the oil was shipped to markets in western Europe. Thus far we have excavated at two such cooperages. Both are located on high ground overlooking but upwind from the ovens and in both cases domestic as well as coopering activities are suggested by the material recovered. Such tools as adzes, a plane iron, a saw fragment and head vises are joined by coarse earthenware and majolica sherds, glass fragments, food bone and personal possessions to give the unmistakable impression of structures in which coopers both worked and lived. The clinching evidence for these structures having been cooperages, however, comes from waterlogged deposits where the refuse from the coopering operation was discarded. Some of the staves, head pieces, hoops, bindings, and discarded "blanks" from which barrel parts were cut appear almost as fresh as the day they were deposited more than 400 years ago. It is from these areas that the bulk of our preserved organic specimens are derived.

Finally, at least some shore stations had "wharves" or "cutting-in stages" built from rock ballast and squared timber cribbing located at the shore immediately in front of the oven. At least two structures have been recorded although no wood has been removed from them to date.

Excavations at five of these stations seem to have confirmed the basic pattern suggested above. We hope, in the next two seasons, to complete these excavations and to excavate completely the largest shore station to determine what other structures and/or activity areas might also be present.

Field Conservation

Frequent reference has been made above to the recovery of preserved organic remains. A number of deposits have been excavated which have produced textiles, leather, seeds and large quantities of wood. The last owes its preservation to a number of circumstances in addition to burial in sterile bog environment created by ground-water or run-off being trapped in various sized basins in the water-worn bedrock of Saddle Island. Some organic materials owe their preservation to the fact that they have been carbonized by heat from the rendering ovens or cooking fires. Others have become infused with oil or fat which seems to act as a deterrent to deterioration while still others have undergone a combination of fat infusion and carbonization. Finally, there are specimens of wood, such as the pea-size wooden beads recovered from the first cooperage to be excavated which were preserved by some fluke which we do not understand. In the case of this particular collection of objects, which apparently comprised a portion of a rosary, some sort of miracle is suspected.

As might be expected of material from such a variety of burial environments, condition varies greatly. Some wood is observable merely as a stain in the soil or peat in which it was encased while other specimens appear as fresh as the day they were discarded. Obviously not all of this material is worth the expense of conservation and a sizable percentage may actually lie beyond the present range of conservation technology.

Decisions must be made daily, therefore,

about what pieces should be retained for treatment and what should be done with those specimens which do not warrant treatment. Such decisions are admittedly subjective and are based primarily upon such factors as the nature of the object itself and its condition. It may seem heretical to discard any archaeological specimen, for even the most humble wood chip probably bears some information about the past. However, the cost in both time and dollars makes the complete conservation of every specimen impractical (in the case of dollars) and virtually impossible (in the case of time). In attempting to make such decisions, therefore, the amount of new information an object is likely to yield is balanced against its condition and the feasibility of conserving it.

Objects not brought to the field laboratory for preliminary treatment are recorded in the field using one or more of a number of techniques. They may be recorded in notes, drawings, photographs or any combination of the three. Occasionally the importance of an object in extremely friable condition warrants unusual recording techniques. One such object was the collapsed wooden working platform behind the oven wall. The individual planks which comprised the platform were in such an advanced state of deterioration that all concerned agreed that they probably could not be removed, let alone conserved. Therefore, the usual recording techniques were employed, following which Canadian Conservation Institute staff took a polysulfide rubber and plaster of Paris impression of a 1.5 metre section of platform, rock wall and one of the firepits on the opposite side of the wall. The mould itself and the cast made from it will not only provide a three-dimensional record of this feature, which, despite being sandbagged, may well be destroyed by winter storms, but will provide a reproduction potentially useful in interpretation either on the site or elsewhere. Although such a technique is difficult, time consuming, and somewhat expensive (not to mention messy) I believe that the preservation of unique and informative features more than compensates for the costs involved.

Once recording has been completed the objects may be reburied (either in place or, if their removal is necessary, in some similar environment) or may be discarded. Even if objects are to be discarded samples are usually retained either for wood identification, conservation research, or both.

Objects about which some doubt exists - i.e. can they be conserved? are they worth conserving? - are usually removed to the field laboratory for further study and recording. In many cases these must be block lifted since they are too fragile to withstand mechanical pressure of any kind. I suspect that an unsuccessful block lift will make some decisions for us but thus far we have had good results. This is particularly true of lifts using melted PEG 4000 for very degraded wood rather than plaster of Paris, paraffin, or some other substance. This technique seems to have several characteristics to recommend it. It is cheap and at least some of the PEG is reusable. It is also an inexpensive technique in terms of time, for the PEG sets rapidly and the object can usually be removed within an hour of the application of wax. PEG can also be applied directly to the object, without an isolating layer of foil or some other substance. This seems to provide a great deal more adhesion between fragments of disintegrating material and may account for our success in removing friable objects. PEG is also much more easily removable than a number of other substances as it can be dissolved in water or melted at low heat. A matrix of PEG also permits radiography much more easily than such substances as plaster of Paris. Finally, in the case of extremely degraded wood, PEG probably starts the conservation treatment immediately - at least it does not need to be removed before treatment begins.

Once in the laboratory a final decision regarding conservation can be made. In any case analysis of the object can be conducted under more or less controlled conditions and additional recording (e.g. studio photography) can be done if necessary.

If a decision to conserve the object is made it is stored in fresh water until it can be transported to the place where final conservation is to take place. Memorial University of Newfoundland has developed facilities for PEG immersion and freeze-drying of wooden objects. For the most part this has worked very well although the absence of an archaeological conservator causes us to rely heavily on the Canadian Conservation Institute. Since they are located about 2500 km from St. John's, delays and communication difficulties are inevitable.

Conclusions

Despite the positive experience we have had with the Saddle Island material I would like to conclude by mentioning some of the problems we have faced. Some are technical problems which are clearly better addressed by professional conservators. Others are of a more general nature and can be addressed by a number of scholars including both conservators and archaeologists.

I shall begin with a technical problem. It is not a complaint but simply echoes what a great number of papers at this symposium have also said. The treatments we are employing (immersion in PEG 400 and subsequent freeze-drying) has provided excellent results insofar as the retention of shape, dimensions, and even colour of the specimens is concerned. For research purposes this product is more than adequate. However, for some other purposes it may be somewhat lacking. Even the most durable objects have a soft surface and a very light weight which leads me to believe that they will not stand up to even limited handling by researchers, museum curators and others who may have occasion to do so. A surface coating with a higher molecular weight PEG may be the answer to this particular problem. I doubt, however, whether any grade of PEG has the adhesive qualities necessary to conserve some wooden objects. Objects which are cracked or split, even though the two pieces are in physical contact, will not be bonded when the PEG treatment is completed. This is acceptable when dealing with a few objects broken into a relatively small number of pieces but when fragmentation is extreme or a large number of objects is to be treated some substance with greater adhesive qualities than PEG would be desirable. Although my own definition of a properly conserved object (i.e. one into which you could drill holes and then use as a cribbage board) probably cannot always be achieved, there seem to be some disadvantages to the PEG treatment which could be overcome. I will limit my comments on technical matters to those raised above largely because I am not qualified to discuss them further and they are probably naive, anyway.

Several other more general problems might also be mentioned by way of conclusion. The first involves what should be conserved and to what state objects should be restored or stabilized. I have already mentioned this above but the solution reached at Red Bay is probably not satisfactory to all parties. Our decisions have been based largely upon the amount of information likely to be yielded by an object as well as its condition and the cost (in time and/or dollars) to conserve it. This decision has been arrived at by consultation between conservators and archaeologists. Notably lacking from this process are considerations involving display and on-site interpretation as well as the opinions of curators or museum directors who will ultimately be responsible for the curation of the collections. I suspect that their opinions would be at variance with our own and might range from "conserve everything completely" to, for example, "don't bother with all those nails, we don't want to store them." I cannot offer a solution to this dilemma but can only make a plea for more cooperation among conservators, archaeologists and those institutions legally responsible for the collections resulting from archaeological excavations. Perhaps at some future meeting of a group such as this the opinions of museum personnel might be expressed.

My second point also involves a group of people who have been left out of a process vital to the cooperation between conservation and archaeology. Those of us who attended this conference are probably all convinced of the need for cooperation between conservation and archaeology throughout the duration of any field project. This includes having conservators involved at the planning stages and continuing their involvement until the final report is completed. Many more archaeologists need to be aware of this, but I think the balance is now tipping toward the positive side. Archaeologists are becoming aware of the benefits of having professional conservation staff involved in field operations. If this trend continues we will soon be faced with a shortage of conservators. This shortage, in Canada at least, will not result from a lack of capacity to train conservators. It will result from a lack of places for them to find employment. Archaeologists may be able to provide 2 to 3 months contracts for summer field work but few research grants are generous enough to permit the hiring of a full-time conservator. This is where the "left-out" third party comes in. Not only must archaeologists be made aware of the conservation profession but the people likely to be able to hire conservators must also be made aware of the benefits of full-time conservation staff. In many cases the same people responsible for collection curation are unable or unwilling to employ conservation staff. This is true not only of museum personnel but of university administrators, as well. In the university setting conservators could not only provide the usual stabilization and restoration services but could also enrich considerably both undergraduate and graduate curricula in archaeology. Many departments of anthropology and archaeology teach specialized courses in soils analysis, zooarchaeology, geology, and so forth to archaeology students, but, to my knowledge, no university course in Canada pays more than lip service to conservation. Again, I cannot offer a solution to this problem but will suggest that it will be a major one before the decade is out.

To summarize, then, our experience at Red Bay has been a very positive one. It probably would not have been possible, and certainly would not have been ethical, to undertake these excavations without professional conservation assistance. As a result of this experience I am convinced more and more archaeologists will seek the assistance of the conservation profession. I am also convinced that unless something is done to make others aware of the vital role conservation can play in the study of archaeology there will be a desperate shortage of conservation personnel before the end of the 1980s.

Questions

John Coles: Thanks very much Jim for that paper. My first comment that I wrote down is, "here is an archaeologist who speaks to conservators and who actually listens to what they say," and from your pictures and from your comments, although you end up with well-founded cautionary notes, it would seem to me that this is a good example of collaboration and perhaps something that many other archaeologists could adopt.

Victoria Jenssen: I want to comment about your remarks about the surface of treated wood. Really, in your conservation treatment you have successfully coped with the hard part, that of stabilizing the wood. Coating the wood after treatment is something which many people do, but is not something which is often discussed. Things which have been used are shellac, waxes and PEG 4000.

Colin Pearson: What we have found out this morning is that the requirements of archaeologists may vary considerably. Whereas one may be satisfied with an object in a relatively poor condition another would like to be able to eat his dinner off an object! This makes it very difficult to arrive at standards for conservation. Should the conservator produce something which is satisfactory or aim to produce an ideally conserved object?

James Tuck: For analytical purposes the end product we are getting is just fine, but for the purpose of our excavation permit we are required to return these objects "fully" conserved. Which means to the people to whom they are returned that they are going to last for a thousand years and stand up to handling by school children and museum curators etc.

Howard Murray: I have found that archaeologists want material to be in either one of two conditions:

(i) Suitable for photography and drawing only. An object for which the archaeologist has no further use and can go to a museum. This can be treated to a conservation standard rather than a handling - archaeological standard.

(ii) Other pieces may be required for school services and need to withstand much handling. You can treat these by epoxy resin impregnation.

John Coles: What do you mean by "conservation standard;" as opposed to a "museum standard."

Howard Murray: I mean a process which has the potential to be reversed, and which does not involve a surface treatment. I prefer that the object should not be handled.

WATERLOGGED WOOD FROM THE CITY OF LONDON

Suzanne Keene
Museum of London,
London Wall,
London England,
EXZY 5HN

Introduction

In this paper I want to talk first about the kind of waterlogged wood excavated in London and then about the collaboration that we have set up between conservators, archaeologists and museum curators, and finally give a brief summary of the conservation methods we use.

Sites and Artifacts Encountered

As you know, London is situated on a river, and several tributory streams also flow through the city, though nowadays these run underground. This means that many excavations in the City of London include waterlogged levels, and of course many wooden artifacts have been found.

The excavated deposits range in date from iron age and roman through medieval to post-medieval and even modern. Types of structures reflect most human activities: the medieval and Tudor Baynards Castle; industrial structures such as hearths and so on; dwellings; streets; drains; and even a sort of pond. But London excavations are probably best known for waterfronts, that is to say quays and related structures, buried during past land reclamation. In spite of our country's economic plight there is a great deal of redevelopment, and up to 10 sites may be excavated in a year. Though some of them are very small, many do cover quite large areas. The excavations are largely funded by the developers, who treat the excavations as part of the development costs. 1% of this goes to conservation, and we can spend this on equipment and materials.

A feature of these excavations which makes life difficult for conservators is their unpredictable output in terms of finds. It's easy to tell which sites will include waterlogged layers, but of course we can't be sure what structures they will contain, let alone what objects will be present.

To illustrate the unpredictable output of sites, the late medieval rubbish dumps at Baynards Castle, uncovered in an excavation in 1972, produced hundreds of items of wood, but a further Baynards Castle site excavated this year produced only 15. Admittedly this site was much smaller, but there was plenty of waterlogged soil. Forward planning is therefore not easy.

Next year the site of the Billingsgate fish market is to be excavated. It has been a trading centre since Saxon or even Roman times, and is right down on the waterfront. About half a million pounds is to be spent on the excavation, which will last between one and two years, and a modest proportion of this is earmarked for conservation. We will have an on-site laboratory, a conservator, and if requirements exceed resources I am fairly confident that extra cash will be forthcoming.

The excavations produce a great variety of wooden objects, from little wooden pins right through to boats (we have two or three awaiting treatment) and even wooden quays. Artifacts are made from many species of wood: native species such as fruitwood; maple and box for delicate items; oak for structures; and imported wood like pine for roman writing tablets and even mahogany in the case of an eighteenth or nineteenth century box.

Though most wood has obviously survived because it was in continuously waterlogged layers, some is particularly well preserved and looks fresh and natural, rather than black and soft. This may have come about through exclusion of oxygen rather than waterlogging. Burial in heavy impervious clay seems to cause this excellent condition.

Organisation of the Excavations

The excavations are carried out by the Museum of London's Department of Urban Archaeology, which has 52 people, including its own Finds and Environmental Section. In addition to this are the actual excavators. There are curatorial departments for various periods: Prehistoric and Roman; Tudor and Stuart; and Modern. The Conservation Department is another separate department and serves the whole museum, and includes an archaeological section with three conservators; and we like to have an intern as well.

It is an important function of the Conservation Department to serve as an information exchange; and ensure that both curators and archaeologists take part in important decisions about finds. For example, each month, all the finds recovered during that time are inspected by a committee of curators, finds-assistants, and conservators, and the probable importance of each object is assessed. In this way, we conservators can find out about special features we might miss, assess likely problems, keep an eye on the post-excavation care of finds, and make sure that precious conservation time is directed to the items which need it most. It's tricky to extend this assessment to wet wood, but we certainly try.

We talk also to the excavators themselves, because conservation has to begin at the moment of excavation. We offer seminars and visits to the conservation laboratory. We have a lot of contact with the Finds Section of the Archaeology Department. Most sites have a finds-assistant working on the spot, and a conservator visits each site once a week to help with any problems regarding finds, to keep an eye open for the unexpected, and to get people used to the idea that conservation is part of excavation. So we can feel reasonably sure that wooden artifacts are well looked after from the moment of discovery, which may take place under very difficult conditions. Moreover, we generally avoid the well-known Friday syndrome, in which it is announced on Friday morning that a large and awkwardly shaped object, which has been left in situ since its discovery some weeks earlier, absolutely must be lifted and removed from the site by Monday morning.

With this close contact, finds-assistants can deal with preliminary cleaning and recovery of wooden artifacts by themselves. Finds are then sealed into individual polyethylene bags, together with labels of polyethylene fibre and a small amount of sodium orthophenylphenate as fungicide. This packaging helps the object to attain equilibrium with its future storage environment. The bags are stored in stacking polythene boxes. In this way finds survive well for years; the longest storage period to date has been for artifacts from Baynards Castle; these were conserved nine years after excavation. Some of the finds had small dark spots in places, due to fungus growth, but in general they were in perfectly good condition.

By the time they arrive in the laboratory for treatment, most finds have had the wood species identified and been carefuly drawn by the Finds Section. When treatment is planned, finds are sorted into groups of objects of comparable thickness. Except for poplar and willow, which seem to require special treatment, objects are not sorted according to wood species. Objects are thoroughly washed, and conservation record cards are completed. Each find is enclosed in a bag made from heat-sealed Terylene net, and an identification tag of non-woven polyethylene, written in soft pencil, is attached with a non-rusting staple.

Treatment Used

Most of the standard treatments are used: the normal polyethylene glycol method, freeze-drying, acetone-rosin to a limited extent, and slow drying.

Polyethylene Glycol:

To consider first polyethylene glycol treatment, so far only fairly small items have been treated in this way. We use small tanks of high density polyethylene, or plywood lined with polyethylene sheet, and put them inside ordinary laboratory drying ovens. Objects up to 1 metre long can be treated like this. The treatment time is decided on according to the thickness of the wood and is not less than 6 months; the amount of solid PEG to be added each week is calculated; evaporation is controlled by a floating lid of bubble pack polyethylene; and water is added to keep the liquid level constant. The concentration is taken up to about 90% PEG, and the temperature to 50°-60°C. PEG 4000 has been chosen in the past, but wood treated with this tends to be a bit dry and splintery, and so we are adopting PEG 1500 (i.e. not 540 Blend). When PEG treatment is complete, great care is taken to remove excess wax. Objects are washed in hot water while they are still hot, and then blotted well with paper towelling. They are wrapped in fresh towelling, and allowed to cool and dry slowly over about a week. If there are dark patches on the surface then the objects have a final rinse in industrial methylated spirits (i.e. technical grade ethanol containing 10% methanol) and water.

The results of this treatment are good. Objects are fairly heavy, and brittle, but there is usually little or no shrinkage. The colour is not unacceptably dark and fine detail is well preserved. The thorough removal of excess PEG means that breaks can be repaired using cellulose nitrate adhesive.

We are in the process of scaling up our PEG facilities. We plan to use a 380 litre (100 gallon) fibreglass tank with a polyester/glass fibre immersion heating panel. The tank measures 122 cm x 61 cm. (This set-up may not work, but it was quite inexpensive!)

Freeze-Drying:

We have much less experience with freeze-drying, but early results are promising. As pre-treatment we have tried various grades and concentrations of PEG in water. We felt that PEG 400 left the wood rather too light and fragile, and we prefer results from 40-50% PEG 1500 or 4000. (ed. Note: It is unwise to take the concentration too high because the PEG - water system has a pseudo eutectic point at about 55%.) An unforseen problem with PEG 400 was that objects treated with it, smell terrible. Also, it caused cellulose nitrate adhesive to soften; perhaps it acted as a plasticizer. Polyvinyl butyral seems to be an effective substitute as an adhesive. At the present state of our art, freeze-drying seems to be less reliable than complete bulking with PEG; we have had some unexpected failures.

Slow Drying:

Slow drying has also worked well. We know that even the most enormous and apparently very solid wood cannot be allowed to dry naturally without damage. Some enormous timbers from a Roman quay were retrieved in order to cut sections for dendrochronology, and left undisturbed afterwards. The sawn section was fresh and solid. Six months later, the decayed outer layers had broken up and the solid core had split drastically. But two elm water pipes were also both slow dried. One was dried over 15 months; there were considerable problems with fungus, but no dimensional changes occurred, and cracking and checking were minimal. The other was partly treated with PEG, to 30% by immersion, and by subsequent brushing. It then dried over about 4 months. In spite of the presence of PEG, it shrank a good deal more than the first pipe, and developed worse cracks. A third item, an oak Roman water pipe, is in the process of drying. It was wrapped in polyethylene sheet with slashes at intervals and left. Now, ten months later it seems to be progressing well. On one side, where the polyethylene sheet did not cover the wood properly the surface is damaged, but most of the remaining surface has fairly small cracks. We want to try treating another similar piece now: we will ensure that there are two well secured layers of polyethylene sheet over it, and aim to dry it over at least 18 months.

Acetone Rosin

I admit to a "love/hate relationship" with this method! Colleagues have had superb results with it, especially for metal objects with wooden handles, which are found in quite large numbers on the Thames foreshore. But I have had some disappointing results, furthermore its long-term virtues are uncertain. The hazard to people who handle hot acetone solutions is a matter of concern, moreover there is great difficulty in finding a secluded place in the City of London where the treatment can take place. I prefer to use other ways to preserve wood.

Conclusions

Treatment is, we hope, only the beginning of conservation. We try to make sure that treated objects are protected from mishandling and environmental excesses. We are lucky in having a large and helpful technical department, which will make a crate to fit any individual item. The museum also has a large unheated warehouse in the City, which has a naturally cool and equable climate. Internal stores and displays have air conditioning and although not without problems, they have a quite stable environment.

To conclude, I feel that we have been reasonably successful up to the present in treating waterlogged wood. We have also laid solid foundations of good communication and shared responsibility. We have mainly dealt with small items, and as everyone knows these are not really the most difficult problem. The future holds much more serious challenges, such as the medieval Blackfriars boat now and perhaps even portions of quays.

Obviously the problems of treating these are far from solved, but this conference has raised many practical possibilities. Above all, I realize that we can't leave the development of new techniques entirely to the scientists. Conservators must not underestimate the part they should, and can, play in this. The ideas often come from us, and development of techniques is almost entirely in our hands, but we do need to make our observations more systematically, and it is absolutely essential that we then share them.

Questions

John Coles: I think that one of the important things you brought out was the simple statement that conservators in the City of London are treated very much as a part of the excavation process. This is something that we have been battling with for most of today, and in many parts of the world, this has not yet happened.

Colin Pearson: What was the adhesive you used for waterlogged wood treated with PEG?

Suzanne Keene: For the standard PEG treated wood we can use cellulose nitrate. However for freeze-dried wood, this adhesive just softens and does not adhere. We have found polyvinyl butyral to be a good replacement.

David Grattan: The wood technology literature suggests that rescorcinol glues are good for PEG treated wood. Does anyone have any experience with them?

Richard Jagels: The problem with them is that they are a dark red colour, which will make the glue line very obvious, but they are very resistant to water.

Jamie Barbour: They are also non-reversible.

David Grattan: When you air dry material do you follow the weight loss?

Suzanne Keene: No.

Victoria Jenssen: From where do you obtain non-corroding staples?

Suzanne Keene: These are of nickel/silver, and are supplied by J.H. Rosenheim & Co., Quay Road, Rutherglen, Glasgow, G73 1RN, Scotland, and they can fit any stapler.

Victoria Jenssen: In North America, Swingland standard size (desk model) Monel staples are available in boxes of 5000 from: Talas, Technical Library Service, 104 Fifth Ave., N.Y. N.Y. 10011.

General Discussion: Conservation for Frozen or Wet Land Archaeological Sites

John Coles: Summing up: We've had by my count, seven papers today. Five by archaeologists and two by conservators. The two conservators, both from Britain, seemed to have developed a very good dialogue with the archaeologists with whom they are working, which is a pleasing development. The five archaeologists have obviously had a mixed series of contacts with conservation authorities, good contacts and dialogue for Jim Tuck, but not so good for other people. There are however, problems in other areas. Dr. George MacDonald set it out quite clearly in his introductory paper by calling for a greater development of contacts, the establishment of various standards of investigation involving conservation specialists and archaeologists. And he pointed to the great problems that are pressing upon us in many parts of the world where these waterlogged sites are known to exist and known to be deteriorating. The essential thing, I think, for this particular session is to make sure that not only in this room, but in the back of our minds for communicating to other people, we accept the necessity for this common dialogue and the development of agreed aims between archaeologists and conservation specialists in their particular areas. It's too much to expect that we are going to get a common agreed set of standards applicable to the whole of the Northern latitudes, but it is important that each of us can develop this set of contacts we have been talking about in his or her own area. The vocabulary has been mentioned (common standards). I think that conservation people assume that archaeologists know more about conservation than archaeologists actually do. As an archaeologist myself, I have not understood all the terms that have been bandied about, as throw-away lines by some of you, but I'm learning! I was very puzzled by one of Dr. Per Hoffmann's tables, when he produced one with "O.D." on it; according to him that is "oven dried," but to me that is "ordnance datum," and he also had "A.D.," "air dried," but according to me that's not air dried but the year of our Lord! And the only thing that was lacking was "B.C.," before conservation I suppose! So I think we should really get together on our vocabulary.

Now I would like to hear comments from conservation specialists about their requirements from archaeologists. If archaeologists are going to be calling on the services of these experts, they should really know, in advance, what conditions are likely to be imposed by conservators before they are prepared to undertake this work. And then I think the archaeologists could respond by explaining what they want.

Charles Hett (Conservator): The first requirement should be that the conservator be involved from the beginning, at the survey stage. This is probably particularly needed in Canada where there are serious logistical problems and you have to plan very carefully. If one thing gets left 2000 miles behind, the lack of it can very well bring matters to a halt. So I think, bringing in the conservator at an early stage in the planning of an excavation is extremely important. And continuing that, as planning through the field season continues, a fairly steady level of communication is required in order to achieve the best results for both parties.

On a remote site it is possible for the conservator to act as a kind of registrar, taking overall responsibility for the welfare of objects, for suitable packing materials and the procedures that are going to be used in the field. Before the excavation starts, all this should be worked out and agreed; the distribution of work should be agreed upon, and carefully organised. The conservator should at least have some input into how things are recovered from the field and how they are handled. Particularly when objects are to be shipped.

Lorne Murdock (Conservator): Once the conservators' profession and abilities are recognized and they are asked or permitted to go on-site, much of the work load can be taken off the shoulders of the archaeologists by them.

Colin Pearson (Conservation Scientist/Educator): I think that this does infer that you have got the availability of an on-site conservator, which means you either have a service facility perhaps like CCI (which happens in very few places), or you can put the money in your budget to get an on-site conservator. One would like an on-site conservator to be employed in every excavation - if there is a need for conservation. But in the situation where there are insufficient funds to employ a conservator, what should the archaeologist do?

Howard Murray (Conservator): You are going to need some sort of adequate legislation to stop archaeologists digging without conservation! That ought to solve part of the problem.

I can see the possibility of a test case, with an archaeologist charged with "Cultural Vandalism." (Laughter)

John Coles (Archaeologist): Then I suppose, an archaeologist could take a conservator to court for spoiling his finds! (Laughter)

Stephen Brooke (Conservator): I think what I would ask the archaeologist is that we be treated as professional equals, so that we are brought in on the planning, and on the distribution of funding from the beginning, so that we are taken into the field so that we may do our job and at the same time enable the archaeologist to have better data and results from his field programme.

Howard Murray (Conservator): In a lot of sites that I have worked on, the conservator is a sort of technician who "tarts-up" things so that the archaeologist can take them round to lectures. There is a grey area in the distribution of money. The funding is all there for excavation, and there is nothing provided to look after the finds from the site to the museum. There is little understanding that conservation may cost much more than excavation.

John Coles (Archaeologist): Do you think if archaeologists did build funding into their estimates, that there would be enough conservators available, in view of the lack of training facilities?

Howard Murray (Conservator): I don't think so, at the moment.

Victoria Jenssen (Conservator): I think things have changed over the last ten years. I believe there are now many conservators available; we at Parks Canada are overwhelmed with requests for internship. I'm sure it is the same at CCI, with the Museum of London and many other places.

George MacDonald (Archaeologist): On the question of the availability of conservators. There are constraints there. I agree entirely with this kind of joint work, as I mentioned earlier today, between the conservator and the archaeologist. The problem in Canada seems, that it is only sites which have been declared of National Historic Interest and so on, that can have that kind of cooperation from the beginning. At Parks Canada, the conservation is restricted to sites that have gone through The National Historic Sites and Monuments Board and been approved. In the case of Red Bay, that must have been a record breaking declaration. As soon as the site was found it was declared! But the real problem lies in the fact that the archaeologist so often doesn't know how rich a site he or she might be getting into. It is impossible in Canada at the moment, for every archaeologist who might encounter wet-site material, to get a conservator involved in the excavation programme right away. Almost everyone working in the Arctic runs a high risk of encountering preserved perishable material. Most archaeologists on the West Coast, if they start a shell-midden, they might very soon be into a wet-site, without ever having planned that in the first place. So, it is certainly a good idea to begin the planning with the conservator, but I see it as being far from a practical reality. I might just add to the earlier comment, that conservators might seek funds themselves. This is the problem that I have always found to date, if the archaeologist is keen on going ahead with a particular wet-site project, he is also responsible for finding all the money to set up a full conservation programme. But I think there could be a lot of fruitful dialogue on this point. If conservators could come up with some of their own money, I'm sure archaeologists would be overjoyed to have them in right at the very beginning.

Suzanne Keene (Conservator): I'd like to ask of archaeologists, that they pay more attention to the finds when they interpret a site. Because, at the moment it is far from clear that the finds are of real importance in interpreting of sites; at least in England. I think that finds-researchers should also be upgraded, so as to make the value of the artifacts clear, because it is only then that conservation begins to seem really worthwhile.

Jacqui Watson (Conservator): I'd like to explain how we deal with this at the Ancient

Monuments Lab. in London. When a site is proposed for excavation, we first of all have a large meeting to discuss it. The excavator, the finds-assistants, all the environmentalists, and the head of conservation as well all attend. We discuss what is likely to be found, based on previous work nearby. We all explain what we are interested in, and what we can do and where we can be contacted. When excavation is finished and material comes to the laboratory, we all sort it all out. The excavator explains to us at this point what he or she wants for the publication. We aim to assist in that, we liaise with all concerned, the draughtsmen, excavators, finds-assistants etc. Then we consider the objects which have to be dried for the convenience of storage, or for cleaning to find out information, and we only conserve those which really must be done. The remainder are stored. But, as far as we can, we try to work together.

Stephen Brooke (Conservator): I think conservators have the view that the archaeologists are frequently more interested in the data derived from the artifacts, than the artifacts themselves. I can say that in my country (ed. The U.S.A.) this has been the case in the past, and continues to be so today. We have large collections of anthropological and archaeological material in museums which receive no care and no budget. I think that part of the problem is that archaeologists must recognize the need for conservation. When archaeologists become spokespeople and demand the training programs, more conservators can be trained to work with them.

Colin Pearson (Conservation Scientist/Educator): To come back to the big problem of funding, for instance in the laboratory where I worked, in Western Australia, ten archaeologists were fully employed doing maritime archaeology, and there were 15 conservators fully employed, treating the objects. That cost half a million dollars per year, to employ people to work on underwater archaeology, which is a lot of money. And if somebody applied tomorrow to establish this system, of legislation, excavation and conservation (a beautiful but not holy trinity) the money would not be forthcoming because people cannot conceive that it costs this enormous amount of money to run -each year. Thus I can understand the problems, it is often easier to get the funds just to do the excavation, and hope that the archaeologist can sneak in the conservation.

One thing that I would also like to plead for with the archaeologist is, do they always need to raise, 5000 of this and 220 of that? I'm sure that anyone who has been involved with shipwrecks has worked on thousands and thousands of silver coins or something similar. Do they all need to be treated? The amount of time and money involved mass treating lots of objects is colossal. Does the archaeologist need them all in the conserved state? I honestly don't think so, since most can be analyzed or drawn before conservation.

John Coles (Archaeologist): I think that is right, archaeologists do not require all, but I think you will find that most archaeologists who are dealing with large quantities of wood, are already making that choice in the field. From my own experience at a site which I didn't mention today, it had about 200,000 pieces of worked wood. We exercised our choice and saved 500 of them. The others are gone, there was just no capacity for dealing with them. It is a valid point to make.

Mary-Lou Florian (Conservation Scientist): It seems to me that when archaeologists go into the field, they should do a good environmental study of the area and the soils, etc. prior to excavation. Much of this information may be valuable in terms of conservation and treatment. I feel that in the papers that we have had today, there's been a lack of environmental description, which as an analyst, I feel the archaeologist should be giving to the conservator.

John Coles (Archaeologist): But how are you going to educate archaeologists, there are about five of us here at present? How are you going to get across to all the others the need for this?

Mary-Lou Florian (Conservation Scientist): There is a trend towards the interdisciplinary approach in archaeology now, and that there are often biologists who are brought in to do analyses. Perhaps this is all it would take?

Victoria Jenssen (Conservator): It is not only for our information. The message can also be brought home by pointing out that it is for the benefit of the archaeologist who is interested in the preservation of sites; a subject which we are discussing at length at present. We are interested in preserving the Red Bay wreck site in Parks Canada. Information gathered during excavation would be most useful in this. The information is not really very useful for conservation treatments yet, since we don't yet know how to apply it, but from the point of view of the preservation of sites it is very important.

Janey Cronyn (Conservator): I would like to ask the archaeologists here, what they really think of objects. So often I get the feeling that we are fighting a battle to save objects, that the archaeologist should be fighting. The onus is put on the conservator by the archaeologist asking "is this conservable or not?" - and this answer is used to decide if attempts should be made to keep it for the future. When a conservator is not asked, we have to interject "think of the people in the future who may wish to research your finds." Perhaps the archaeologist would prefer objects self-destructed after publication, since after that they don't matter. Do archaeologists think it important that artifacts should be kept? Do they just think that it is a bit sad that objects are lost? Or, do archaeolgists truly believe that long term conservation techniques should be found? Are we fighting a battle for ourselves and our general public for display of the objects?

John Coles (Archaeologist): A good question, who would like to answer it?

George Macdonald (Archaeologist): I can't really answer it, but I would point out, that you really do have two kinds of archaeologist; at least in North America. Those who have a museum base, who have a big stake in the artifacts because part of their career depends on the exhibits and re-use of the artifacts, thus it is very important to them that the artifacts be maintained. They thus depend heavily on the conservator.

I would also comment, that I think to pose the question to the conservator, "can this be preserved?" is perfectly valid, because the archaeologist cannot answer this. I don't think archaeologists should ask conservators to make the decision of what to preserve, but they can say, "I like this object, think it is important, can it be preserved?" The other kind of archaeologist is in some other kind of institution. In North America we have for example archaeologists employed by private consulting groups, who once they've finished with the artifacts, out they go. The finds may go into some institution where they may never see them again. These are professional archaeologists making a living in archaeology. You also have university based archaeologists, who maybe are associated with a university museum and therefore have an academic interest, and also are interested in the artifact because they take their students into their museum for classes etc.

There are now many archaeologists in North America, who really have no follow-up interest, or very little; and there are others who are very dedicated. So you must always take this wide spectrum of attitudes into account; a simple characterization is not possible.

Janey Cronyn (Conservator): But you do see our problem?

George MacDonald (Archaeologist): Oh yes, the problem is that responsibility is being put on you.

Janey Cronyn (Conservator): That we are trapped into having to make the decisions, and we are often not given the level of decision making to do this, we may have lower status than the archaeologist and yet we make the decisions on preservation.

George MacDonald (Archaeologist): I don't think the archaeologist should rely on you to do that.

(Tape Change)

I think that there are budget problems. Out of my own budget I have to service five or six projects, in that case the responsibility is automatically shifted to the conservator. The archaeologist says, "do as many for me as you can." Archaeologists are used to that sort of compromise approach with radiocarbon labs and in other situations too.

Barbara Purdy (Archaeologist): Let me just say that I think there has been a trend in North American Archaeology for ten to twenty years, to use artifacts as explanations for human behaviour and changes in human behaviour, and we have moved to such a high theoretical level, that (I think you are correct) sometimes the artifacts are not supposed to mean anything other than what you can tell about human behaviour from them. So a beautiful mask or totem pole, is supposed to be just the same as a piece of detritus, because in each case you are actually working with objects representing human activity. I wish that I were more theoretical and I try; I try to think of artifacts in that way, but part of me still likes to go to museums and see things. I really believe very strongly, that the mental well-being of people is dependent upon archaeologists supplying material for museums which are just as important as musicals and art galleries etc. So I believe that American Archaeology is losing a lot of "clout" with the public, because we get too sophisticated and can't relate to the ordinary person. I think that archaeologists should go ahead and be interested in human behaviour in the past because this is their job, but I believe that all of the really beautiful artifacts should be preserved, so that they will last for thousands of years. I think, that is what we want the conservators to do for us.

Janey Cronyn (Conservator): But what about all the non-beautiful objects, which would be re-interpretable by other archaeologist in the future? Would you see these as being worthwhile for conservation? Is this valid?

Barbara Purdy (Archaeologist): I think that a sample of these should be preserved. One of the recent trends which is good, is that we know how to get a representative sample, so that we are not losing that. John Coles mentioned that he saved 500 objects out of 200,000, this is what we have to do. If we have a system which takes every tenth specimen, and number five which is scheduled to be discarded happens to be a beautiful mask, we must be able to be flexible enough to keep it.

Suzanne Keene (Conservator): I am glad to hear museums brought into the discussion, because I don't think it should be archaeologists who should decide on the fate of finds. It is museums that have the most knowledge about artifacts, and it is the museums that have to store and "take care of the baby", when the archaeologists have finished with it. So I think museums should be brought in at a much earlier stage. No excavation should start without a conservator and a final home for the finds and a date when the finds will be put in that final home.

Barbara Purdy (Archaeologist): You can almost relate that to some of the bad things going on in shipwreck archaeology at present. Where the archaeologist might say that we are for instance losing all the evidence about 16th-17th century behaviour from a shipwreck, because the underwater salvager is going in and taking out only the dramatic artifacts. The archaeologist will therefore take issue with your attitude, because as anthropologists we think we are the only ones who can really decide. Attitudes are polarised on this issue.

Howard Murray (Conservator): In my opinion, many wet wood or underwater excavations are not really archaeological in nature. They are really conservation excavations with an archaeological input rather than vice-versa. It really has become a question of lifting and salvaging fragile objects, rather than archaeological interpretation of the site. Some of the sites should therefore be entirely run by conservators with an archaeological advisor, rather than the other way round.

(Laughter)

Richard Jagels (Wood Biologist): As neither a conservator nor an archaeologist, I am going to step into a very sensitive area which probably nobody really wants to discuss, but I'm going to say it anyway. Objects from dry sites which need little conservation survive for a very long time, enabling future workers to re-interpret them and shoot holes in the theories of earlier archaeologists. In the wet-site area however, if you don't conserve things you don't have that problem later on. That may be perhaps overstating something which may not even be in the minds of archaeologists. I'm being somewhat facetious about this, but the importance of the interpretation of the site and the objects versus the value of the objects in the future is what we are really arguing about. I'd like to hear the response of an archaeologist to this.

Barbara Purdy (Archaeologist): What you are saying really is that at a wet site (and this is why I disagree) that conservators should go in and remove objects and the archaeologist be there to help. At Hontoon there was almost complete correspondence between the resources utilized and those available. So we are really able to study the behaviour. Would the conservator be interested in conserving all the plant remains and fungal remains for instance? We are very interested in that now. We would want to examine everything before taking random samples of artifacts.

Victoria Jenssen (Conservator): Well, we are here to provide a service!

Stephen Brooke (Conservator): In work on objects form the Defense, we have been asked to conserve, potato peelings, remains of food, as well as the artifacts from the wreck itself. We are also in a conservation facility at a museum, which has legislative authority (ed. Maine State Museum) to issue or withold permits for archaeologists excavating on State owned properties -which in Maine, includes all the navigable waters, inland, coastal, and out to sea for 3 miles. So that the Museum does have some means of control of the collection it inherits.

John Coles (Archaeologist): I'd like to comment on this discussion, which I have noted down here as "Throw away archaeology," because it seems to me that there is some little confusion about why people are saving things and Barbara Purdy has put her finger on it when she talks about ancient behaviour. Archaeologists are very interested in this and in the development of early societies. And you will know many archaeologists and anthropologists today believe very firmly that patterns of behaviour did exist in the past, and that they can be retrieved from the surviving material culture on sites. Many theoretical models have been produced on the basis of the distributions and the character of the surviving material culture from various sites, and you are well aware of many of them.

You also, I'm sure, are well aware from historical documentation and ethnographic literature, that many undeveloped communities relied very heavily on wood - up to 90-95% of their artifacts, materials, fences, houses, barns, boats and so on, were all of wood, and yet the theoretical archaeologists, the model makers, are looking at sites where that aspect of society has decayed. They are looking at dry land sites, they are therefore looking for patterns in 5-10% of the material. And they are building their models on them. It seems to me that archaeologists who are dealing with wet sites and conservators who are concerned with them must get together, not only to retrieve these wet sites, but also to conserve that evidence, to allow the theoreticians (if you like) to look for patterns in that range of material culture, which would be increased ten fold by our retrieval and conservation of these organic elements. And how much better (in my opinion) would these theoretical models about behavioural patterns be, if archaeologists were able to work on the wet sites; and it can't just be the wet sites that were excavated by me, Barbara Purdy or anybody else, results published and everything destroyed or wrecked, we have got to conserve enough range and quantity and quality of this wet material so that not just ourselves but theoretical archaeologists can come in and see (a) how inadequate their models have been in the past, because they have been based on 5% of the evidence and (b) how much more accurate or reliable a model can be produced by looking at the type of evidence that can be produced from Florida and other places. And that is why I think we are not talking about just the beautiful objects, but a wide range of the common objects, as well as the exotic objects, that have to be not only retrieved but they have to be conserved. The publication is not the end of it for thinking archaeologists; it is necessary that they see the material through the conservation and into the museum or the storage stage, and be satisfied that it is there, not just for them but for the next generation of theoretical or practical archaeologists who can use this material. So that I think, is the real importance of wet-sites.

Jamie Barbour (Wood-Biologist): One thing that has not been discussed is that there are different degrees of conservation, that a potato peeling or a wood chip etc. doesn't have to be taken care of to the same extent that a mask does, or something that's going into a museum. A lot of our artifacts are for example conserved but never cleaned. We simply treat them in PEG, and don't remove the excess of wax after treatment. This keeps the object in good condition, they stick to the trays better (Laughter) and should someone wish to work on them at a future date, it is very easy to remove the excess PEG. This is something also that people should think of when faced with 1000 pieces of worked wood. Maybe the conservation is not that big a problem, and the conservators are going too far - they don't all have to look like museum pieces.

Richard Jagels (Wood-Biologist): To add to that, as biologists we keep specimens for great periods of time in liquids-fixing fluids. Potato peelings don't have to be dried and preserved in some shape, they can be placed in a bottle of fixative solution, and kept possibly for hundreds of years. In fact they are probably more useful in that state for further analysis than if treated in PEG, for instance.

Jacqui Watson (Conservator): Returning to the point of the complete conservation of objects. Some of the objects that I have encountered, which have been drawn before they were treated, have been found to have fine surface detail only after the freeze-drying treatment had been completed. One example was an oak gaming board, which did not look like anything much before treatment, was clearly seen as a Nine Men Morris board after treatment. Now, we simple pencil draw objects at the wet stage, so that the drawings can be amended after freeze-drying.

Barbara Purdy (Archaeologist): We had an similar experience using an infra-red scanning scope on a piece of unpreserved wood excavated 90 years ago from the Key Marco site, and we wanted to see if the scope would actually work. Using this we found a little image, that nobody could see with the naked eye. I wonder therefore if sometimes we miss designs!

Victoria Jenssen (Conservator): We had a wooden navigational instrument, with designs all over it, and it was impossible to record these satisfactorily visually. Treatment was going to take a long time, so the object was photographed immersed in water with infra-red sensitive film, and the result was excellent, all the design was revealed.

John Coles (Archaeologist): I will sum up very briefly.

First of all a point of information and I hope interest. In the U.K. there has been established a dialogue between the academic archaeological societies, and the grant giving body, the Department of the Environment, and major national societies have been asked and have provided priority documents for the development of archaeological funding, research, and rescue priorities in the U.K. I can only speak for the prehistorians because I happen to be involved in that particular society, and was involved in the preparation of the document, which has now gone in to the Department, which gives out about 3 million Pounds in direct grants for archaeological units, plus a lot of back-up. That document, (26 pages long and I won't read it all) says two important things, high in the list of priorities. At the top of the list is the comment "We wish to draw attention to areas of particular prehistoric concern. These are situations where biological and archaeological remains are likely to be particularly rich and well preserved, they include waterlogged deposits. We hardly know anything about the organic elements of prehistoric economies, structures and equipment, and any sites likely to yield such evidence from peats, muds and silts should receive support." That document is now in the hands of Her Majesty's inspectors, and we firmly believe (and are led to believe) that it will result in a substantial injection of funds into wet site archaeology in Britain. The second comment we make is "That few regions in Britain are equipped to deal with existing quantities and characters of material requiring conservation; either immediate or long-term. This matter is well known and is a disgrace to both practising archaeologists and the public. And it is a matter of great concern that so few regional units are equipped to deal with anything but the most elementary procedures." I am not at all certain in my own mind that this comment will result in a massive injection of funds into conservation, but I do feel that as for the first priority i.e. the first quotation, if the archaeologists who are concerned with wet sites play their cards in the right way (and you have to do that!) then it will not only involve funding for wet site archaeology, but concomitant would be development of conservation facilities to go along with it. I think the point is well made today, the need to involve conservation specialists from the beginning. Finally, after a century of attempts at conservation, and some of these pioneering attempts have been talked about in the last few days, it seems to me that we are only now beginning to get standardisation of aims, and methods, communication between conservators themselves, and between conservators and archaeologists (I'm sorry that there are not many museum people here). We are beginning to get a full range of experiments, some of which have been described, and a much better comprehension of wood characteristics and rates of decay. These are all good developments that I have heard about this week. Archaeologists are going to be asking conservators in the near future, if not already, that they should bring to their projects aspects such as interest, initiative; the expertise we assume will be there. We expect firmness in the decisions that the conservators make on-site, we expect them to deliver the conserved material; and in other words be honest in the promises for completion and make real attempts to not only conduct the ordinary day-to-day conservation but attempt constantly to research methods of conservation of this waterlogged material. It doesn't seem to me that there are very many aspects that we ask of conservators, and I'm sure from what I have heard and seen today that all of that is available. I think that conservators should get from and expect to get from archaeologists elements such as consultation (a point very well brought out) time, accessibility, facilities (on and off site), funding and probably sympathy! I think that all of these things are important for archaeologists and conservation specialists to share. I am only concerned that we may convince ourselves here of the need for all of these elements, and we may agree on them but it is essential now that these types of agreements, (or whatever else is devised) should be communicated. And it has got to be communicated by both conservation specialists and archaeologists. Because I feel very strongly, that although there are bad patches of the world and bad episodes, where things don't seem to have worked out too well (and I don't refer to anything mentioned today, but in general), there are great opportunities for wet site archaeologists, for waterlogged wood etc., and really if archaeologists don't take their chances now and conservators don't go along with them and help them and gain much for themselves from it, then there is little point in this conference and there is no hope for the development for a greater understanding of ancient societies, and that is what the archaeologist is really after.

SURVEY OF ROEBUCK PREHISTORIC VILLAGE SITE GRENVILLE COUNTY, ONTARIO

Wednesday, 16th September 1981

MAJOR JAMES F. PENDERGAST, LL.D.

JAMES V. WRIGHT, Ph.D.
Chief,
Archaeological Survey of Canada

GEORGE F. MacDONALD, Ph.D.
Senior Scientist Archaeology,
National Museums of Canada

Transcripts of on-site description

Dr. James Pendergast: Who were the Iroquois?

If you can imagine in 1535, almost on the anniversary, the 3rd of October, Jacques Cartier walked into the village of Hochelaga, which was a palisaded Iroquoian agricultural village, and saw, as the first European, an Iroquois, the Iroquois, any Iroquois. And then when Champlain came in 1603 he went by the sites that Cartier had reported in his 1535, and 1536 visit and they were gone. So the question arises who were they? Where did they come from? And where did they go? And this Roebuck site looms large in that story.

In 1854, W.E. Guest from the Smithsonian Institute, was examining sites in the Watertown and Ogdensburg area, on the New York side of the river, and he heard about this site, its fame was already that well known. He came over here and he examined it. He dug into it, he took away bones and pottery, etc., and very interestingly, a walrus tooth. Not too unusual if it was a fossil, because this is the bottom of the Champlain Sea here. He went back and published a report in 1856, thereby, commencing the first of a long series of dissertations on the Roebuck Site.

In 1912, and again in 1915, W.J. Wintemberg, Guest's assistant archaeologist in the National Museum, under Harlan I. Smith, came down here and dug extensively (**Figure 1**). And it is his report, published in 1936 by the National Museum, that constitutes the pillar and case for St. Lawrence Iroquoian studies, and in many ways, all Iroquoian studies. For as Stenton has said, Wintemberg didn't just write a site report, he wrote an ethnography of the Iroquois, and to this day it stands. You can read it four or five hundred times, and each time, you learn something new. Today Jim Wright and I had a discussion: "Was the wooden bowl from the swamp?" I said, "Oh yes, I have read about it many, many times!" He said, "Oh no, it's from a grave!" Every day, you learn something about this site!

The next thing that happened was, a large number of sites came to light to the east of here, excavated by an amateur, and when the ceramics and the artifacts were examined they were seen to be quite different than the Onondaga-Oneida which R.S. MacNeish had attributed to this site. (Onondaga-Mohawk, Onondaga-Oneida, Onondaga etc., any spread of that triangle, part of the East Group of Iroquois.) And there was quite a discrepancy between their artifacts, and the artifacts found in the Iroquois villages found along the St. Lawrence. It emerged, that there were two clusters of these villages, one about 60 miles east of here in the vicinity of Cornwall which seems to be earlier than the other, in one of which we are right now. (There is another one about 5 or 6 miles from here, there are in fact, about 6 of them within 12 miles of this village.) The characteristics of these people are that they were farmers. And you might ask how they became farmers? Well it seems, and it is day one in this story, it seems that they were an indigenous Late Woodland people (Late Woodland Period: approximately 1000 to 250 years ago) of hunters and gatherers who frequented the shores, islands and the lakes along the St. Lawrence. Agriculture was introduced into that economy bit by bit, commencing around 1200 AD - 1250 AD and it seems to have come around the end of Lake Ontario from both sides. Of course when you look at the geography of Lake Ontario, looking from west to east, you see the Adirondack Mountains on one side, leaving very little distance between the shore and the base of the mountains, the same is true on the north side with either the Frontenac axis, which makes the Thousand Islands when it cuts the St. Lawrence at Brockville, or, indeed the base of the Gatineau Hills or, as they come down to Montreal, the Laurentians. And into that funnel came agriculture around both ends of Lake Ontario. On the south side probably linked up with a group of Iroquoians were proto Iroquoians, who eventually emerged to be at least one or two of the Onondaga, Oneida, Senecas, Cayugas, five nations and six nations group. And around the other side, were the Pickering-Middleport-Ontario Iroquois group. So you have New York agriculturalists, early proto agriculturalists and Ontario feeding agricultural traditions into this area, and emerging from that Late Woodland hunting/gathering base, came these Iroquois. Roebuck is almost the culminating point. Probably, Hochelaga itself is the next site or next but one, of the whole sequence. Now, I mentioned all these villages, if you took them all to be contemporaneous, you might see a large population. It is unlikely, however, that there was a large population. It was a small population that moved around in the Iroquois fashion. Everything that you see here would have been corn and the forest would have been driven back bit, by bit, by bit, as they struggled to live in the 47°C below zero temperatures we had around here last Christmas. So you can see that the wood would disappear quickly. So if they were contemporaneous there would be a large population, but because of the wood supply and the fertility of the ground on which they grew the corn, they sought new villages every 12 to 20 years. So it was probably the same population and their ancestors, moving around. That is not to say that any one village wouldn't be large. It is not unlikely that this village with 40 long houses, by extrapolation and examination, on the portion of the site from that large bushy topped maple tree to the white house on your left, would support anywhere from 1600 to 2000 people. But, there wasn't another village with 2000 more contemporaneous over there, or another over there. It was the same population, sometimes coalescing in the one village and sometimes, in hard times during feuds etc., they broke up into smaller groups.

Well that tells you the story of where they came from. Now where they went? Well it seems that when the Europeans came into the mouth of the St. Lawrence, into what was then called Canada, at Hochelaga and Stadacona, these people were the middlemen, between the inland tribes where the furs came from, and the Europeans, to whom they were traded. But the people trading the furs to the Europeans didn't need the middlemen, and these Iroquoians were the middlemen, and they disappeared, sometime around 1580. Their ceramic remains, which are diagnostic are found on the treck among the villages of the people who were initially the fur providers, and who probably did away with the middlemen. So they came from an indigenous Middle to Late Woodland base into which agriculture had been introduced, and they left because they happened to be the middlemen between the fur suppliers and the Europeans who wanted the furs.

The Previous Excavations of Roebuck

Dr. James V. Wright:

(The most recent archaeologist to conduct excavations who was instrumental in the purchase of the site by the Province of Ontario).

A number of years back, a contractor bought the land to take out sand; and you will see a large barrow pit as you start to walk across the site. When he started to take out the sand he found skeletons. The St. Lawrence Iroquois, unlike their kinsmen to the west, buried their dead in the village. So it is almost impossible to dig a St. Lawrence Iroquois village without coming across burials. Well this upset the contractor, so I got in touch with Bill Cranston in the Province of Ontario and raised this matter. Things got speeded up when the contractor said that he was selling the land to another individual who; "Wouldn't care who's body he was digging up - not even his mothers! The sand will come out no matter what." So Bill said, "Good, but can you get in there and do a quick assessment of what's left of the site? I believe its been pretty potted and damaged." So I finished off some work up north and came down with a crew. The original contractor had conveniently shoved all of the plough zone off the site, and you will notice the unnatural lips on the edge of the marshy part of the land, and indeed it probably covers part of the marsh. So I didn't need to remove the plough zone to start coming up with long houses. And I think we exposed in a very short time some 14 long houses averaging 100 feet in length - these are the typical matriarchal structures of the historic period, where, they were controlled by the old woman of the house. (Where if you got

Figure 1. W.J. Wintembergs original excavation in the main spring area between midden 9, 10 & 11 at the Roebuck Site. No pumps were used and the area had to be abandoned despite its yield of well preserved organic artifacts. Photograph with kind permission of Archaeological Survey of Canada, National Museums of Canada.

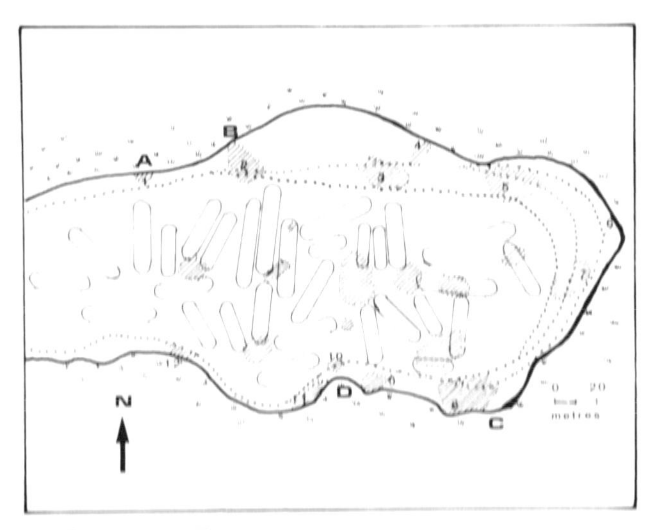

Figure 2. The map of The Roebuck Site was taken from: Wright, J.V. La Préhistoire du Québec (Ottawa: Musées nationaux du Canada, 1980). Annotations were made by the Editor.

into an argument with your wife, she could tell you to pack up your pipe and weapons and go home to mother, and that was it!)

By overlaying the settlement pattern data that Reid came up with, on Wintemberg's information, and extrapolating on the total extent of the site, we come up with some 30 long houses, averaging about 100 feet in length, which could have housed about 2000 individuals.

Only one thing to mention before we start coring, Wintemberg ran his excavation trenches out into the surrounding swamps, around a spring which is still extant today, and which we will see later, and he came up with some of the preserved organic material you see in the small exhibit at CCI. There was quite a bit more of the material, things like carbonized squash seeds and pieces of cut and whittled wood. And this is what of course, the whole process today is, how much of the marsh that rings basically the entire site is capable of the potential of producing preserved organic artifacts. Which is an area we know virtually nothing about in the Iroquois, except what we get from the ethnographic information.

On-Site Sampling

George F. MacDonald:

I would like to make a few comments about what we intend to do now. We will start our auger probes on the north edge of the site. This is the beginning of one midden that has come over the edge of the sand ridge on which the village is located, and as you can see it is quite marshy in this spot. We will put in a simple auger core, we are not using any fancy sampling techniques. The idea was that we would bring you here when we first tested the site so you will be part of that experience of seeing whether it is worthwhile to go ahead with. We had thought originally that we would do the test in February when the land was frozen, which would have allowed us to take out a frozen core, and the chance then of studying the exact features and the amount of deterioration would have been much better. But, Dr. Jan Terrasmae of Brock University, who is an expert on bogs, informed me that the bog would not be frozen, as bogs around here just don't freeze, even in temperatures which read 46°C below, unless such conditions continue for a very long time. When this conference was being planned, we felt that it would be appropriate for you to come around with us as we pull the first cores out of what we think are water saturated deposits around the edge. If it is successful and lives up to our expectations, and we do have the bits of wood recovered by Wintemberg, the plans are, tentatively for six weeks beginning next May, 1982, that we would do some excavation with a crew of students. Also the CCI has indicated that it might be able to do some of the conservation of the material. We will have to be careful not to remove too much! It was mentioned yesterday that when archaeologists get going on a wet site they swamp the conservators and it becomes a problem. So we figure that 6 weeks will be quite adequate to get the kind of sample we would need of Iroquois organic material culture, and further work might be continued in succeeding years. That would be enough to give us at least a peek at what the full range of organic prehistoric Iroquois culture is.

So we will walk around the edge of the marsh and core adjacent to the edge, with the idea that the middens are running down into the swamp, and that the bog had continued to form over them. So we would expect a certain amount of sterile, boggy material underlain by something recognizable as midden. Hopefully, there will be some samples of preserved wood or organic structures that will be worth adding to the samples of preserved imperishable cultures that we already have from the site (Editor's Note: core samples were taken at several locations around the edge of the site, starting on the north side and working around eastwards to the southern edge. The main cores were taken at positions A,B,C and D on **Figure 2.**)

Discussion Following the Roebuck Site Visit led by Dr. George MacDonald:

Today we are again dealing with the Roebuck Site and after hearing the comments yesterday afternoon in particular, it came across clearly that conservators would like to be in on the plan with the archaeologist right from the very beginning. This was generally what was in the back of my mind when I suggested taking people out to the Roebuck Site for the initial test. We held back from doing more extensive probing and testing earlier at the site so we could have you there at the very beginning. The aim is to have an ideal approach worked out to aspire to in a full excavation, interpretation and conservation programme at a site like Roebuck.

You have all seen the site, so you have an idea of the factors we are talking about (**Figure 2**).

Adjacent to midden no. 1 is the area of a small spring, and as far as I'm concerned the simplest thing is just to put in a probe to find out the depth of the wet deposits. All along the north side of the site there appeared to be very shallow lenses of sand, as deep as 20 to 30 cm lying at the base of the bog deposits. The material, although wet, does not appear to be wet enough considering the height of the water table this year, to ensure organic preservation. I am sure it dries out periodically and that there will be very little preservation here, except for very large objects that might have sunk down into the bog. If the palisades run into the swamp at this point, I would expect the tips of the palisade poles to go down even further into the sand below the bog, and be preserved. That is what my expectation would be for the north side of the site.

Along the south side we are getting into the major stream drainage and there are a lot of minor feeder streams as well. The more you approach the larger spring, surrounded by middens 9, 10 and 11, the more water saturated the deposits become, until you are into an area of total saturation. We don't yet know how the depth of the organic deposit thickens or thins towards the stream, but we do know it is continuous to the stream. We therefore have a potentially productive area all along the southern flank of the site. There have been some activities from the farm that have affected the deposits along the southern flank of the site adjacent to the bog, particularly the barnyard refuse and effluent has damagd these areas over the past century. There is also the gravel pit operation which we are not too worried about since the dry land site has been salavaged as far as is necessary. The part of the site that we walked over is owned by the Province, and it is close to a relatively major highway, linking the Trans Canada Highway with the Nation's Capital. There is the potential that if the site produces interesting remains it would be ideally suited for some kind of a regional interpretation centre for the public presentation of St. Lawrence Iroquois research. Of course, this would be up to the Province of Ontario to pursue once the project was done, but it is well located from almost every aspect and it could be developed into an excellent interpretive programme. The first question for discussion is therefore, what would you conserve of the material that came from the site?

On the Roebuck Village map (see **Figure 2**), the major spring we were talking about, lies between middens, 9, 10 and 11. The map shows roughly the position of the long houses on the site. The middens, or concentrations of refuse, are concentrated along the margins of the spring. There is a lot of evidence of cultural material and what appears to be good environmental conditions for preservation. In this area, the augers encountered a solid mat of wood. There was too much wood to get the auger through. Presumably this is made up of dead-fall trees and pieces of tree trunks, at the lowest level. It isn't recent material. It is at least as old as the occupation and some of it is undoubtedly older. Mixed in with the solid mat of wood would be a lot of cultural material.

One of the things that has been suggested for a number of years is the need for a thorough environmental survey of the area before we enter into any archaeological work. Mary-Lou Florian has made comments to me on what would be useful or even ideal, in an archaeological project to do in the way of environmental testing. I would like to call on Mary-Lou to suggest the kind of environmental factors we would look for in an initial test of the overall site area. We hope by this discussion to determine something of the characteristics that would indicate good preservation of organic material.

Mary-Lou Florian: Well, I think there are really two things that have to be done in terms of environmental descriptions of archaeological sites. You have to take an overall environmental description, and this is important after the excavation in terms of biological conservation of the site to know what type of eco-system you have; to know the delicacy of the system, the amount of traffic it can handle in terms of setting up interpretive centres, the stability, whether it is a tremendously disrupted kind of environment or whether it is an interesting biological environment, in that it may be a climactic forest or an ideal Prairie etc. So this has other implications other than conservation or preservation of the organic material that comes from the soil. What has to be done is soil analysis, and this just hasn't been done, in terms of matrix or artifact material. We keep talking about, "it must be an anoxic, or anaerobic environment, to preserve wood," but we are just talking, we don't really know. So now is the time to find out. It seems logical that in the region where the organic material is found that it is an anaerobic situation. When the cores come up, we have to examine them and measure certain specific parameters. It really is very simple to define the type of environment you are dealing with. You have oxidizing and reducing agents in soils and these are two chemically different situations, and you get precipitation of different chemicals and different chemical activity. These can be described by the E_h or redox potential (Editor's Note: Redox, or oxidation - reduction potentials (E_o, E, E_h, etc.) measure the potential drop involved with oxidation and reduction processes. For example, the conversion of ferrous to ferric iron (oxidation) or sulphide to sulfate (reduction). Redox potentials are measured against a standard electrode potential, and show the tendency of different oxidation and reduction reactions to occur) So what you need to do as you go down through the depths of the soil, is to obtain redox potentials and oxygen levels. The other interesting thing would be the pH, and this is going to change as you go down through the depths also, and this is what we are really interested in, in terms of conservation. We are always saying that this or that has been conserved because it has been in an acidic environment, but I haven't seen any data, I haven't seen any pH measurements quoted. So if we are going to take a scientific approach to these things we just have to take down the initial data. When you plot the E_h and pH you can go over a territory such as this and you can plot the various parameters which describe a bog situation, then you can be sure that this is going to be an area where you are going to get maximum organic preservation. There are also indicators that suggest that tannic acid or tannins are acting like biocides and protecting these organic materials. So perhaps this is another thing that should be characterized. I do not think we should be stating things without facts, and I think that the value of this type of work is to identify these parameters so that the next time round we know that we were right: there was organic material; it was in this kind of state etc. So that next time we can simplify our job.

George MacDonald: Translating that into an archaeological programme. If you took core samples, could you get measurements of the tannin, pH and the E_h levels? The first survey should be an environmental one for the redox potential, the basic nature of the soil, and tannin levels from core samples. I would probably approach that by a grid over the entire bog around the site and taking core samples at 5 metre intervals. I would then map the pH etc. on a number of levels over the bog. Presumably that would give us an isobar map of likely areas of good preservation around the site. Is this how you would visualise it? We would be mapping the swamp on the level of isobars, and see where the optimum factors for organic preservation coincide.

Mary-Lou Florian: It would be a three dimensional picture too. Also moisture would be very useful.

George MacDonald: It does present some problems in terms of the extent of the area and the depths of deposits. There are very shallow deposits in some areas, and very deep ones in others. You would probably generate a whole series of maps, possibly at 10 cm levels through the deposit, and have a series of 4 different index maps at every 10 cm level. For some areas, there is more than 1.5 metres of organic deposits, so that the contour interval might be spread apart somewhat. Ultimately, these maps should tell us where to concentrate our efforts to locate the optimum preservation environments. Does anyone have other suggestions that would be useful in the initial environmental survey?

Charles Hett: Water flow.

George MacDonald: Water flow. So in a bog how do you measure water flow?

Jamie Barbour: Dr. Sparling, who was at the University of Toronto, has done work on measuring the hydraulic conductivity through peat bogs; and he has also done work on bogs which have lots of organic material in them. Perhaps he would be the person to contact.

George MacDonald: I might say that one of the things that appeals to me about the Roebuck Site is that it is so close to the Geological Survey of Canada and all the other research laboratories here in Ottawa. It is not like taking an expert 3000 miles to the west coast, that costs in excess of $1000 for travel alone. Here we can bring down a variety of scientists for the day by car to look at the site.

Mary-Lou Florian: As an archaeologist, do you have any feelings about core size? You were talking about a 5 m grid over the site.

George MacDonald: Well that is a good question for debate. You want to take a core only once, if at all possible, not having to keep going back and taking more. In that case you have to isolate the second core from the first, and may have to set up a subsidiary grid system. So the best thing is to devise a sampling technique that will give you enough material in the first core, that, if well preserved, and able to be divided into parts, will be sufficient for the range of tests. That raises a whole list of questions, for instance if you are interested in the redox potential, and you take a core and let it sit around for awhile will it oxidise rapidly and become useless. Perhaps the ideal would be to take a frozen core and store it in the frozen condition, then cut it up and use it when you needed it. For that you would have to go into liquid oxygen or some other super-cooling medium, and use a jacket and cover that would freeze the core fully and keep it refrigerated until it was analysed. There we are getting into high cost factors, since we need quite a big core and quite a lot of coolant.

John Coles: I'm not sure what size of core you are thinking of taking, but the size might have implications for the site. After all you are puncturing the site in a mutitude of places, and this could introduce a lot of surface water and air down into your holes.

George MacDonald: One thing one could do is to use plastic tubes, if it is stable, of the same diameter as the auger and fill them with a neutral sand material that would fill the hole permanently, and at the same time not introduce any changes to the microenvironment of each hole, but simply provide space fillers. Is that a possibility?

Barbara Purdy: The other solution would be not to take them every 5 metres. I think 5 metres may be excessive - depending upon the total area of course.

George MacDonald: Yes - a 10 metre grid is probably more realistic.

Barbara Purdy: Then if you just use a 1 to 1-1/2 inch core, you will probably find you don't have too much to fill.

George MacDonald: I don't want to devote too much time to the core problem, although it is an important one, but I don't feel that 1-1/2 inch core is enough. It seems to me that you have to take at least a 3 inch core, because you are probably going to be subdividing that core. Would somebody like to comment?

Robert Barclay: I wonder do you need to take a core for pH and redox measurements? Can these not be made in-situ more conveniently?

George MacDonald: Presumably using a probe. Yes, that is a good suggestion. If you build everything into one core sample, you are going to need a very large one. If some of these tests can be split off, that would help solve one problem. However, you must work out a pattern. Do you have to do all of the tests at the same time of the year, so that you have the same water level throughout? In fact it would be very difficult, unless you have a sizeable crew, to do a large area at one time to avoid these fluctuations. So this is another problem to consider.

Jamie Barbour: What about inorganic salts, are these parameters that you will consider?

George MacDonald: I hadn't thought of it. What information would that provide?

Jamie Barbour: Well I was thinking of the artifact when you take it out. Would it perhaps have some bearing on the treatment process? As far as I know people haven't looked at it, but it might be important in using the PEG treatment. It seems to me that a lot of the salts that could be found at a site could get in the way of that treatment, and it would be interesting to see some data on that.

George MacDonald: Perhaps in that case you could take a few initial samples and see if you are likely to have any problems at that site.

Jamie Barbour: I don't think you need to go overboard, but, it would be interesting to look at iron and sulphate content of the soils and see what they are.

George MacDonald: The next topic for consideration is the excavation technique. And as an opener I would say that, near the major spring, I would focus down and get more environmental detail. From what I saw the other day, in the area of the main spring, it might be necessary to divert the spring, because it is going to be adding water all the time. You might be able to use that water for hydraulic excavation, but on the other hand, you don't want to alter the water content of the soil radically even during the course of the excavation. I would be tempted to let the bank form the natural barrier on the north side, and depending upon the seasonal conditions and rainfall, perhaps consider the idea of using a sandbag loading for a southern perimeter. It doesn't need to be very high to create some loading on top of the bog so that you don't get a lot of uncontrolled run off flowing back into your excavation area when you start to lower that water table very rapidly with the pumps. I would build a carrier system to gain control of inflow into the excavation area. If we have a controllable excavation unit, a fairly traditional approach in wet site archaeology in North America, would be to use a water excavation technique. That is to use hoses with variable nozzles like fire hoses, so that the delivery can be changed from a cutting, hard spray to just a mist. There has been a lot of experimentation with such techniques over the past decade and they work well. Small size garden hoses can feed off the main lines from the pump to provide full control of the amount of water that is being pumped off and used. If the excavation is done with hoses, you would have to plan that the drainage system came from one corner of your excavation unit. From there, you would probably use a sump pump to bring out the slurry. There would be a lot of particulate soil matter in the water which is, primarily how you remove the soil from the excavation. The artifacts, pieces of wood etc. don't get displaced. No metal tools should be used in uncovering artifacts because metal does cut into the wood and soft organic materials. Instead water is used to excavate. It is rather like a dentist performing oral surgery, where water is brought into and out of the mouth by tubes to clean out debris whilst they are working. Although the archaeologist's pumps and hoses are bigger they are doing the same thing; that is, cleaning the working surface and removing unwanted debris.

It is necessary to pump the slurry back into a settlement tank. You could put this settlement tank, just a plastic lined pool, in the old gravel pit. It might be necessary to use two, so that you can switch from one to the other when one gets loaded with sediment, as frequently happens. Do any of the archaeologists have comment on this kind of a system? How does this fit with you John?

John Coles: Well I don't know the composition of the material you are going to be excavating, but you do know, and presumably you know the water system will work. The sediments are presumably loose enough, and fine enough that they will be carried away. Does it mean that you are going to move the positions of a certain number of artifacts?

George MacDonald: Only very slightly. If you can work the material away, you don't even need to disturb the smallest pine cone. The problem is that it goes very slowly, but if you want very detailed information you have to put up with that, and work very, very slowly. You may also have to use small hand held water evacuator lines in some places.

John Coles: The only other comment I wanted to make is, if you think this is the best area for the good conservation of artifacts, why would you start excavation there. Why not in a secondary spot?

George MacDonald: It is my feeling now that the major spring area is the best spot, however, an environmental survey might show othrwise. However, from looking at the other sites, I know that the inhabitants would have had to come here for the water supply. Therefore, the chances of them throwing water containers and other examples of material culture in this immediate environment is very high. These deposits are the wettest of any part of the site that I know.

John Coles: I would have thought it at least worth considering, from the point of archaeological experience and conservation experience, starting on another part of the site first. So the conservators know what to expect when you get onto the better preserved areas. That way you can make your mistakes and they can make their mistakes on another part of the site, rather than going straight in on potentially the best part of the site.

George MacDonald: That is a good point, I would say of course that this isn't the first wet site we have done. As you know the tendency in archaeology is always to try and devise the most economical way of getting at what you really want, and sometimes you take short-cuts. It would be a short-cut to start in the optimal area. But I think your suggestion is good, and maybe I should start in a subsidiary area with the trial excavation.

However, the archaeological funding is unfortunately considered as risk capital by many sources of funds, particularly municipal or regional museum boards, and even more so by private foundations. Negative evidence or careful preparatory work may lead to loss of support. There is constantly pressure on the archaeologist, from the media and the public, to make significant finds. I could cite a case from the Prairie Provinces, where private foundation funds were withdrawn from a sound research project because the finds were not considered to be significant enough to the public for continued support. Ultimately, all archaeological funding comes from the public, and will respond in this fashion. Nevertheless, it should be done at least once, and I shall try to take your suggestion of preliminary tests in peripheral areas at the Roebuck Site this spring.

The inevitability of making mistakes, which you say must be provided for in the research design, applies to many features of wet site archaeology.

Positional recording of evidence is a case in point. Most wet site projects have had difficulties maintaining three dimensional control due to the unstable nature of the ground, the fluctuations of water table at different times at the site etc. Traditional bench marks set near the work area and maintained by tapes and levels is not sufficent.

Optical or reflected beam (sonar etc.) surveying instruments have developed rapidly and are now accurate within a few centimetres and can be hooked up to data tape recorders.

Now if we assume that we have batches and batches of material, lets first talk about sampling and then preservation. The sampling design really boils down to what we keep of all the structures and small finds. There will be lots of plant material, I can see that at the Roebuck Site already. There will be lots of little pine cones, for example, and needles, twigs, roots etc., that the core brought up. Are there any suggestions that we keep everything that is found in the core. For example, when you start moving down through the deposits you will probably throw some of the stuff away - a lot of unmodified sticks etc. that have fallen naturally into the bog as it accumulates.

David Grattan: I was just going to suggest that when you do come across unmodified sticks etc. that you don't want, that to someone like me that is tremendously valuable material. Because I can use the stuff from that environment to do my experimental work on.

George MacDonald: My immediate response is, "How many tons do you want?" I think we have probably taken for granted having an on-site conservator to make some of these judgements, and perhaps, once the archaeological and curatorial needs are fulfilled, then you would get rid of some of the larger pieces of wood, or store them, or tag them and keep them in a nearby pond. What do people think of that idea? If we are into a large wet site with a lot of big structural remains, and you think, "I can't possibly try to preserve all this." What do we do? Do we pile it all up and let it dry out, or do we build plastic lined tanks and put fungicides into them, or do we find a nearby lake, as Barbara Purdy did, and get permission to put everything onto the bottom of the lake? Actually, in Canada I assume that would work quite well because we have such cold lakes. If we could find one up in the Gatineau Hills, where the dug out canoes came from, and we know they preserved the dug out canoes well, so we just put the material in appropriate containers and put them on the bottom of a cold lake. Any comment on that approach.

Judy Wight: Yes. Next summer at Red Bay, Labrador, we will be digging a section which contains great quantities of waterlogged wood; too much to handle in the on-site laboratory. We are therefore, planning to store it in suitable wet spots on the bog, and do the cataloguing on a couple of days each week through the summer. Then we will be able to treat it all together.

George MacDonald: OK. The idea of a holding pond and a holding tank, perhaps.

Suzanne Keene: I was talking to Dick Jagels about this and we thought the best possibility could be to excavate a trench in the bog or moorland site, rather than use an actual water site, and possibly cover the artifacts with fine clay, as this does appear to have exceptional preservative properties, and then fill in the trench.

John Coles: But, you then have to be very sure that that bog is not going to be drained. This has happened in the past.

Rosemary Ravindra: Would it not be better to leave it where it is?

George MacDonald: You have a practical problem - here are a mass of pieces of wood - you can imagine Star Carr for example -you suddenly have a complete platform, how do you get to the layers underneath?

Rosemary Ravindra: Or you could move it to another area of the same bog which is not being excavated? Because you then have them in an environment with which they are in equilibrium, in which they have survived, but when you move them and place them in another environment, say tap water, you can have problems. It may work very well, but there is always the chance that a change will cause new degradation to occur until it reaches a new equilibrium. If the artifact is very fragile to begin with, that extra bit of degradation might be too much!

George MacDonald: So you favour the idea of leaving it in a similar environment, even though John Coles' point about the drainage, also applies?

Rosemary Ravindra: Yes, there are things that happen to lakes too.

George MacDonald: I suppose you could build a plastic lined tank, at more cost, and then you are always going to be assured of having water.

Victoria Jenssen: If you stick to thinking about timbers, and here we get back to small and large again, at the beginning you should be asking the people who evaluate these timbers about their drying capability. I understand that these bogs are very variable from one place to place and wood preserved in one area might get very punky in another. But depending upon the site at which you are working, you may find that some of this wood can be dried without storage.

George MacDonald: I think that could be an option here in Canada, especially since the success with the natural outdoor freeze-drying has been so encouraging. We might find if we got into a lot of large timbers, and this is a possibility with the west coast sites where you could find whole, collapsed plank houses, then the idea of freeze-drying a complete building would be practical.

Suzanne Keene: The second possibility would be, instead of storing all artifacts, you could adopt the scientific approach and simply bottle the smaller artifacts in preservation fluid at low concentration.

George MacDonald: I agree. One of the points that came out of yesterday's discussion, that hit me immediately was that we should start putting an awful lot of our finds into some kind of fixative material. But I know for a fact that archaeologists, even wet site archaeologists, are abysmally ignorant about what kinds of fluids to use. So they would have to go to people in the natural sciences, who are doing this regularly, and say, "Look here, I have some plant material, some bark etc., what fixative should I use?" So, this would mean that you could be quite liberal with your sampling and if you think someday somebody will want to study cones, then you take some from each stratigraphic level and keep them in jars. And you would want to keep core samples too.

Jamie Barbour: I have a couple of things I would like to say. Firstly, with these big timbers you really don't need to treat them as they have been treated in the past, and this can save you a lot of money. This, was the whole point of the paper I gave at the IIC-

CG meeting in May (Annual Conference of IIC -Canadian Group, held in Victoria, B.C., May, 1981). For a cubic metre of western red cedar, if you just replaced the water in the cell walls with PEG, the cost I think was 96 cents per pound and it would cost you approximately $138.00. However, if you filled all the spaces in the wood it would cost you about $1600.00. And this is one thing I think archaeologists and conservators should start thinking about - cost. You need to know what you have to do, so that you don't do too much. I think that in a lot of cases things can simply be air dried as was suggested. Another point, as regards burying material back in the same soil, if you were going to do that, you would have to do it in such a way that the soil was very tightly packed in around the objects. Because once you disturb these things you are going to get a lot of oxygen that is going to be moving around and the environment is going to be quite a bit different and there will be a lot of oxygen trapped in the soil as you fill it up. Now I don't know how fast this is going to be used up, but I do know this much, that we have a lot of lignin left over in these woods. Now, whilst nobody, as far as I know, has published any information relating to an anaerobic step in the degradation of lignin, the oxidative steps in aerobic conditions will deteriorate lignin very quickly. So once the oxygen gets mixed in, I wonder how long the wood can sit around before deterioration gets going. I wonder, with material stored in lakes and bogs, how much deterioration occurs during storage.

Barbara Purdy: One thing you do not want to do, is to try and retrieve the material when the ground is frozen.

George MacDonald: Yes, we tried to retrieve a canoe from Mississippi and that was frozen solid. They had dug a trench, put the canoe in and filled it with water, but they didn't take into account that the soil was very porous and so the water level fluctuated a lot. When Barbara and I went to look at it, the water level was halfway up the gunwhales, it was frozen solid and we had to break the ice around the canoe before we could lift it.

Barbara Purdy: The bad thing was that when they first put the canoe in the water it floated, and to keep it down they put great big cement cylinders on top. When we looked at it, the cylinders were frozen solid to the canoe.

Herman Heikkenen: There is something involved here in your question of sampling design. You have to know your objective and with a lot of papers I have heard here, I was not really sure what the objective was. If it is collecting for collectings sake, fine.

George MacDonald: I don't think we set one single objective. The objectives for collecting range from pure research to pure display and public availability.

Herman Heikkenen: Then if you have multiple objectives there is no way you can come up with sampling design.

George MacDonald: Well I think you could probably isolate sets of objectives and then set up sampling designs for each one. For example, the sampling design for public need is going to be very simple. For example, they are going to want to see the very best things you encounter in your whole excavation.

Herman Heikkenen: For example, if your objective is to find out which tree species was being used for these canoes, then you have solved that. How many more samples do you need to know that they used pine.

George MacDonald: Except that it was cypress in Mississippi. There are going to be odd ones. But testing what the canoe is, is a very simple test so that there is no need to be elaborate with your sampling design in that case. But, you are right, that is a very good question and one we probably won't resolve here, but it certainly merits further discussion.

Victoria Jenssen: The only comment I would make is that we are taking almost the same point with the environmental survey as with the archaeology. If we knew what we were going to get we wouldn't need to do it. The reason you excavate is to test hypotheses, and the reasons for the environmental survey is to evaluate the environment of the site. There are certainly lots of examples to go by.

George MacDonald: There are lots of good research hypotheses regarding this site. For example, was there a shift in aboriginal water traffic with the introduction of metal tools? Here is a site that could have metal tools, the later occupation is right at the proto-historic period, not at white contact. A lot of these smaller settlements on the lower St. Lawrence could be coming up the river with metal from Basque fishermen shortly after 1500 AD. It would be important to elucidate the period between totally prehistoric and proto-historic. It will generate a whole set of sampling questions.

Barbara Purdy: You have mentioned total sample, maybe some people may not realize what you mean by total sample, how you get it, what you use it for, etc. I think it's really invaluable on a wet site, from each zone as you are going down.

George MacDonald: By total sample I imply two different kinds of samples and that should be clarified. Some remains occur in relatively small quantities and can provide much significant information, so should be collected in toto. Examples of this might be structural timbers, or perhaps hafts and shafts of tools and weapons. The opposite might apply to cut or burned sticks. In a large site you could encounter millions of wood chips and it is virtually impossible to preserve all of them. The other kind of sample is a bulk sample whereby a standard volume of material is removed for total analysis in the laboratory. The difficulty of taking a bulk sample at a wet site using hydraulic excavation techniques, is that you have to separate the collecting of the bulk sample from the normal excavation procedure.

Barbara Purdy: I think especially when you are going to excavate this way, despite the fact you are going to have good control, there is still the chance of mixing. If you clean off the profile and then take a pure profile sample, there is less of a problem.

George MacDonald: I think that splits down very clearly too. If you have good controls in regard to cultural material, but in regard to the environmental materials yo do not. No site known to me, using sluicing techniques, has really maintained control during this kind of excavation. Yo take all your bulk samples with coring devices before you start your sluicing. Most of your environmental information comes from the bulk samples.

Herman Heikkenen: You mentioned pine cones, permit me to mention a conference that was held at Virginia Tech. on dendrology, at which there were some historical dendrologists who spoke of our tree species as "scampering across the continent." That is, when you consider the ice and temperature ages over time, these species have been moving all over. The work that is coming from this area of science is absolutely essential for those people.

George MacDonald: In terms of trees, of course, we are dealing with a very narrow time slice at the Roebuck site. But one of the critical problems has to do with corn agriculture and the bog should reflect the introduction of corn pollens. In the polynological test we should get some idea of the level at which corn pollen comes in.

May I summarize the discussion on "what you keep" and suggest that there is a certain amount of material that you would keep traditionally for the museum collection. You might also want to build a holding tank where you could control water levels, perhaps in a spare building somewhere. Beyond that, rather than throw away things that might have some potential but are too bulky to be afforded high class treatment, the option would be to either put it into a cold lake, a bog, or outdoor freeze-drying and storing it dry.

End of Discussion.

Further information about Roebuck, the Iroquois and the archaeology of this area, can be found in the following:

1. Wintemberg, W.T., Roebuck Prehistoric Village Site, Grenville County, Ontario: Bulletin No. 83 (Ottawa: National Museums of Canada, Facsimile Edition, 1972).
2. Wright, J.V., The Ontario Iroquois Tradition: Bulletin 210 (Ottawa: National Museums of Canada, 1973).
3. Trigger, Bruce G., Volume Editor., Handbook of North American Indians: Volume 15: Northeast (Washington: Smithsonian Institute, 1978).
4. Wright, J.V.; Ontario Prehistory (Ottawa, National Museums of Canada, 1972).

SESSION IV

PROGRESS IN THE TREATMENT OF WATERLOGGED WOOD

Colin Pearson
Session Chairman

SOME PROBLEMS OF USING TETRAETHOXYSILANE (TETRA ETHYL ORTHO SILICATE: TEOS) FOR CONSERVATION OF WATERLOGGED WOOD

Kirsten Jespersen

National Museum of Denmark,
Konserveringsafdelingen for Jordfund
Traekonserveringslad,
Brede - 2800 Lyngby,
Denmark

Abstract

Partial dehydration and subsequent conservation with TEOS has been carried out with 67 wood samples, representing 12 different species. The age of the samples covered a period from 5000 B.C. to 1300 A.D. Acetone and t-butanol were used for dehydration. The water content in the samples after dehydration varied from 4-23% and they were submerged in a bath of TEOS from 2-9 days. The results have shown that it is possible to stabilize well preserved coniferous wood with TEOS and also in some instances deteriorated deciduous wood.

Introduction

For some years we have been considering tetraethoxysilane; usually abbreviated to TEOS; as a possible agent for the conservation of waterlogged wood.

Ever since this method was presented by Irwin and Wessen (1), I suppose that many of us have tried to impregnate wood with TEOS. Without asking critical questions we threw ourselves into the method simply following Irwin and Wessen's description. The results from these initial attempts were, however, most variable and the great optimism was somewhat reduced. And yet, we did see wood-samples preserved with TEOS in Neah Bay, which appeared to be fairly good, so what was wrong?

To answer the question it was necessary to test the method more seriously, and on our archaeological wood. Experiments were therefore undertaken at our laboratory from 1978 to 1979.

The Method

Although the method is probably familiar to you, I shall give a brief description of it. The objects are dehydrated in acetone, which removes only the free water. This pretreatment leaves sufficient water for the formation of microcrystals of silicon dioxide, when the wood is immersed in TEOS or when TEOS is injected in.

The following equation expreses the chemical reaction between TEOS and water:

$$Si(OC_2H_5)_4(Sol.) + 2H_20(Sol.) \rightarrow Si0_2(c) + 4C_2H_5OH(g)$$

According to Irwin and Wessen the main advantage of using TEOS in conservation is the very short treatment time. For example, for a large object of say 1 m long by 25 cm diameter, 13 days; for medium sized: 25 cm long by 5 cm diameter, 6 days; and for small objects: 20 cm long by 5-10 cm diameter, 4 days.

Furthermore, the treatment is cheap and no cleaning of the wood before conservation is required. The excess TEOS evaporates within 8 days and the treated wood will be resistent to moisture, fungi and micro-organisms.

Important information, such as the kind of wood used and its degree of deterioration was lacking in the report. Specifications of weight, dimension, volume and water content of the samples both before and after conservation were also missing.

Aims

The aims of this investigation were:

(a) To make a microscopical examination of the wood used in the testing programme, in regard to the species and condition.
(b) To determine the water content of the wood before treatment.
(c) To measure and compare dimensional changes when the untreated wood was air dried at room temperature.
(d) To study the extent of dehydration with acetone and t-butanol.
(e) To alter the water content in the dehydrated wood before treatment with TEOS, so that the effect could be studied.
(f) Conservation with TEOS.
(g) To study the treated wood's drying time.
(h) To regenerate used TEOS.

The suitability of this method was tested on wood frequently found in Denmark of different ages, from different localities and with varying degrees of deterioration.

The preserved wood was tested with regard to:

(i) The change of dimension.
(ii) The surface treatment necessary.
(iii) Suitability of materials, which could be used for joining breaks and for filling gaps.
(iv) Whether the treated wood could be reshaped.
(v) The behaviour of the wood to variations of relative humidity (RH).

Experimental Work

The experimental work was begun by mixing TEOS with different percentages of water varying from 2-20% in 25 ml test tubes. After 24 hours it was found that the water was still in a clear phase at the bottom, but after briefly shaking it the solution turned to jelly. After 4 days it was taken out and dried for 48 hours. The crystals formed were all different in appearance, ranging from milky to very clear, and from small powdery and spherically shaped to very large size.

A test mixture of 10 ml TEOS with 10 drops conc. hydrochloric acid produced one big and extremely hard crystal of 14 mm in length and a diameter of 9.5 mm. A similar test was made with sodium hydroxide; no crystals were formed, but instead a fine white powder was produced.

Before the dehydration with acetone and t-butanol, the specific gravity variation with concentration for acetone-water and t-butanol-water mixtures was plotted. Experiments were also carried out in order to check the dehydration time by measuring the rate of diffusion of water into acetone and t-butanol. It was found that the dehydration with acetone was 5 times faster than with t-butanol, (it took 3 days with acetone and 16 days with t-butanol).

The Wood Used

The wood which was used for all the tests, came from different sites and 12 species of wood were represented. The age ranged from 5000 B.C. to 1300 A.D.

Tests

(a) Microscopy

For the microsocopial examinations, samples were impregnated with paraffin wax and slices of 15 microns in thickness were cut on the microtome.

(b) Air Drying

For the comparison between treated and untreated wood, 14 samples were left to air dry at room temperature and at a RH of 55-61%. The loss of water after 14 days ranged from 65%-93% based on the wet weight. Dimensional change was recorded, and varied greatly from sample to sample.

(c) Water Content

The determination of the water content for 16 pieces of wood was determined by drying wood samples at 105°C in an oven until a constant weight was reached. The water content ranged from 77.4 to 94.8% based on the wet weight. (ed. 342% to 1824% based on the dry weight.)

Material was also kept in water as a reference.

Pre-Conservation, Water Content of Wood

Irwin and Wessen (1) said that conservation with TEOS requires that some water be left in the wood, but not how much.

It was therefore decided to preserve the 67 samples with different percentages of water remaining in the wood.

To obtain the chosen percentage of water after dehydration of the sample, the measured water content was used as the basis for the calculations. Since the amount of waterlogging water was known, and the amount of acetone used in dehydration could be varied, by adding a known amount of acetone to a known amount of wood which would contain a known amount of water; after interdiffusion of the water and acetone was complete the amount of water remaining in the wood could be calculated.

In this way samples of wood were dehydrated in acetone in a seven stage process, to give wood containing between 4 and 23% water; similarly for dehydration in t-butanol, a six stage process was used to give water contents between 8 and 21%.

The immersion in TEOS was carried out in sealed polyethylene jars, and the periods of immersion used were 48, 72 and 96 hours. On removal, samples were allowed to air dry at room temperature. All samples were measured for shrinkage and loss of weight during the drying period.

Results

Four of the preserved samples represent the general trends of the results:

Sample No. 1, medieval pine, (pinus silvestris) suffered only small and acceptable shrinkage after conservation. The best treatment for this sample was obtained with dehydration with t-butanol, 8% residual water in the wood, and a soaking time of 51 hours in TEOS. The interior of the wood was still intact, but the exterior was soft and friable.

Sample No. 2, ash (fraxinus) from 5000 B.C. also showed small dimensional change. However, the samples dehydrated with acetone possessed innumerable cracks, suffered internal collapse and after drying the wood shattered into pieces. The three pieces of sample 2 dehydrated with t-butanol had no cracks on the surface. But, once dry, they were cut into two, excessive shrinkage/collapse of the interior was seen. The high water content of 19-21% after the dehydration, gave a very hard external surface, which prevented it from splitting. The drying period was about 45 days.

Sample No. 20, a piece of elm (ulmus), dating from 2500 B.C., consisted of an extremely hard core and a completely deteriorated external layer of thickness 2 to 3.5 cm. The deformation when left to dry at room temperature was drastic. One piece, dehydrated with t-butanol to give 12% residual water and immersed in TEOS for 6 days was not acceptable, whereas a slice which was dehydrated with acetone for 5 days to give a residual water content of 23% and immersed for 8 days in TEOS gave an almost acceptable result, but even then the shrinkage of between 9 and 10% was much too high. The drying time was more than 2 months.

Sample No. 22, was a completely deteriorated piece of poplar (populus) with a water content of 94% based on the wet weight. The TEOS treatment had no effect at all. The tangential shrinkage after conservation ranged from 25 to 49% and the results were quite unacceptable.

Of the 67 samples tested only 8-10 could be regarded as partly acceptable. These included three samples dehydrated with acetone giving a residual water content of 23% followed by immersion in TEOS for up to 9 days, and 5 samples dehydrated with t-butanol giving a residual water content of 12% followed by 2-4 days immersion in TEOS.

It was very difficult to find any relation between the differences in treatment and the results, or to discover factors which could have been responsible for the best results obtained. Since 8 out of the 10 best results had a high water content after the dehydration, it was important to examine the relation between the water and the amount of silicon dioxide absorbed. The degree of filling of SiO_2 in cavities inside the wood was calculated for all samples and it was found that:

(i) The % water after dehydration had no influence on the amount of deposition of SiO_2 in the wood,

(ii) that the amount of silicon dioxide, which the wood was able to absorb, was not proportional to the time of immersion in TEOS. This means that prolonged soaking in TEOS will not create a higher silicon dioxide content and

(iii) that there was no connection between a greater SiO_2 content and successful conservation.

By comparing the best results to the remainder it can be demonstrated, which factors, such as; the species of wood, the degree of deterioration, the water content after dehydration and amount of SiO_2 absorbed; lead to good results.

It was found that well preserved pine could be impregnated fairly well without special consideration as to the solvent used for dehydration, the residual water content and the amount of silicon dioxide absorbed.

For the other sorts of wood it was found that dehydration with t-butanol was better than with acetone, and that the water content after dehydration should be more than 10%.

Surface Treatment

The surface of TEOS treated wood is normally porous and very light coloured making a surface treatment necessary.

The most suitable were found to be (i) shellac, (ii) Hoescht PA 520 wax (in a 60% PA wax/paraffin mixture dissolved in xylene) and (iii) a 1/2 dammar resin/beeswax mixture dissolved in 3 parts xylene to one of the wax mixture.

Many types of adhesives such as acrylic, polyurethane and cellulosic were useful for repairing breaks. For gap-filling a moisture hardening polyurethane glue or plastic wood were suitable.

Aesthetics

From the aesthetic point of view TEOS treated wood is not pleasing, it is discoloured and spongy and very light in weight. The specific gravity is very often less than that of cork (0.25). For some of the samples it ranged from 0.18 to 0.21.

Conclusions

A few of the samples from this investigation showed that TEOS could be used for the conservation of waterlogged wood, but the uncertainty with regard to exact water content in the wood, the extent of deterioration, and hence the difficulty in deciding the immersion period in the TEOS bath etc. limit the use of this method for treating museum objects.

Results from our normal freeze-drying process from t-butanol and PEG 4000 exceed in all respects those obtained with TEOS.

In spite of a few good results obtained with TEOS I am convinced that this method should not be used for the treatment of waterlogged wood.

It does however seem to be more promising as a middle step in the conservation of rope and textile. This research is at present taking place in our laboratory.

Appendix

During this work it became obvious that a way of controlling the reaction between TEOS and the water must be found in order to allow the formation of small crystals; bigger crystals too often cause disruption of the wood.

A supplementary investigation was undertaken by two students from the Technical University in Denmark in 1979/80.

From their results with 26 samples of waterlogged oak (quercus) and (fraxinus) the best solvent for TEOS was found to be ethanol. When 80% of the water in the wood was exchanged with ethanol, and TEOS was added to this bath, then the reaction between TEOS and water could be initiated with a few drops of 0.1 N hydrochloric-acid, which catalyses the process. The creation of crystals and their size has been examined by leaving the wood in the water/ethanol/TEOS bath for periods of 0, 5 and 9 days before the addition of the catalyst. Tests were also carried out to see whether the reaction requires heat. Their best results, in which small crystals were always produced, were obtained with the following conditions: when the reaction was initiated immediately after TEOS addition to the ethanol/water bath and with reaction at room temperature.

References

1. Irwin, Henry, T. and Gary Wessen, "A New Method for the Preservation of Waterlogged Archaeological Remains: Use of Tetraethyl Ortho silicate," in: Pacific Northwest Wet Site Wood Conservation Conference Vol. I, proc. of the conference, 19-22 September 1976, ed. Gerald H. Grosso (Neah Bay, Washington: 1976), pp. 49-59.

Questions

David Grattan: Are you worried about the toxicity of TEOS?

Kirsten Jespersen: TEOS is not toxic, there has never been a report of any toxicity.

David Grattan: I know this, but it seems to me that if you inhale TEOS it could well hydrolyse in your lungs and deposit SiO_2. Which I feel must be a rather hazardous effect. Why shouldn't it lead to silicosis, for instance?

Kirsten Jespersen: We came to the same conclusion and thus we are very careful about contolling the fumes.

(editor's note: see for example toxicology of TEOS:

from Dangerous Properties of Industrial Materials 4th Ed., N. Irving Sax, (New York: Van Nostrand, 1975)

p. 1148: Tetra ethoxy silane & p. 704 Ethyl Silicate

Howard Murray: We have used TEOS to treat waterlogged rope, but it makes it rather brittle.

Kirsten Jespersen: We have been investigating the use of TEOS to treat rope and it does make it very brittle, but we are able to make it flexible, by treating the rope after TEOS impregnation, with an elastomer. (ed. the details of this elastomer are not available at present.) After this treatment even the most fragile and brittle rope becomes soft and quite beautiful and easy to handle. This research is not yet finished, but we have found that the TEOS treatment does retain the dimension of the rope.

David Grattan: Dr James Hanlan and Ruth

Norton of Queen's University, Kingston, Ontario, has been carrying out experiments with Dimethyl diethoxy silane, and methyl triethoxy silane, for the treatment of dry and decayed wood. The latter substance was found to give a flexible glass under controlled hydrolysis with dilute hydrochloric acid, and wood so treated has some flexibility and seemed a large improvement over TEOS treated wood. The results of TEOS treatment that I have seen are very poor. I also had concluded that this technique was unsatisfactory on the basis of a comparative study, which I hope will shortly appear in Studies in Conservation (1982).

Kirsten Jespersen: I have totally finished with TEOS, I will never use it again for wood, it is really destructive. (Laughter).

Mary-Lou Florian: Was there a drastic reduction in the specific gravity of the coniferous woods.

Kirsten Jespersen: Yes there was.

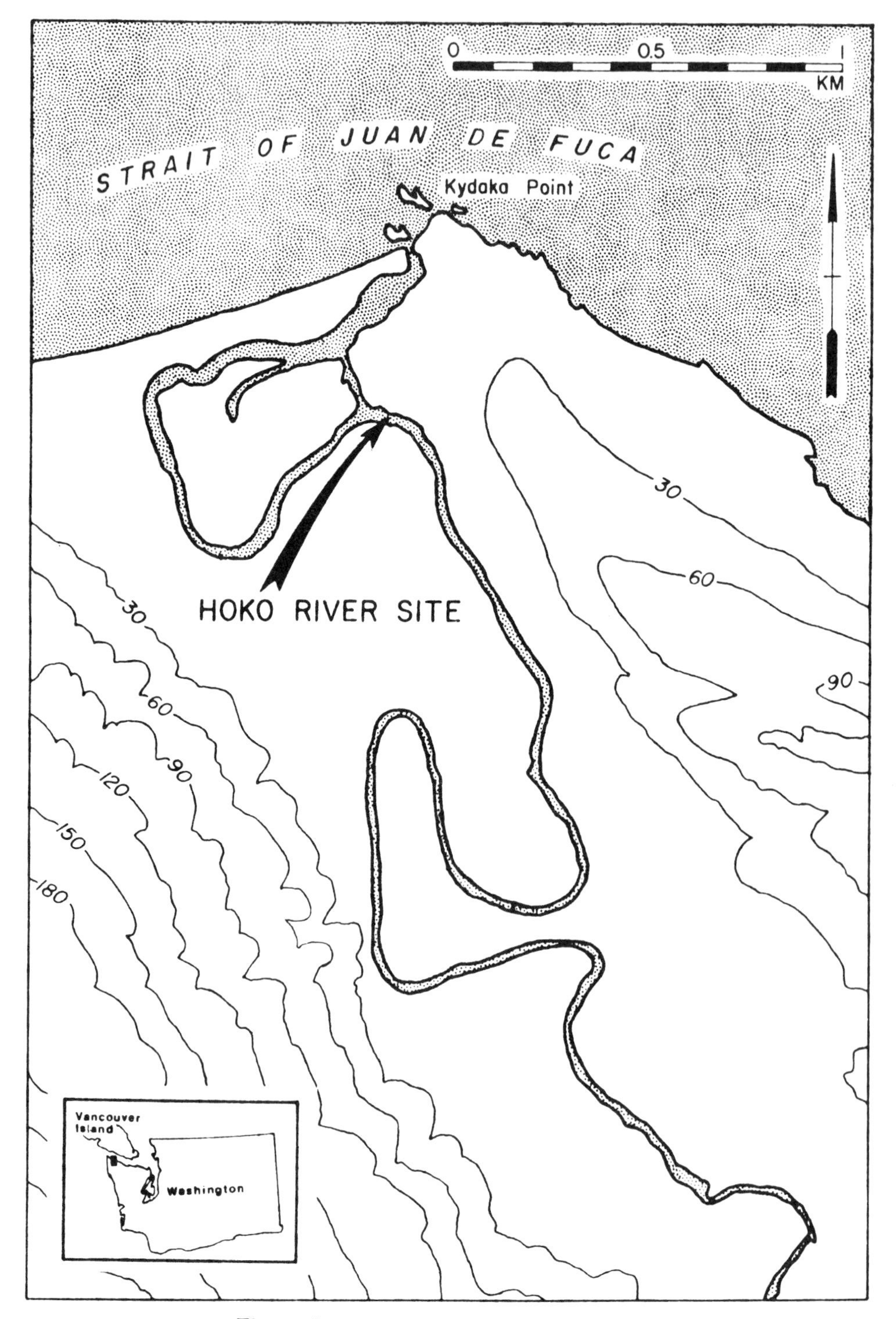

Figure 1. Location of Hoko River site.

SHRINKAGE AND COLLAPSE IN WATERLOGGED ARCHAEOLOGICAL WOOD: CONTRIBUTION III HOKO RIVER SERIES*

R.J. Barbour
L. Leney
University of Washington,
Seattle, Washington,
U.S.A.

Introduction

The Hoko river site is a 2500 year old fishing camp located at the mouth of the Hoko river. The river empties into the Straits of Juan de Fuca which form the boundary between Washington State's Olympic Peninsula on the U.S. side and Canada's Vancouver Island to the North. The site lies about 23 km east of Neah Bay, Washington (**Figure 1**). The Hoko site is characterized by an abundance of preserved "waterlogged" perishable artifacts, including basketry, cordage, fishing equipment and wood working tools (Croes and Blinman, 1980).

Wooden artifactual materials recovered from this site was either discarded or lost during use, found its way to the river bottom, where it was covered by silt and remained until excavation. Further details of this process are given by Blinman (1980) and Stuki (1982). This wood falls into two distinct categories. The first makes up more than 75% of the artifact sample and is made up exclusively of coniferous woods. It is relatively sound and dimensionally stable. Upon drying this wood will warp and check but its shape will not be completely lost. It can easily be stabilized using PEG treatment after which the wood appears quite natural, if somewhat dark.

The second group is made up mainly of angiosperms but does contain a few conifers. It is much more deteriorated. When wet it can be ground between the thumb and forefinger. Upon drying it undergoes a drastic loss of dimension, finally drying to a hard, dark mass. This end product bears more resemlance to coal than the wood it originally came from. It can be stabilized, in the green state with 540 Blend PEG but the end result is waxy and rather unnatural in appearance.

For the past two years the major thrust of conservation research at the Hoko River Project has been to understand the properties of the more highly deteriorated wood. The aim here has not been to solve any particularly pressing conservation problem. Instead it has been to examine the behaviour and condition of the wood in some detail, and to explain that behaviour in terms of the condition, from a wood technological point-of-view.

Hopefully, this paper and those that follow will help to add to the pool of knowledge concerning the properties of waterlogged archaeological wood (The term waterlogged is used here to mean the wood is at or near its maximum moisture content; Skarr, 1972; p. 84). This information should make it easier for the conservator and archaeologist to examine wood samples recovered from test excavations and to make meaningful recommendations as to how the wood should be handled.

The material reported here is a brief summary of the techniques used to describe the dimensional behaviour observed during the drying of a piece of red alder (*Alnus rubra*) recovered from the Hoko site. This work was carried out at the College of Forest Resources, University of Washington under contract from the Hoko River Project, Department of Anthropology, Washington State University.

Drying of wood is a complex problem. During drying, cells change size and shape as moisture is removed from the wood. These movements result in a change in the overall shape of the piece. In this report, qualitative descriptions of this change are termed dimensional behaviour. When quantitative measurements, of the dimensional behaviour, are made the magnitude of these measurements is term-

*Contributions:
I. Croes, Dale R. and Eric Blinman
"Hoko River: A 2500 year old fishing camp on the Northwest Coast of North America," *Reports of Investigations*, No. 58. (Laboratory of Anthropology, Washington State University, Pullman, 1980)
II. Flenniken, John Jeffrey
"Replicative systems analysis: A model applied to the vein quartz artifacts from the Hoko River site," *Laboratory of Anthropology Reports of Investigation*, no. 59, (Washington State University, Pullman, 1981)

ed the dimensional change.

Two causes of dimensional change will be considered here. They are capillary tension and desorption. Capillary tension is the first to occur during drying. It arises as a pressure differential develops when water evaporates from the capillaries which form the permanent void structure of the wood (Tiemann 1915, Hawley 1931, Siau 1971). Capillary tension type collapse is the observable phenomenon that results from the force generated by evaporation from the permanent void structure of the wood.

Desorption is the removal of water from intermolecular association with the cell wall substance (Stamm, 1964). Most authors refer to this water as bound water while the capillary held water is referred to as free water (Skarr, 1972). As bound water is removed from the cell wall substance the molecules making up the cell walls draw together. The observable phenomenon that results from this coalescence of the cell wall substance is known as shrinkage (Stamm, 1964). However, the term shrinkage has, in the past, been applied rather loosely to any reduction of dimension occurring during drying of a piece of wood. This loose usage has distorted the meaning of the term. For clarity, cell wall shrinkage is used here to mean dimensional change resulting from the force created during desorption of water from the cell wall material.

While capillary tension and desorption begin the sequence of events leading to dimensional change, they are not the only influence on its outcome. As they occur stresses are generated within the wood. These stresses must either be tolerated or dissipated through a realignment of the wood structure. The way these stresses develop and how the wood will react to them depends largely on the characteristics of individual cells, but also on the general wood structure.

For purposes of this report these stresses are divided into two broad categories. The first category comprises the intercellular stresses. These develop within individual cells in response to capillary tension and desorption. If the cell is not strong enough to resist the stresses, deformation will occur. The amount and direction of the deformation is determined by the anatomical characteristics of the cell. Kuo and Araganbright (1978) describe collapse in two hardwoods and two softwoods. Although they do not use the terminology given here they describe intracellular characteristics that influence collapse. In three of the species studied the latewood showed collapse that was thought to be related to the small radial diameter of the cells. Similarly Thomas and Erickson (1963) attribute collapse of earlywood tracheids in western red cedar to the thin walls of these cells.

For sound wood Boyd (1974) contends that the anisotropic nature of cell wall shrinkage can be linked to intracellular characteristics. He recognizes two "dominant determinates." The first is the difference in the lignification of radial and tangential cell walls. The second is the cell form factor or differences in the cross-sectional shape and the difference of wall thickness between early wood and latewood cells. These determinants, as Boyd calls them, act to dictate how individual cells will either dissipate or tolerate the stresses generated within them.

Since every cell does not act in perfect unison with those around it, as they dry, stresses also arise between cells. These intercellular stresses make up the second broad category of stresses. Deformation due to the intercellular stresses can be thought of as the result of differential drying and (or) nonuniform dimensional behavour of different cells, groups of cells or regions within the wood.

Intercellular shrinkage stresses have been studied in sound wood, where they are commonly called shrinkage stresses (Skarr, 1972) or drying stresses (Kauman, 1958; Stamm, 1964). they are used in the explanations of phenomena such as casehardening, reverse casehardening, and checking (Skarr, 1972) as well as a specific type of collapse (Stamm and Loughborough, 1942). The intercellular shrinkage stresses reported by these authors are of the type caused by differences in the cell wall shrinkage in the different layers of the wood due to moisture gradients, within the piece.

Another example of deformation due to intercellular shrinkage stresses is that reported by Erickson (1958), Pentoney (1953) and others, where the greater shrinkage of latewood deforms earlywood cells. This is a result of the cell form factor discussed by Boyd (1974). In this case the intercellular shrinkage stresses arise in response to differences in the magnitude of cell wall shrinkage of different

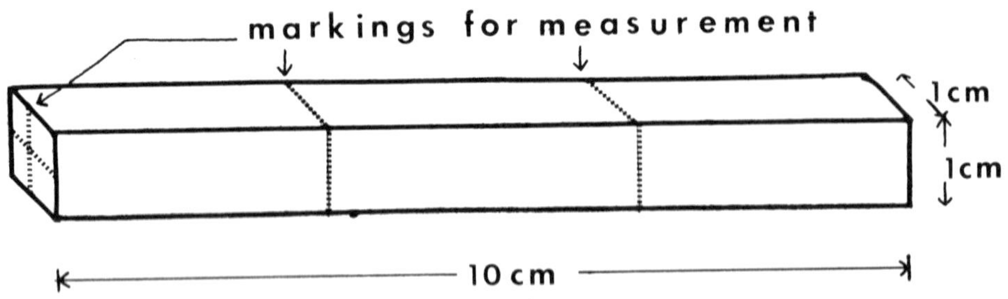

Figure 2. Schematic of sample blocks use to evaluate dimensional change.

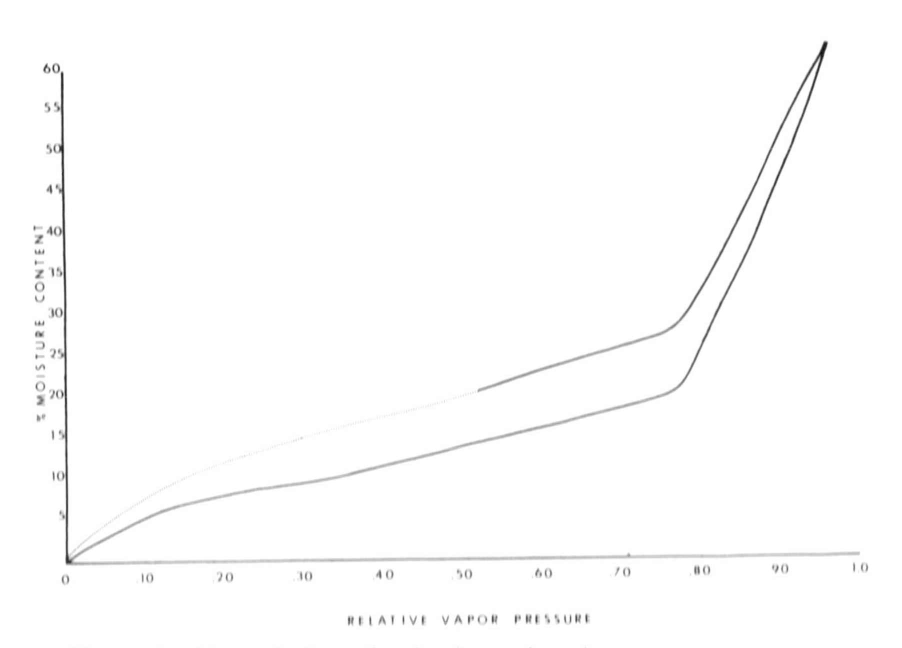

Figure 3. Hysteris loop for isothermal moisture sorption at 21°C. Dashed portion of desorption isotherm represents incomplete data.

ell types, rather than a variation in cell wall shrinkage caused by moisture gradients.

While intercellular shrinkage stresses due to cell wall shrinkage are of great importance in sound wood they may be of secondary importance in waterlogged archaeological wood. This is because the wood, which is waterlogged must dry through the moisture content range where capillary tension type collapse will take place before cell wall shrinkage can occur. Therefore, deformation due to both intracellular collapse stresses and intercellular collapse stresses would develop first. In addition, the magnitude and potential for unequal dimensional change are greater with capillary tension collapse than with cell wall shrinkage (Kauman, 1958). Thus, stresses developed due to capillary tension collapse are very important in the dimensional behaviour of waterlogged archaeological wood.

Procedures

I. Material collection and storage:

Wood was collected during July and August of 1980. The red alder used in this study was located in a silty deposit adjacent to the Hoko site.

The wood was removed from the site, sealed in plastic bags and shipped to the College of Forest Resources at the University of Washington in Seattle. Upon arrival, it was placed in a cold room at 2°C where it remained until needed.

In order to obtain material free from post-excavation infections by micro-organisms, ten centimeters of material were removed from each exposed surface before samples were cut. To further ensure against infection, cut samples were kept in distilled water in a refrigerator and were used without delay.

II. Evaluation of dimensional change:

Sample blocks were cut slightly oversized using a bandsaw. They were then carefully trimmed to final dimensions of 1 cm radially (r) by 1 cm tangentially (t) by 10 cm longitudinally (l) as in **Figure 2.** The final trimming was done using a hand held vibritome constructed from electric hair clippers, similar to that described by Brazer and Knight (1980). This procedure provided smooth surfaces for accurate measurements. The transverse surfaces were left quite smooth by the saw blade and were not modified.

When the sample blocks were cut to the desired length (10 cm) the pieces cut from each end gave two end matched samples which were used to determine the density (ovendry weight, green volume basis) for each sample block.

On each sample block, measurement lines were drawn on the transverse, radial, and tangential surfaces, (**Figure 2**). These lines marked the positions where all measurements were made during the experiments.

The sample blocks were then measured using a hand held caliper, accurate to 0.1 mm, weighed and their volumes determined. From here the blocks were placed in an Amico-Aire controlled environment cabinet at a relative humidity of 48% and temperature of 21°C. These conditions were chosen to simulate those expected in many laboratories.

The blocks were periodically removed from the cabinet, weighed, measured and returned to controlled conditions. This process was repeated until weight loss was minimal between weighings. Then the blocks were ovendried at 103 ± 2°C and their constant ovendry weight recorded. Moisture contents were calculated based on this ovendry weight.

III. SEM observation of collapse:

Sample blocks for this study were prepared exactly as those in Section II. In this case the average density, of each sample block, was used along with its volume to calculate the ovendry weight of the sample block. With this information, the moisture content of the block could be calculated for any known weight. The samples were then dried, as in Section II, to a weight which would give a desired moisture content. After measurement for dimensional change, a five centimeter piece was cut from the center of the sample. This piece was frozen in liquid nitrogen and fractured into a number of fragments to expose fresh surfaces. These fragments were then freeze-dried by the method described by Erickson et al. (1968). Unused parts of the sample were ovendried to accurately determine moisture content.

After freeze-drying, the samples were stored over phosphorous pentoxide until they were mounted and sputter coated for observtion in the SEM.

IV. Evaluation of cell wall shrinkage component.

To evaluate the cell wall shrinkage component, shrinkage samples measuring 1.5 cm(r) x 1.5 cm(t) were cut with a band saw. They were then frozen in the freezing compartment of a refrigerator. While frozen a transverse surface was cut on a sliding microtome. A sample 0.5 cm in the longitudinal direction was then taken from the freshly cut end, using a micro saw. This shrinkage sample was weighed, marked for measurement, numbered, and its volume determined. After a photomacrograph was made of the transverse surface, shrinkage blocks were freeze-dried as previously described.

After freeze-drying, the blocks were weighed, photographed and placed in a controlled environment established over saturated salt solutions. Air was continuously bubbled through the salt solutions to ensure equilibrium conditions in the environmental chamber. The temperature was controlled by keeping the apparatus in a controlled temperature room at 21°C.

The samples were taken through six moisture content steps to the fiber saturation point. They are now in the process of being returned to zero moisture content over the same series. At each step the blocks were weighed on an analytical balance. When at equilibrium, they were photographed to record their dimensions. In this way sorption isotherms as well as shrinkage curves have been drawn for the shrinkage blocks.

To evaluate cell wall shrinkage in the longitudinal direction, sample blocks like those described in section II were prepared. They were only 7 cm long because of the limited size of the freeze-drier. These sample blocks were first measured unfrozen, then when frozen and again after freeze-drying. Then they were dried over phosphorus pentoxide and their total cell wall shrinkage determined. Since freeze-drying does nothing to inhibit cell wall shrinkage (Erickson et al., 1968) reasonable estimates of cell wall shrinkage can be made in this way.

Results and Discussion

I. Moisture sorption:

A hysteris loop for isothermal moisture sorption is presented in **Figure 3**. The dashed portion of the desorption isotherm represents the portion of the curve that has not been completed to date. Upon inspection it can be seen that the shapes of the adsorption and desorption (sorption) isotherms are quite similar in shape to those for normal wood. That is, they are sigmoidal or type 2 isotherms as described by Stamm (1964). There are, however, significant differences between the isotherms presented here and those reported elsewhere, (Skarr, 1972; Stamm, 1964) for normal wood.

The isotherms pictured in **Figure 3** show that at any relative vapor pressure the equilibrium moisture content (E.M.C.) of the wood studied here is about twice as high as that expected for normal wood. This phenomenon has been discussed by Scheffer (1973) who points out that EMC values may be either higher or lower in deteriorated wood. From the data presented here and those of Noack (as reported by Rosenqvist 1975) it appears that the EMC of waterlogged archaeological wood can generally be expected to be higher than that of normal wood. The fiber saturation point (FSP) follows this trend. Its value is 60%, for the wood studied here, fully twice that reported for normal wood in the Wood Handbook, (Anonymous, 1974).

Another interesting difference between sorption isotherms for normal wood and those reported in **Figure 3** is the variation in shape of the curves. Isotherms for normal wood display an abrupt increase at a relative vapor pressure of about 0.90. According to Stamm (1935) this can be attributed to capillary condensation in fine permanent capillaries which are present in wood. Namely, those with effective diameters of less than 10 nm. For the wood studied here the rapid upturn in the isotherms begins at a relative vapor pressure of 0.80 not 0.90. However, no equilibrium moisture contents were measured between these points. Additional equilibrium moisture contents should be measured in this region. No explanation will be offered until these additional data are available. The phenomenon is mentioned to bring notice to the fact that differences may exist in the shape of sorption isotherms of waterlogged archaeological wood and those for normal wood. Hopefully, this will be the subject of future research; as it could be rather important in understanding and predicting the sorption behaviour of this wood.

II. Cell wall shrinkage:

Figure 4 presents data collected for cell wall shrinkage of the shrinkage samples. These are preliminary results, as the moisture

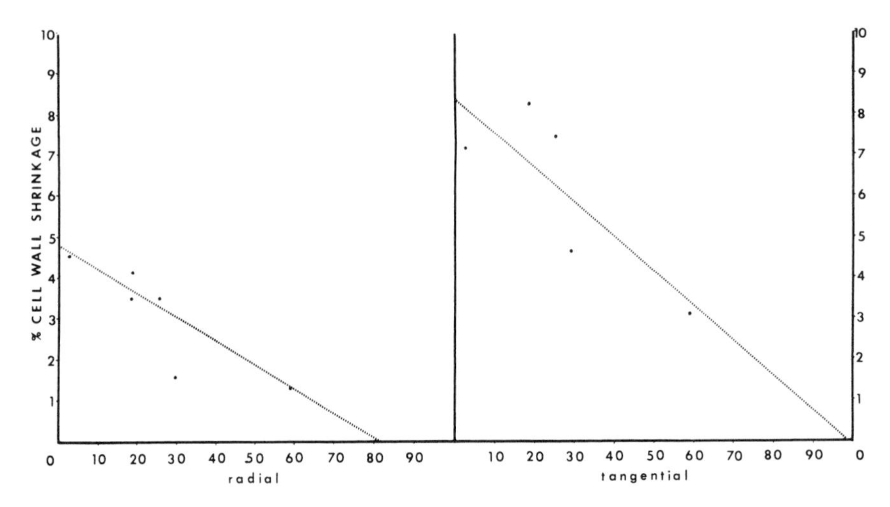

Figure 4. Cell wall shrinkage for radial (r) and tangential (t) direction when conditioned to various moisture contents from near fiber saturation to 2.5 percent over salt solutions. The regression lines for the data points are shown. The r^2 for (t) is 0.67 and for (r) is 0.94.

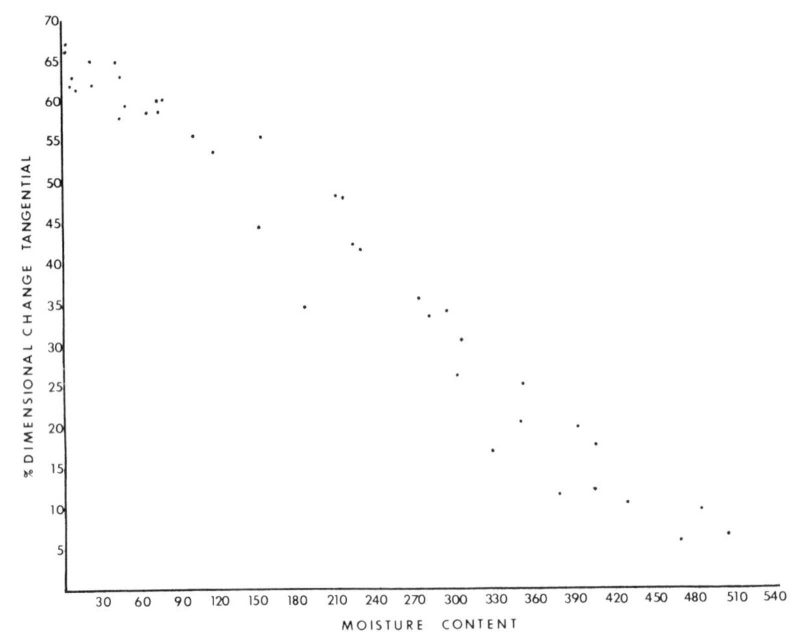

Figure 5. Tangential dimensional change of sample blocks. Dimensional change plotted against moisture content. Drying conditions 21°C, 48% R.H.

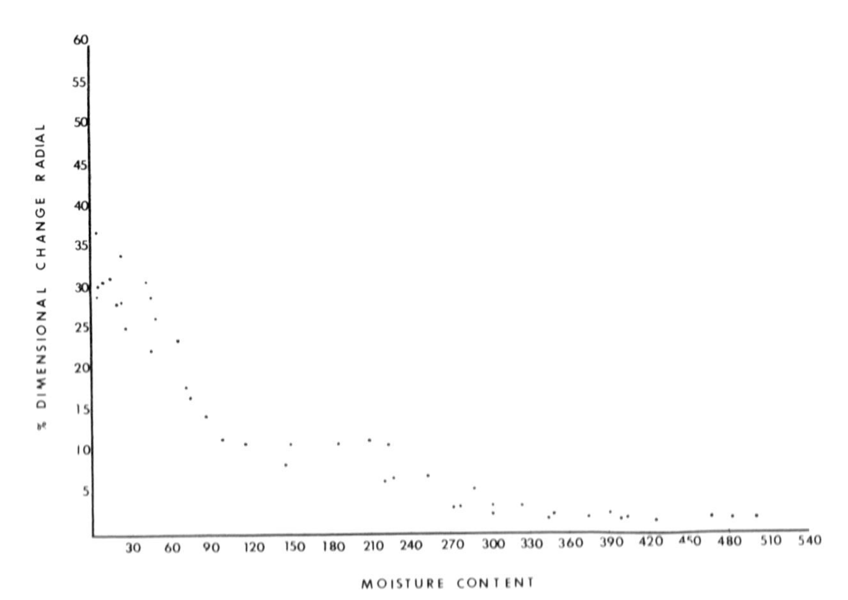

Figure 6. Radial dimensional change of sample blocks. Dimensional change is plotted against moisture content. Drying conditions 21°C, 48% R.H.

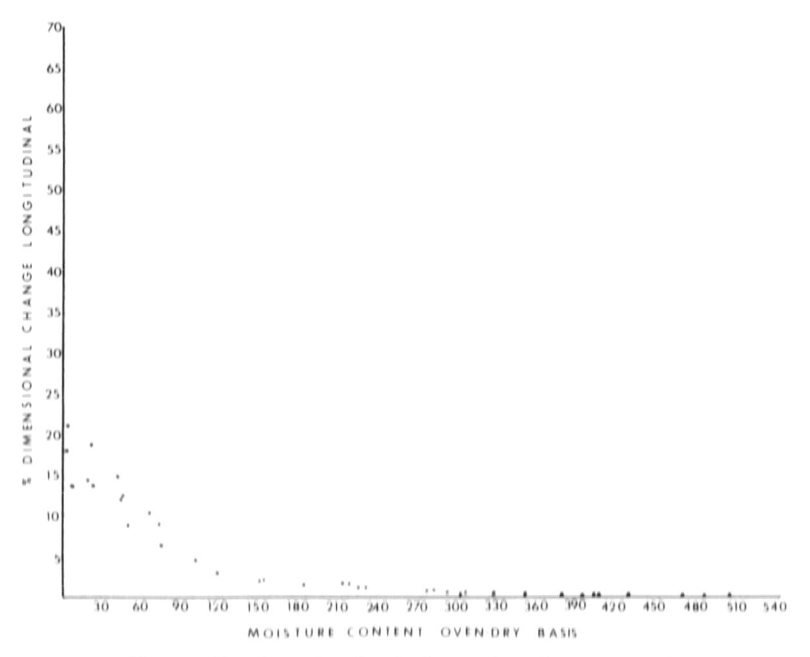

Figure 7. Longitudinal dimensional change of sample blocks. Dimensional change is plotted against moisture content. Drying conditions 21°C, 48% R.H.

content cycle has not been completed. However, data available thus far are useful in understanding the overall dimensional behaviour of the wood under study.

In **Figure 4** regression lines are drawn for the data; actual measurements are represented as points. The linear regressions show good correlation between moisture content and cell wall shrinkage. The r^2 value for radial data is 0.94 and 0.67 for tangential data. The total cell wall shrinkage predicted by the regression lines is 4.8% in the radial direction and 8.4% in the tangential direction. The Wood Handbook (Anonymous, 1974) gives values of 4.4% for radial shrinkage and 7.3% for tangential shrinkage for normal red alder.

The cell wall shrinkages observed here are higher than those expected for normal wood. A 9.1% increase in total cell wall shrinkage is seen in the radial direction and 15.1% in the tangential direction. This means that if a piece of waterlogged archaeological wood were dried, without collapse, it would not come to the same final dimensions, as it would have done before waterlogging and subsequent deterioration.

These findings support the contention that the goal of conservation should be to stabilize the archaeological wood in the size and shape it is found in rather than attempting to regain its original pre-waterlogged dimensions. This position is strengthened when data for cell wall shrinkage in the longitudinal direction are considered.

Estimates of cell wall shrinkage in the longitudinal direction give a value of approximately 10%. This value was arrived at by measuring the dimensional change observed after freeze-drying and drying over phosphorus pentoxide. Using this value for longitudinal cell wall shrinkage, volumetric shrinkage can be calculated at 21.5%. This is nearly twice the 12.6% volumetric shrinkage given for normal alder by the Wood Handbook (Anonymous, 1974).

The amount of cell wood shrinkage observed in the longitudinal direction is extremely high, between 50 to 100 times as great as the 0.1% to 0.2% expected for normal wood (Anonymous, 1974). Differences of this magnitude cannot be ignored, conservation treatments chosen for wood that is known to react in this manner must be designed to prevent cell wall shrinkage or they will not be successful.

III. Analysis of overall dimensional change:

The data collected for overall dimensional change (**Figures 5-7**) represent the cumulative effect of cell wall shrinkage and capillary tension type collapse. Using this information and that gained during observations in the SEM (**Photos 1-9**) and from cell wall shrinkage experiments (**Figure 4**); reasonable conclusions can be drawn about how the forces, generated during drying, act upon the wood.

Tangential dimensional change (**Figure 5**) begins as soon as drying commences and continues linearly to the fiber saturation point. In this range good correlation between moisture content and dimensional change is seen. These data have r^2 equal to 0.96. Below the fiber saturation point the slope of the line decreases and the correlation falls to r^2 equal to 0.42.

Radial dimensional change (**Figure 6**) is much more erratic than tangential change. There is an initial loss of dimension followed by a long period of relative stability. Below 300% MC the curve increases irregularly to below 150% MC. From this point dimensional change increases linearly to the dry condition.

A third pattern is seen for longitudinal dimensional change (**Figure 7**). There is very little dimensional change until the moisture content is reduced below 150%. If a linear regression is run on the data for dimensional change below this moisture content very good correlation is seen. These data have r^2 equal to 0.83.

A photographic record, of transverse dimensional behavior, is given in **Photos 1-6**. Together with the graph described above this allows a more complete understanding of how the wood reacts during drying.

Tangential dimensional change (**Figure 5**) appears to be the easiest to explain. Eight tenths of the total tangential dimensional change has occurred by the time the wood reaches 120% MC. This is twice the fiber saturation point (60% MC see **Figure 3**). The dimensional change occurring here can safely be attributed to capillary tension type collapse. Therefore, tangential dimensional change is mainly a product of capillary forces and their resultant stresses, not cell wall shrinkage.

Examination of **Photos 1-6** offers insight as to why capillary tension type collapse is expressed so strongly in the tangential direction. During the early stages of collapse there is

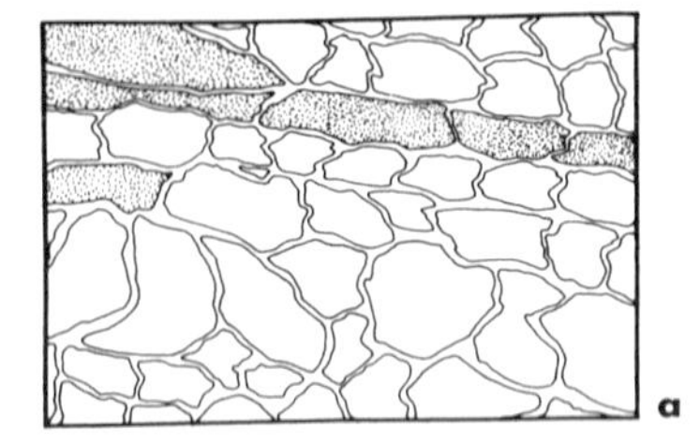

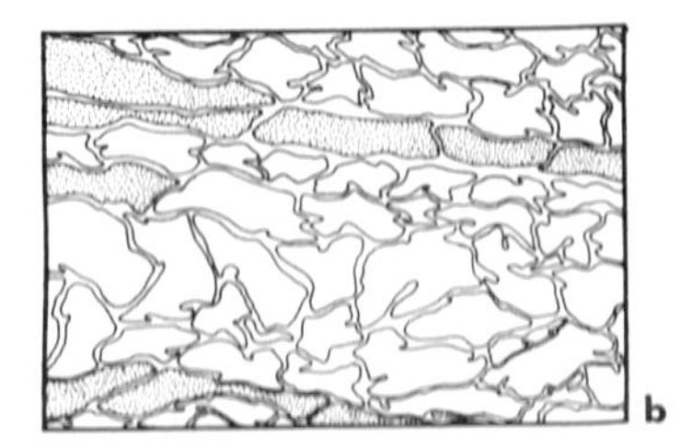

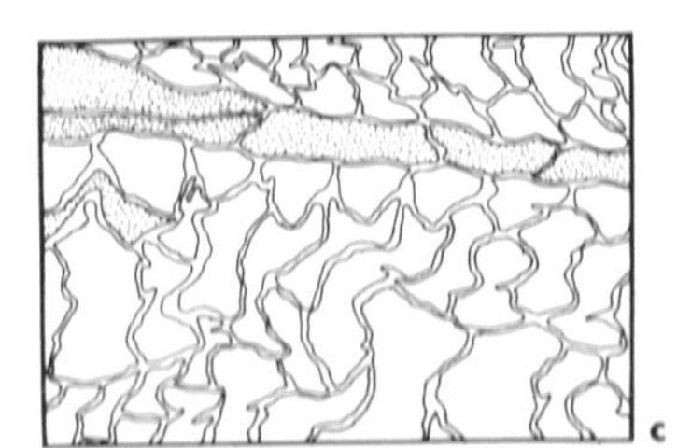

Figure 8. Artist's conception of collapse in a sample block.

A. Fibers much as they appear after freeze-drying.
B. How the same cells actually collapse tangentially.
C. How the cells would appear if they collapsed radially rather than tengentially.

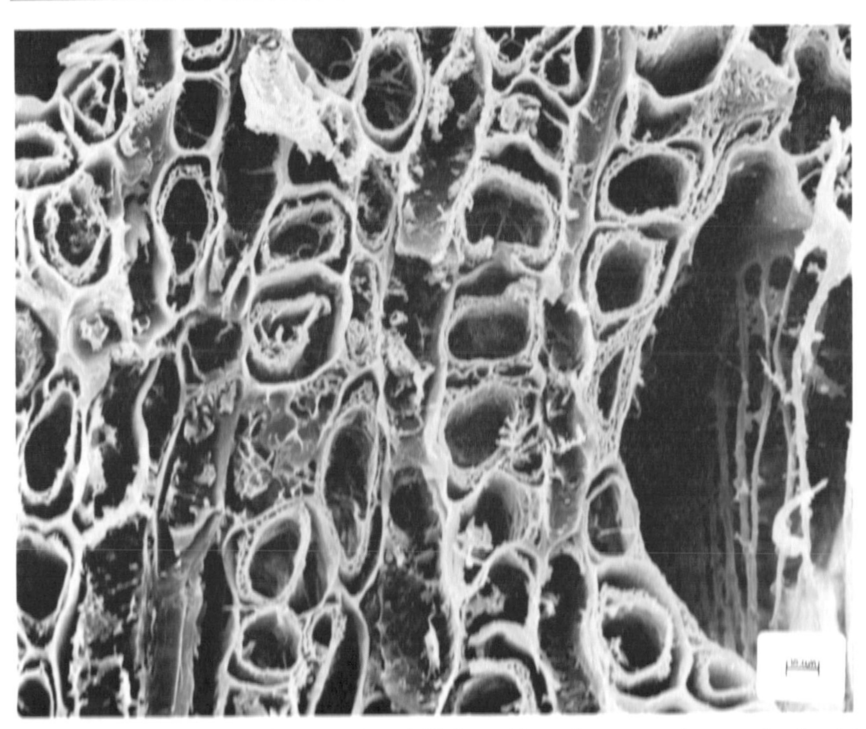

Photo 1. Scanning electron micrograph of Hoko alder that was frozen in liquid nitrogen, then freeze-dried from the waterlogged condition. No drying took place in this sample. Moisture content was approximately 600%. (The scale bar indicates ten microns)

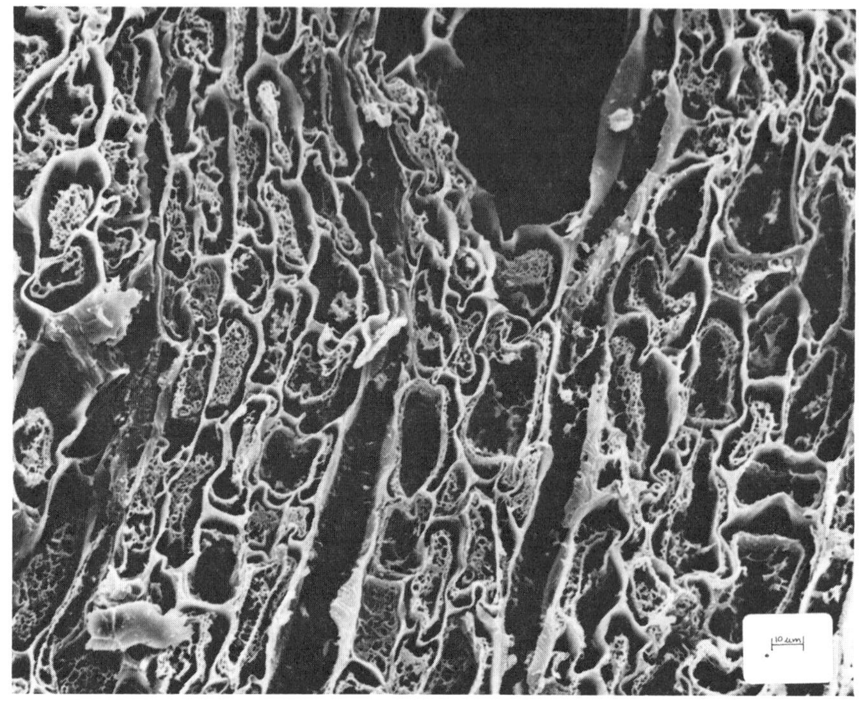

Photo 2. Scanning electron micrograph of Hoko alder conditioned to 300% MC, then frozen in liquid nitrogen and freeze-dried. (The scale bar indicates ten microns.)

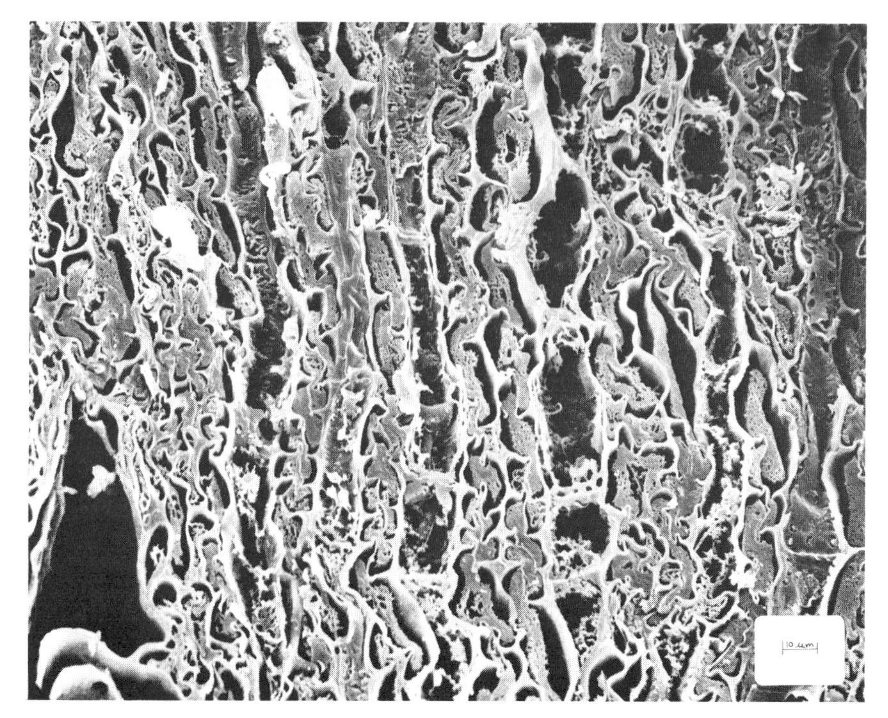

Photo 3. Scanning electron micrograph of Hoko alder conditioned to 255% MC then frozen in liquid nitrogen and freeze-dried. (The scale bar indicates ten microns.)

considerable buckling of the tangential cell walls but relatively little in the radial walls (**Photos 1 and 2**).

The behaviour of the radial and tangential fiber walls offers one possible explanation of why the wood shows a predisposition to collapse in the tangential direction. The compound middle lamella seems to be the only wall layer that remains intact in this wood (**Photos 1 and 2**). It is known that this layer is considerably thicker in the radial walls than in the tangential walls (Skarr, 1972) For this reason, a uniform load placed on the walls would be expected to buckle the tangential walls first. Following the terminology given above this would be considered an intracellular deformation brought about by the characteristics of individual cells.

Two other factors which could direct collapse tangentially are intercellular in nature, as they arise due to interactions between cells. The first of these is radial restraint. This is where the rays provide a means of support for the radial walls. **Figure 8** depicts how the rays would help to restrain radial collapse while encouraging tangential movement. **Figure 8a** shows the fibers much as they appear after freeze-drying. **Figure 8b** is an artist's conception of how the same cells would actually collapse. This figure is modeled after **Photo 2**. **Figure 8c** shows how the cells would appear if they collapsed radially rather than tangentially. In order for this to happen, the cells adjacent to the rays must become almost triangular, and there is considerable buckling of the radial walls. The type of deformation depicted in **Figure 8c** was not routinely observed in this study.

The second intercellular collapse stress that may contribute to the tangential orientation of collapse is created by the behaviour of the vessel elements and the fibers surrounding them. The vessel elements show only a small amount of distortion at higher moisture contents and are commonly relatively collapse free when the wood is oven-dried (**Photo 6**). Their lack of collapse acts as a restraint to radial movement and encourages tangential movement in much the same way as the rays. However, in this case the action is slightly more complicated.

Fibers are located on both the radial and tangential walls of the vessels. Fibers located on the tangential walls of the vessels would be forced, by intercellular collapse stresses, to collapse radially. Those adjacent to the radial vessel walls would be expected to collapse tangentially. This phenomenon can best be seen in **Photo 4**. **Photo 9** shows a sound piece of alder. Upon inspection of this photo it becomes immediately apparent that more fibers are located on the radial vessel walls than on their tangential walls. Thus, the net effect would be for the vessels to restrain collapse in the radial direction to a greater extent than in the tangential direction.

Unlike tangential dimensional change, radial and longitudinal dimensional change occur about equally above and below the fiber saturation point. Therefore, cell wall shrinkage must play a major role, along with capillary tension type collapse, in bringing about the final dimensional change observed in these directions.

The graphs for the radial and longitudinal directions (**Figures 6 and 7**) show a slight initial loss of dimension at very high moisture contents. This can be ascribed to a generalized tension placed on the wood at the outset of drying. As soon as the tangential fiber walls begin to buckle the failure orients itself in the tangential direction, at this point radial and longitudinal dimensional change essentially stop. With continued drying dimensionsal change proceeds almost exclusively tangentially until the fiber lumina are closed tightly enough that the collapse must reorient itself as it continues.

After a long period of stability radial dimensional change begins rather suddenly, as the wood dries below 300% MC. Again the photographic record is useful in interpreting why the wood behaves as it does. **Photo 2** is a sample block at 300% MC. The fiber lumina are very distorted but at this point almost all of the distortion has taken place tangentially. The rays are easily discernible and the radial fiber walls are still relatively free from buckling. At 225% MC (**Photo 3**) the radial walls of the fibers have begun to buckle except where the fibers are in contact with a ray. Here, the walls are conspicuously straight. The rays, while not tightly collapsed, have begun to become twisted by the pull of the cells collapsing around them. This type of deformation of rays, during collapse, has been reported by Kuo and Arganbright (1978).

At 150% MC (**Photo 4**) the radial fiber walls

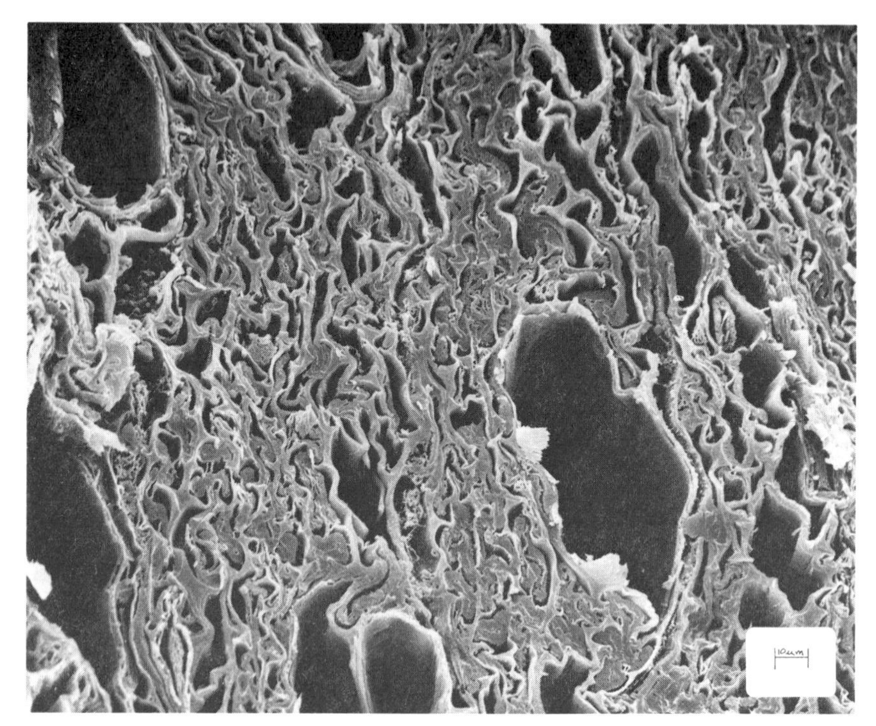

Photo 4. Scanning electron micrograph of Hoko alder conditioned to 150% MC then frozen in liquid nitrogen and freeze-dried. (The scale bar indicates ten microns.)

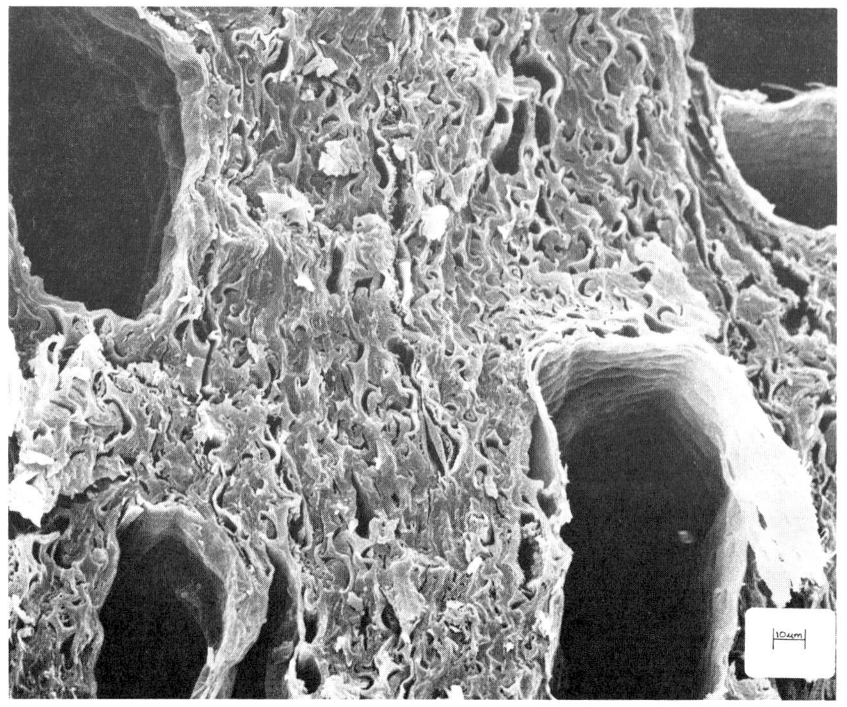

Photo 5. Scanning electron micrograph of Hoko alder conditioned to 100% MC then frozen in liquid nitrogen and freeze-dried. (The scale bar indicates ten microns.)

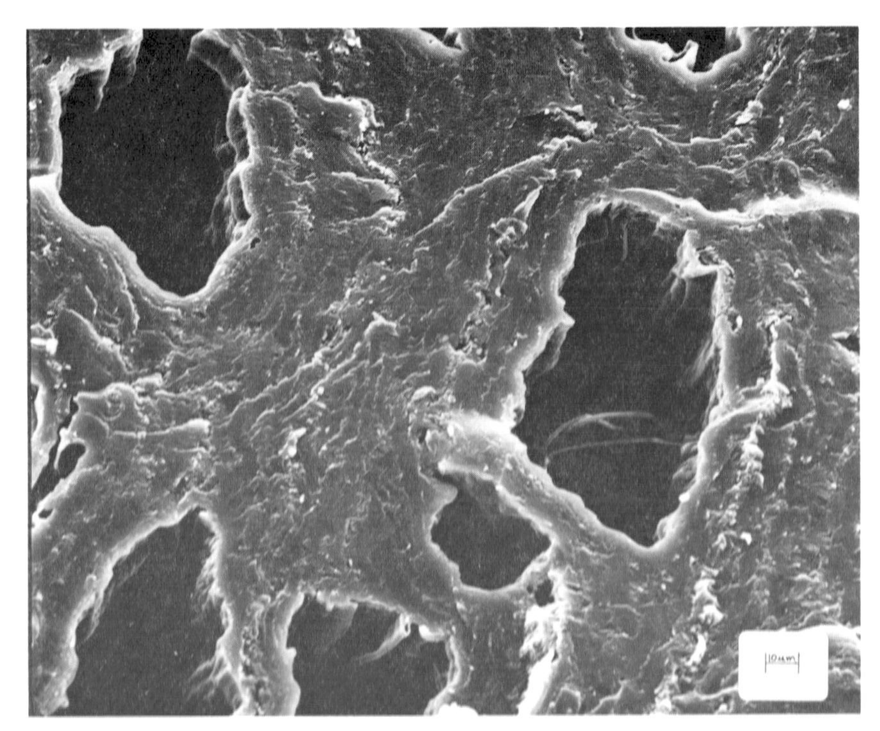

Photo 6. Scanning electron micrograph of Hoko alder conditioned to equilibrium then oven-dried at 103 ± 2°C to constant weight. (The scale bar indicates ten microns.)

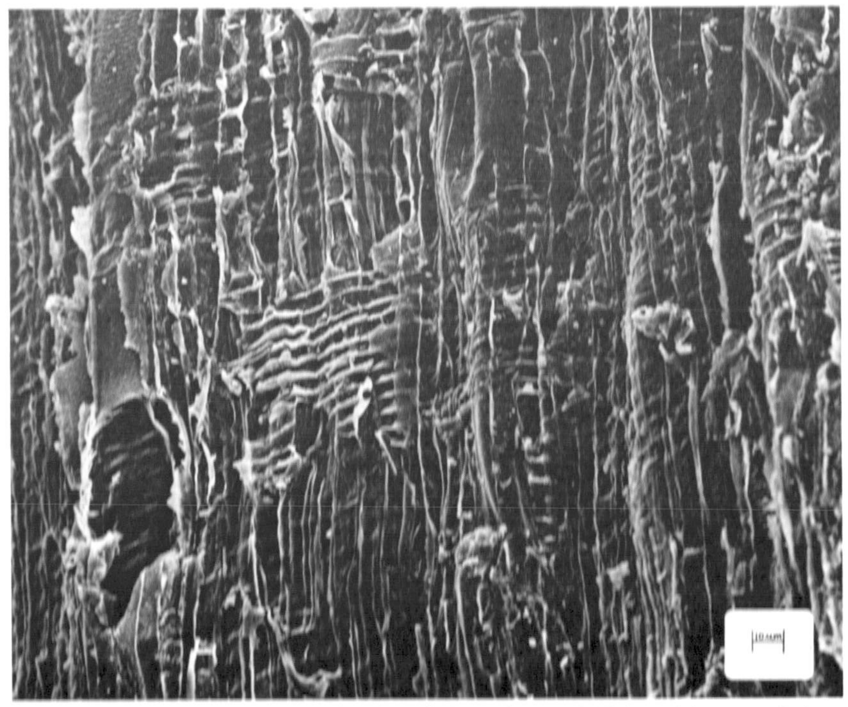

Photo 7. Scanning electron micrograph of radial surface of Hoko alder conditioned to 100% MC; then frozen in liquid nitrogen, fractured to obtain an internal surface and freeze-dried. (The scale bar indicates ten microns.)

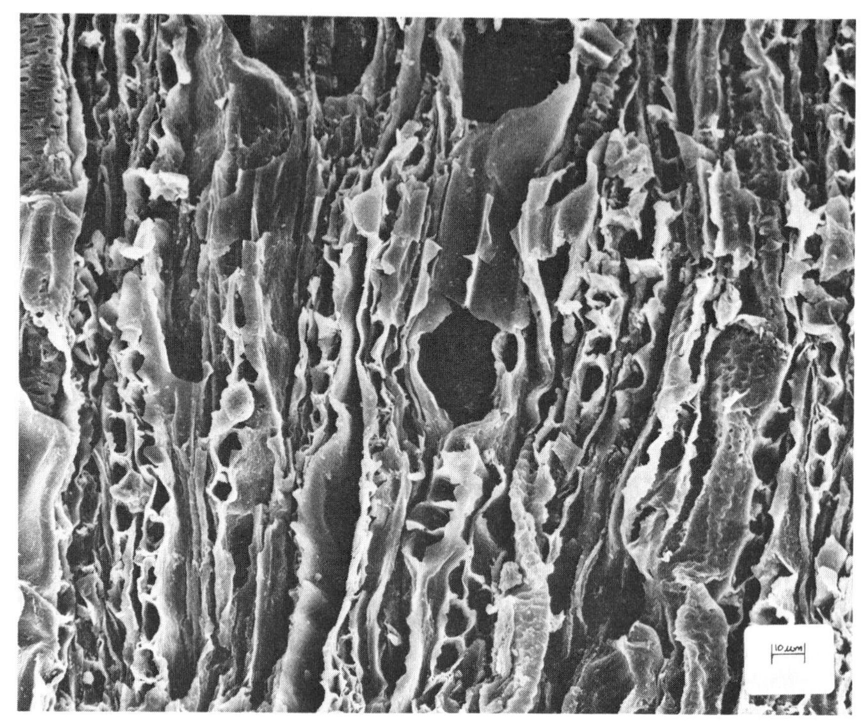

Photo 8. Scanning electron micrograph of tangential surface of Hoko alder prepared as Photo 7. (The scale bar indicates ten microns.(

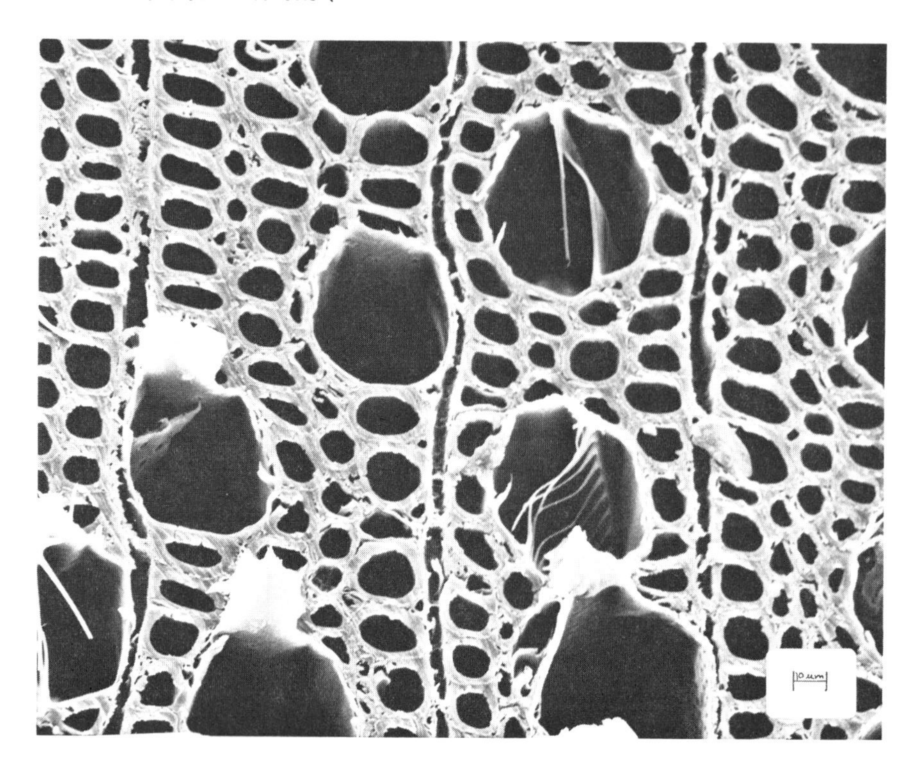

Photo 9. Normal undeteriorated alder. (The scale bar indicates ten microns.)

are very buckled and the rays are difficult to distinguish. At 100% MC (**Photo 5**) the fibers are so distorted that it is hard to distinguish radial walls from tangential. The rays are now nearly invisible. It is believed that this is an optical effect, rather than an indication that the rays have collapsed as tightly as the fibers. **Photo 8** is a tangential view of the same sample block as **Photo 5** (100% MC). The surface was created by breaking a fragment from a piece of the sample block that had already been freeze-dried. From this surface it can be seen that the rays are not as tightly closed as are the fibers, at this moisture level. However, care must be used in assessing the actual dimensions of the ray cells. The method of fracture did not create a true tangential surface nor was it completely smooth. The ray cells were broken at varying angles to their long axes. Thus the diameters of these cells appear to be greater than they actually are. In fact, if the same type of preparation is used to create a radial surface (**Photo 7**) a certain amount of radial collapse is seen. The main point is that the rays are not as collapsed as the transverse surfaces make them appear. They are difficult to pick out in transverse surfaces below 100% MC because: 1) the rays have been distorted by the intercellular stresses, created by cells collapsing around them; 2) the fibers have collapsed more tangentially than radially and they look very much like ray cells; in effect, camouflaging the rays; and 3) the rays have collapsed tangentially to a limited extent (**Photo 8**).

Like radial dimensional change, longitudinal dimensional change undergoes a drastic increase when the moisture content falls below 150% (**Figure 7**). And again about half the total longitudinal dimensional change occurs before the fiber saturation point is reached. The fiber lumina are almost completely closed at this point (**Photo 4**); for capillary tension type collapse to continue the fiber walls must be drawn together even more tightly (**Photos 5 and 6**). This can no longer be accomplished by pulling the walls into lumina. Thus, the remnants of the secondary walls must be crushed, as they are pulled into one another (compare **Photos 1 and 2** to **Photos 4 and 5**). At the same time there is a realigment of the cells that leads to the radial dimensional change described above. The same type of realignment must also be responsible for the longitudinal dimensional change that occurs above the fiber saturation point.

Therefore, the radial and longitudinal dimensional change above the fiber saturation point, is attributed to the distortion that occurs as the lumina close and radial walls begin to buckle. This gives rise to an increase in the tension between cells, or the intercellular collapse stresses. These cause distortion of the rays and realignment of the rest of the wood structure.

As the wood dries below the fiber saturation point; and possibly at higher moisture contents if moisture gradients develop (Stamm and Loughborough, 1942); desorption and the resultant cell wall shrinkage becomes dominant in causing dimensional change. It might be expected that the amount of dimensional change recorded in the sample blocks, below the fiber saturation point, would agree with the values found in the cell wall shrinkage experiments. This was true only for dimensional change in the longitudinal direction.

Cell wall shrinkage sample blocks showed a dimensional change in the longitudinal direction of about 10%. This is in good agreement with the amount of longitudinal dimensional change observed in the sample blocks dried under controlled conditions, below the fiber saturation point.

For the transverse directions a considerable amount of excess dimensional change was seen in the conditioned sample blocks (**Figures 5 and 6**). With oven drying approximately 12% dimensional change occurred in both the radial and tangential directions, below the fiber saturation point. From **Figure 4** it can be seen that cell wall shrinkage in these directions should have been 4.8% and 8.4% respectively. The reason for this discrepancy is unknown. However, one possible explanation is that during collapse the integrity of the secondary wall remnants was destroyed and the compound middle lamellae were so distorted that radial and tangential measurements are no longer completely valid at this point. It could also be, that the cellular material became more dense during collapse and this caused the increase in cell wall shrinkage. Obviously a complete answer to this question cannot be made until further research is conducted.

Conclusion

The aim of this study has been to point out

the complexity of the internal deformation that leads to the observed external dimensional change. However it has also been demonstrated that while these changes are complex they can be interpreted in a logical manner. It is believed that by understanding the cellular characteristics of cell wall shrinkage and collapse, evaluation of treatments will become much more meaningful.

Using the techniques presented here it will be possible to determine how treatments work. By treating sample blocks to varying degrees then repeating these experiments, much could be learned about the exact action of stabilization processes. By changing the size and dimensions of sample blocks or the drying conditions it would be possible to elucidate the effect of moisture gradients and their resultant stresses. In cell wall shrinkage experiments, observations at higher magnifications would allow an assessment of the shrinkage behaviour of individual cells or groups of cells. Much the same as has been done for collapse in this paper.

Combining this type of study with other methods for classifying waterlogged archaeological wood, such as chemical analysis and moisture content data, will also be useful. Providing another means to compare samples will be helpful in establishing a meaningful system of categorizing this wood.

References

Anonymous, Wood handbook: Wood as an Engineering Material, (Washington: U.S. Dept. of Agriculture, Forestry Service, 1974), Agriculture Handbook, no. 72.

Bazen, G.T. and D.S. Knight, "An inexpensive vibrating microtome for sectioning fixed tissue," Stain Technology, vol. 55, no. 1, January 1980, pp. 39-42.

Blinman, Eric, "Stratigraphy and Depositional Environment," Hoko River Archaeological Project Contribution, no. 1, Washington State University Laboratory of Anthropology Reports of Investigations, no. 58, June 1980, pp. 64-87.

Boyd, J.D., "Anisotropic Shrinkage of Wood: Identification of the Dominant Determinants," Mokuzai Gakkaishi, vol. 20, no. 10, October 1974, pp. 473-482.

Erickson, H.D., "Tangential Shrinkage of Serial Sections within Annual Rings of Douglas-Fir and Western Red Cedar," Forest Products Journal, vol. 5, no. 4, April 1955, 241-250.

Erickson, A.D., R.N. Schmidt and J.R. Laing, "Freeze-Drying and Wood Shrinkage," Forest Products Journal, vol. 18, no. 6, June 1968, pp. 63-68.

Hawley, L.F., "Wood-Liquid Relations." United States Department of Agriculture, Technical Bulletin, no. 248, June 1931, pp. 1-35.

Kauman, W.G., "The Influence of Drying Stresses and Anisotropy on Collapse in Eucalyptus Regnans," C.S.I.R.O., Division of Forest Products Technological Paper, no. 3, 1958, pp. 1-16.

Kuo, Mon-Lin and Donald G. Arganbright, "SEM Observation of Collapse in Wood," International Association of Wood Anatomists Bulletin, vol. 2, no. 3, 1978, pp. 40-46.

Pentoney, Richard E., "Mechanisms Affecting Tangential vs. Radial Shrinkage," Journal of the Forest Products Research Society, vol. 3, no. 2, February 1953, pp. 27-32, 86.

Rosenqvist, A.M., "Experiments on the Conservation of Waterlogged Wood and Leather by Freeze-Drying," in: Problems in the Conservation of Waterlogged Wood, proc. of the symposium, 5-6 June 1973, ed. W.A. Oddy, (Greenwich: National Maritime Museum), Maritime Monographs and Reports, no. 16, 1975, pp. 9-24.

Scheffer, Theodore C., "Microbiological Degradation and the Causal Organisms," in: Wood Deterioration and its Prevention by Preservative Treatments, vol. 1, Degradation and Protection of Wood, ed. D.D. Nicholas, Syracuse Wood Science Series 5, (Syracuse: Syracuse University Press, 1973).

Siau, John F., Flow in Wood, (Syracuse: Syracuse University Press, 1971).

Skarr, Christen, Water in Wood, (Syracuse: Syracuse University Press, 1972).

Stamm, Alfred J., "The Effect of Changes in the Equilibrium Relative Vapor Pressure Upon the Capillary Structure of Wood," Physics, vol. 6, no. 10, October 1935, pp. 334-342.

Stamm, Alfred J., Wood and Cellulose Science, (New York: Ronald Press Co., 1964).

Stamm, Alfred J. and W.K. Loughborough, "Variation in Shrinking and Swelling of Wood," A.S.M.E. Transactions, May 1942, pp. 349-386.

Stuki, Barbara R., "Fluvial processes and formation of the Hoko river archaeological site

(45CA213) Olympic peninsula, Washington," Master Thesis, (Pullman: Washington State University, 1982).

Thomas, David P. and H.D. Erickson, "A Study of the Permeability of Western Red Cedar and its Relationship to Collapse and Drying Rate in Lumber Seasonings," Final Report to U.S. Forest Service, University of Washington, College of Forestry, March 1963.

Tiemann, Harry Donald, "Principles of Kiln Drying Lumber," Lumber World Review, vol. 28, no. 2, January 1915, pp. 23-26.

A REVIEW OF STORAGE METHODS FOR WATERLOGGED WOOD

John E. Dawson
Rosemary Ravindra
Raymond H. Lafontaine
Canadian Conservation Institute,
1030 Innes Road,
Ottawa, Ontario, K1A 0M8

Introduction

An ideal storage method for waterlogged wood should:

(a) Prevent decay, or at least not accelerate the existing decay rate.
(b) Be compatible with subsequent treatment or examination methods.
(c) Be easy to maintain.
(d) Be inexpensive.
(e) Be safe (i.e. non-toxic) for those people maintaining it.
(f) Leave the wood in a safe (i.e. non-toxic) condition for subsequent work.

Of greatest importance is that the wood always be kept water saturated; from the time of excavation until conservation treatment commences.

Storage Methods

Numerous methods have been employed:

(i) Enclosure in polyethylene, either in sealed bags or wrapped in film. Before wrapping the wood may be sprayed with an aqueous solution of a biocide.

(ii) Immersion in water, with or without the presence of a water soluble biocide. Materials for use in tank construction must be carefully chosen to be compatible with any biocides or possible impregnants, (Oddy et al., 1978; Blackshaw, 1975.) (Methods (i) and (ii) are most commonly used, and have often been mentioned in the literature, (Christensen, 1970; Dowman, 1970; Muhlethaler, 1973; Ambrose, 1976 and Oddy, 1978).

(iii) Spraying with water. Biocides may be added to the spray solution. Ensuring complete coverage may be difficult with this method, and pockets of micro-organisms may develop; (Van der Heide, 1972; Baynes-Cope, 1975).

(iv) Keeping wood buried, and maintaining a high groundwater table and excluding oxygen, (de Jong, 1979).
This method was used to store a shipwreck discovered in the clay soil of Ysselmeerpolder in the Netherlands. Around the wreck, subsurface draininge pipes were removed; plastic sheeting covered the horizontal and vertical surfaces and rainwater was collected to keep the groundwater as high as possible.

(v) Use of storage lakes or ponds. Excavated material is frequently placed in a convenient body of water from which recovery is simple. The Sjvollen ship was stored in this way, but deterioration rate increased, (Rosenqvist, 1975); the Brown's Ferry Vessel was stored in a similar fashion, (Singley, 1981). Ideally, the water in recovery and intended storage environments should be tested and compared, (e.g. oxygen content, pH, mineral content) because wood subjected to drastic changes is likely to experience an increase in deterioration rate.

(vi) Duplication of the environment of burial, by artificial means. This method is planned for storage and display of two armed schooners the Hamilton and Scourge; (Nelson, 1981). This depends on detailed analyses of the lakebed environment of the wrecks.

(vii) Reburial after examination. This method is being adopted by Parks Canada in their excavations of a shipwreck in Red Bay, Labrador; (Stevens, 1981).

(viii) Wrapping in damp sacking and cool storage, with periodic re-wetting; (Hague, 1977; Dowman, 1970).

(ix) Sand burial. This method can be used to store and support waterlogged wood during a drying process; (Dowman, 1970). However, in large scale use in the Machault project, it failed; although biocides were used the wood developed mould growth, and did not dry out; (Jenssen and Murdock, 1981).

(x) Frozen storage; this has been discussed frequently and there are reports of it as a successful method; (p. 59, Oddy, 1975).

However as mentioned at this meeting, problems of salt deposition or cracking can arise; (Watson, 1981). Pretreatment with PEG before frozen storage seems however to be a more successful approach (Murray, 1981).

Storage of Untreated Waterlogged Wood

A study was carried out to evaluate the capability of five different techniques of tank storage for preventing biological growth on waterlogged wood. These were: (a) filtration and circulation, (b) germicidal U.V. sterilization, (c) ampholytic surfactant (An "ampholyte" is an amphoteric electrolyte, that is one which behaves as an acid in alkaline solution and vice-versa. An ampholytic detergent therefore behaves amphoterically), (d) cooling and (e) biological control, with (f) as a reference or "control" tank with no attempt to limit micro-organism growth. Each system was tested in a tank containing five litres of wood and twenty-nine litres of water.

Presented here is a detailed summary of a report, shortly, to be published (In Journal of the IIC, Canadian Group, the information given here is reproduced by kind permission of the editor).

During prolonged storage, waterlogged wood provides an ideal substrate for growth of micro-organisms. Grattan (personal communication) has found that the often produced gelatinous growth can interfere physically with impregnation. Furthermore, the micro-organisms could possibly degrade the wood further.

Initially the aim was to characterize the micro-organism(s) responsible for growth. Microscopic analysis and culturing of the slime revealed that it was being produced by a gram negative, non-sporulating rod form of bacteria capable of degrading wood.

To assess the growth of the bacteria in a storage system, bacterial levels in both the water and on the surface of the wood were determined. The dilution plating technique used in both cases is explained in **Figure 1.**

To compare storage methods, one tank was left undisturbed for the duration of the experiments. Samples were taken weekly from this otherwise undisturbed tank for cell enumeration. The bacterial count obtained gave reference or control values, for comparison of the various storage techniques.

(a) Aquarium Filters:

The aquarium filters used in the study had a filtering capacity of approximately 300 litres per hour. The results obtained are illustrated in **Figure 2** and indicate that although the water in this system remained cleaner than the control, it had little effect on the bacterial count of the surface of the wood. Initially, it was supposed that there would be an equilibrium between the surface and the water levels of bacteria, but this was not found.

(b) Germicidal Ultra-Violet Light:

Six germicidal Ultra-violet lamps (wavelength of 243.7 nm) were arranged 15 cm from the water's surface as illustrated in **Figure 3.** Although the water reservoir kept relatively sterile, this supply of 'clean' water allowed the surface of the wood to reach a level 8 times higher than the maximum for the control (see **Figure 4**). Unfortunately, the U.V. irridiation heated the water to 33.4°C; blowing air over the U.V. lamps did not help. The increase in bacterial growth is a direct result of this increase in temperature.

(c) Tego 51B:

Tego 51B, an ampholytic surfactant was tested because it had been suggested by Mühlethaler (1973) for this purpose, and because it was reported to be very effective (Dawson, 1981). In this trial, Tego 51B was added as a 1% solution (v/v). Initially, there was a marked decrease in the bacterial level in the water but was not evident on the wood (see **Figure 5**). After this decrease, bacterial levels increased rapidly. These results can be explained in part by the existence of a resistant strain of bacteria and perhaps by the inability of the surfactant to penetrate the gelatinous material produced by the bacteria on the wood surface.

(d) Cooling:

Cold temperatures have an inhibiting effect on the rate of growth of most micro-organisms (Carpenter, 1977). A preliminary test established the effect of temperature by inoculating agar plates with the bacteria incubating them at various temperatures and measuring the growth rate. As indicated in **Table 1**, little growth was evident at 5°C and none at -5°C or 44°C. The storage tank was therefore placed in a refrigerator at 5°C. The bacterial level in the water remained below that of the control, and the bacterial level on the surface of the wood remained below that of the con-

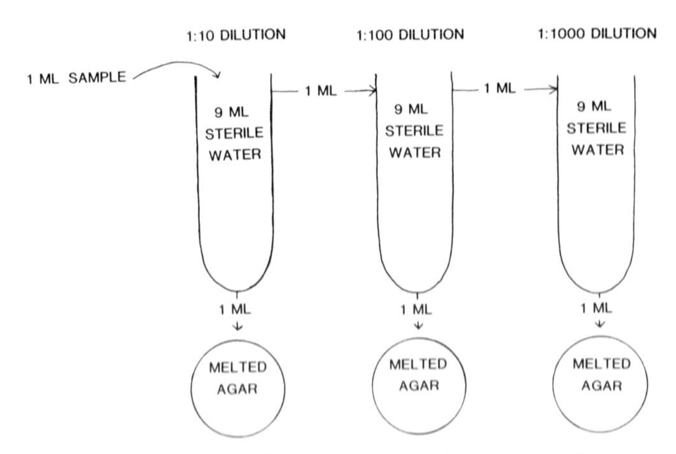

Figure 1: Dilution Procedure for Water Testing. One ml samples of water were removed from the wood storage tank being tested and added to 9 ml of sterile water (1 to 10 dilution). From this dilute solution a 1 ml sample was added to a second 9 ml portion of sterile water (1 to 100 dilution). Further dilutions can be made likewise if needed to obtain a count. From each diluted sample, one ml of solution is added to a petri dish of cooled melted agar.

For wood surface testing the same procedure was used, with the exception that the initial sample was prepared by scraping a 1 cm^2 area of the surface. The material removed was added to 10ml of sterile water and dilutions of 1 in 10 and 1 in 100 were carried out as previously.

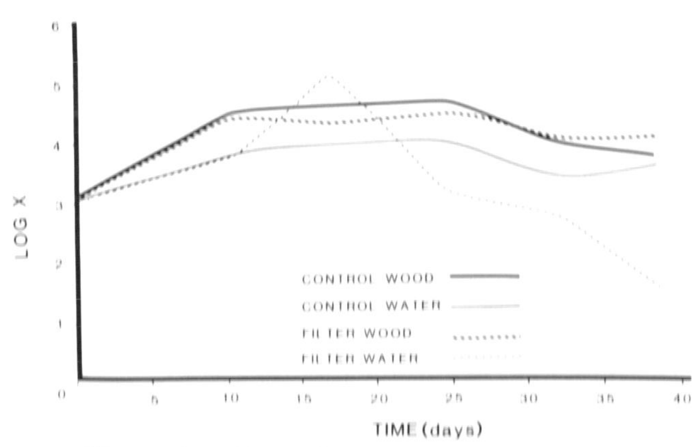

Figure 2: Bacterial growth curves for the control and the aquarium filter system. These data are the average of two trials.

The y axis (i.e. 'Log X') is log_{10} of the number of cells/ml for samples of the storage water, and log_{10} of the number of bacterial cells/cm^2 for samples removed from the wood surface.

TABLE 1

Cooling Test

Plate No.	Temperature, °C				
	-5	5	21	33	44
1	x	x	a	a	x
2	x	x	a	a	x
3	x	s	a	a	x
4	x	x	a	a	x

a - abundant growth
x - no growth
s - sparse growth
Four replicates were used for each temperature.

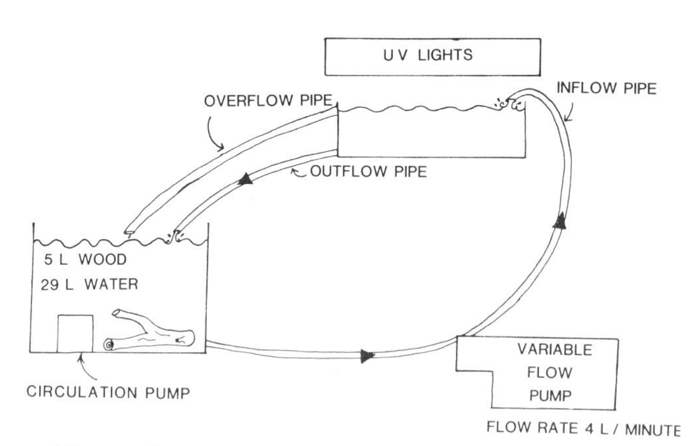

Figure 3: Schematic diagram for the U.V. sterilization test.

To prevent U.V. light escaping, the entire apparatus was covered.

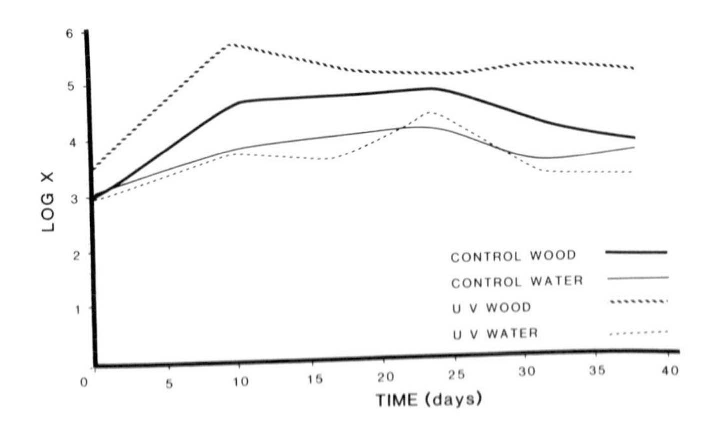

Figure 4: Bacterial growth curves for the control and U.V. irradiated system.

These data are the average of two trials (see legend to **Figure 2** for 'Log X').

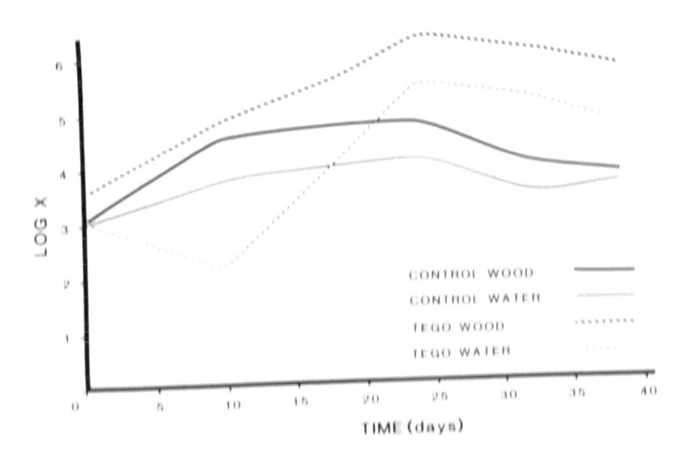

Figure 5: Bacterial growth curves for the control and the ampholytic surfactant system (Tego 51B).

These data are the average of two trials (see legend to **Figure 2** for 'Log X').

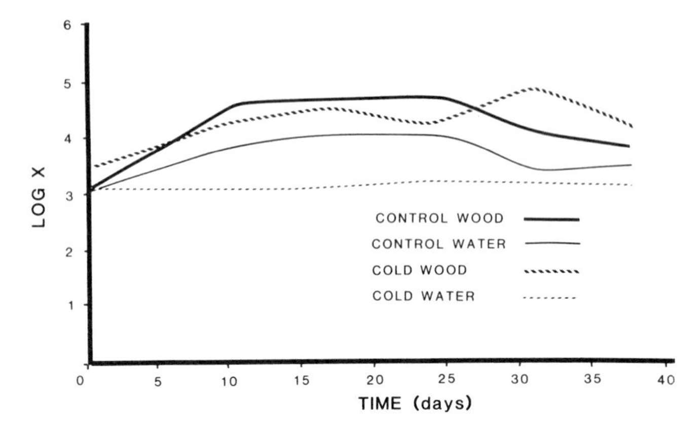

Figure 6: Bacterial growth curves for the control and the cooling system.

These data are average of two trials (see legend to **Figure 2** for 'Log X').

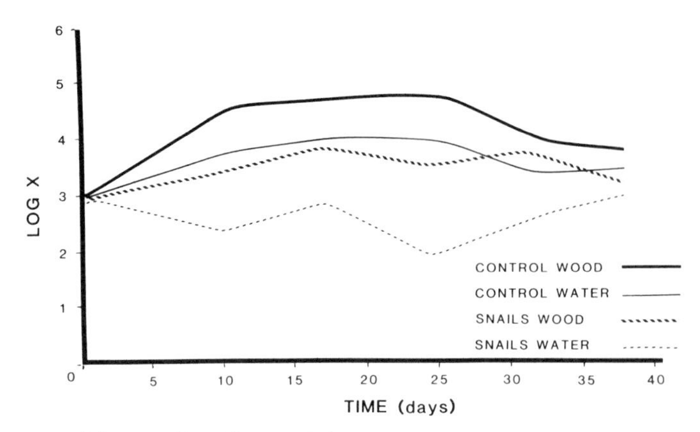

Figure 7: Bacterial growth curves for the control and the biological control system' (snails)

These data are the average of two trials (see legend to **Figure 2** for 'Log X').

trol for half the experimental period (See **Figure 6**). Cool temperatures alone, were not able to prevent slime formation, but did help in retarding development.

(e) Biological Control:

The use of one organism to control a second is referred to as biological control (Andres, 1971). For this study, the common pond snail was well suited for the removal of organisms on the wood surface, because of its great tolerance to various environmental conditions and its feeding habits. Initially, twenty-five snails (Physa sp.) were added to a wood storage tank equipped with an aquarium filter to remove snail wastes. The wood surface and the storage water remained at bacterial levels below that of the control (see **Figure 7**). The system was allowed to continue operating for one year after the conclusion of the test and the wood was still not slimy to the touch. A larger tank of 2,000 litres (approximately one third filled with waterlogged wood) has also been tested successfully with the snail method. Further study is needed to determine how much of the degraded wood is removed from the surface by the snails; it is believed that this is considerably less than the amount that could be removed by brushing the object to remove slime.

Conclusion

Of the methods tested, the biological control method was clearly most successful. Jespersen (personal communication) has indicated that the use of fish, a species of perch, have been very successful in keeping waterlogged wood clean. Clearly methods, other than the use of chemicals can be successfully used.

(The methods are compared in **Figure 8**.)

References

Ambrose, W.R., "Sublimation Drying of Degraded Wet Wood," in: The Pacific Northwest Wet Site Wood Conservation Conference, vol. 1, proc. of the conference, 19-22 Sept. 1976, ed. G.H. Grosso, (Neah Bay, Washington, 1976), pp. 7-15.

Andres, L.A., "The Role of Biological Agents in the Control of Weeds," in: Understanding Environmental Pollution, ed. by M.A. Strobbe, (St. Louis, The C.V. Mosby Co., 1971), ch. 40.

Baynes-Cope, A.P., "Fungicides and the Preservation of Waterlogged Wood," in: Problems of the Conservation of Waterlogged Wood, proc. of the symposium National Maritime Museum, Greenwich, 5-6 October, 1973, ed. W.A. Oddy, (London: National Maritime Museum, 1975), Maritime Monographs and Reports, no. 16 pp. 31-33.

Blackshaw, Susan, "Comparison of Different Makes of PEG and Results on Corrosion Testing," ibid., pp. 51-58.

Carpenter, P.L., Microbiology, 4th ed., (New York: W.,B. Saunders Co., 1977), pp. 63-142 and 217-452.

Christensen, B.B., The Conservation of Waterlogged Wood in the National Museum of Denmark, (Copenhagen: The National Museum of Denmark, 1970).

Dawson, J., "Some Considerations in Using a Biocide," this publication.

Dowman, Elizabeth, A., Conservation in Field Archaeology, (London: Methuen and Co. Ltd., 1970).

Hague, E., "Conservation Problems of Waterlogged Large Size Wooden Antiquities in Dacca Museum," in: International Symposium on the Conservation and Restoration of Cultural Property - Conservtion of Wood, 24-28 November, 1977, (Tokyo: National Research Institute of Cultural Properties, 1978), pp. 59-65.

de Jong, I., "The Deterioration of Waterlogged Wood and its Protection in the Soil," in: Conservation of Waterlogged Wood, International Symposium on the Conservation of Large Objects of Waterlogged Wood, proc. of the symposium, 24-28 Sept., 1979, ed. Louis H. de Vries -Zuiderbaan, (The Hague: Netherlands National Committee for UNESCO, Government Printing and Publishing Office), pp. 31-40.

Muhlethaler, B., Conservation of Waterlogged Wood and Wet Leather, (Paris: Editions Eyrolles, 1973).

Murray, Howard, "The Conservation of Artifacts from the Mary Rose," this publication.

Nelson, D., "The Hamilton - Scourge Project: An Approach to the Conservation of Intact Ships Found in Deep Water," this publication.

Oddy, W.A., and Van Geersdaele, P.C., in: "Discovery, Excavation and Recovery of Remains," The Graveney Boat ed. by V. Fenwick, National Maritime Museum, Greenwich, Archaeological Series no 3, British

Archaeological Reports, series 53, 1978, pp. 17-22.

Oddy, W.A., Blackshaw, S. and Gregson, C., "Conservation of the Waterlogged Wooden Hull," ibid., pp. 321-330.

Rosenqvist, A., Problems of the Conservation of Waterlogged Wood, proc. of the symposium, 5-6 Oct., 1973, National Maritime Museum, Greenwich, ed. W.A. Oddy, (Greenwich: National Maritime Museum, 1975), Maritime Monographs and Reports, no. 16, p. 124.

Singley, Katherine, "The Recovery and Conservation of the Brown's Ferry Vessel," this publication.

Stevens, W., "The Excavation of mid-16th Century Basque Whaler in Red Bay, Labrador," this publication.

Van der Heide, T.D., "Problems of Ship Archaeology and the Preservation of Ancient Ship Remains," in: Biodeterioration of Materials, ed. by Wattens, A.H., and Hueck -Van den Plas, E.H., (proc. of The 2nd International Biodeterioration Symposium, Lunteren, The Netherlands, 13-18 Sept., 1971), (New York: John Wiley, 1972), vol. 2, pp. 376-380.

Watson, Jacqui, "The Appliction of Freeze-Drying on British Hardwoods from Archaeological Excavations," this publication.

Questions

Richard Clarke: What made you think the bacteria were in the water? Did you analyse the water before? and did you know that they were causing the infection of the wood?

John Dawson: Yes.

Richard Clarke: We are spraying some wood with aqueous PEG solution, but we still get the development of slime. Do you know if there are any inherent organisms in the wood before you put them in the water?

John Dawson: Yes, but we don't know if these are the same ones which are causing the problems to the stored wood. The wood in question had slime on it, and there was also a background level of bacteria in the tap water.

Richard Clarke: The tap water at Greenwich even contains the eggs of crustacea.

John Dawson: The important thing about the slime growth was that it had a very distinct colony appearance, so that it could be picked out from other contaminants. This was shown in microscopic and in colony examination.

David Grattan: What I notice in my storage tanks, several of which contain snails, is that the more degraded is the wood, the less is the problem of slime development. It seems to be more of a problem for relatively undegraded wood. Where slimes form, large populations of snails arise and in the other tanks which contain more degraded wood, I have difficulty in keeping them alive.

John Dawson: What David Grattan is talking about is control. If there is plenty of food, the control agent breeds well and hence controls it. As the food source falls off, the control agent will not survive: This is a well known phenomenon, for instance in agriculture where an insect has been used to get rid of a weed; the insect never entirely destroys it. But for our wood, we found that the snails controlled it sufficiently well to keep the wood visually clean.

Victoria Jenssen: One of the reasons why you did this work was the observation by David Grattan and colleagues, that slime development interfered with penetration. Will you now go back and compare penetration rates before and after exposure to snails?

John Dawson: We have unsuccessfully tried to do this. Every time we mounted a slimy piece of wood in the permeability apparatus, the slime floated free. Yes, I think this should be quantified.

Kirston Jespersen: Are there different kinds of slime? Did you analyse your slime.

John Dawson: We analysed the bacterium, but not the slime produced.

Kirsten Jespersen: Well we did, and found out that it was a bacterium, Zoogloea ramigera (saprobic number IV 9, III 1). It appears when the water is heavily polluted, with a high content of organic material. Perhaps you might be interested to find out if you are developing the same slime bacterium. Speaking about cleaning it, you use snails, and I use fish (Perch) with exactly the same result. I use 2 perch in each 7 cubic metres tank.

John Dawson: There is a fish used in North America, commonly called the "algae eater," used frequently to keep fish tanks clean.

Mary-Lou Florian: We should be almost as careful with biological control as with biocides. Some snails are intermediate hosts for liver flukes. -Don't relax!

John Dawson: There is no such thing as the biocide, or the control method. It has to be what will work in your cirucmstances and your type of wood!

Jim Spriggs: One problem of biological controls is the amount of toxic material that the wood itself will introduce into the water; the black material from sulphide products for instance. Presumably the water has to be changed frequently.

John Dawson: I don't know if the snails would have worked for the circumstances you have described. But filtration through activated charcoal, which is what we use, might help.

Rosemary Ravindra: Our tanks have occasionally developed strong sulphury smells, with no apparent adverse effects on the snails!

John Dawson: We tried to keep the water clean, David Grattan just chucked in the snails, David does not change his filter, David does not change his water, David just leave it alone. And the snails thrive in his tanks, at times, better than they do in ours. (ed. No comment!) (Laughter).

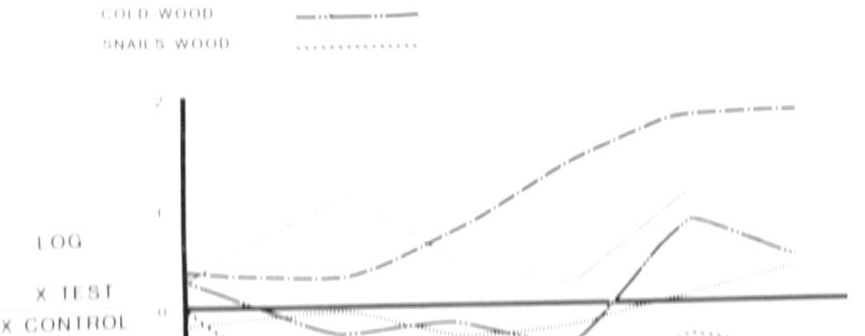

Figure 8: A comparison of the surface growth on the wood for the five storage methods tested.

The y axis is the $\log_{10}$ of the number of bacterial cells/ml or cells/cm^2 over the number of bacterial cells observed for the control.

Above zero values indicate bacterial levels greater than the control, values below zero indicate bacterial levels less than the control.

THE APPLICATION OF FREEZE-DRYING ON BRITISH HARDWOODS FROM ARCHAEOLOGICAL EXCAVATIONS

Jacqui Watson
Ancient Monuments Laboratory,
Fortress House,
23 Savile Row
London W.1., England

Abstract

This is based on work carried out on small artefacts retrieved from multi-period sites in Britain. Freeze-drying seemed to offer the best possibilities for treating these small and sometimes fragmentary items. All work was carried out using an Edwards model EF2 freeze-dryer. Initially all objects were freeze-dried from 10% PEG 400 solution, but the results were widely variable. After studying the microstructure of some of these pieces it became evident that certain factors influenced the success of the treatment. These could be summarized as the species and the extent of degradation of the wood.

Species: for convenience British hardwoods can be divided into four categories by their density. These groups respond slightly differently to the standard freeze-drying treatment, the higher density woods such as Oak providing the greatest problems.

Degradation: Fungal degradation causes thinning of the wood cell walls. Because of the lack of cementing material, wood in this condition tends to splinter and fragment after freeze-drying. Mineral deposition and mineral replacement of the outer layers, usually by iron salts; this phenomenon is common even in prehistoric wood. Buried wood seems to attract and accumulate iron salts, which can ultimately be seen as casts of the original structure. This can result in tool-marks and decoration being retained in a mineral, or part organic part mineral layer. This condition gives concern as to the use of chelating agents on this material, and to the long-term effects of these iron salts on residual PEG in the wood.

Experimental work is under way using various grades of PEG as a pretreatment to freeze-drying.

Introduction

Most of the objects that have been treated are small in size, and in the main composed of indigenous hardwoods. Typical objects have been bowls in a fragmentary condition and small utensils.

Following Ambrose (1), all items were immersed in 10% PEG 400 for 4 to 6 weeks, before freeze-drying, after which bowl fragments were re-assembled. Fragments were firstly consolidated with a resin in solvent, (Acryloid B 72 in toluene) and then repaired with the same resin as an adhesive.

If necessary, gap filling comes next; a basket work support which employs balsa strips is used. This is coated with AJK fibrous dough (2). This gives a very good surface for painting. (Consolidation with resin solutions has now been discontinued because of the unpredictable results, described below.)

All of the treated objects seem to show no dimensional change after treatment, although no very precise measurements have been taken.

They dry to a very light colour, with the wood grain clearly visible; a ladle made from wild cherry wood still retained a slight pinkish colour. The few softwoods that have been so treated have also turned out well. Basic problems with this treatment have been encountered only with oak.

In several instances oak freeze-dried from 10% PEG 400 had surfaces that were badly cross-checked. If there was fine surface detail, such as tool marks which needed to be preserved the object required further consolidation, which darkened the surface considerably.

Examination of oak after freeze-drying from 10% PEG 400 by Scanning Electron Microscopy (SEM), showed deep cracks next to the rays, with distortion and collapse of some of the small fibres.

Grouping of Wood for Treatment

For conservation purposes wood can be grouped by species in order of decreasing wood density (3), as follows:

(i) Oak and Beech:

These are the most difficult to treat, mainly because objects made from these woods are often very large with a thick cross-section. Furthermore, they both have wide rays, and tend to crack in the tangential plane.

Degraded items have been successfully treated with a consolidating solution made up of 10% PEG 400 and 15% PEG 4000. (This is discussed in more detail below.) This considerably reduced the amount of cross-checking on the surface of the object.

Objects in very good condition need only be treated with 10% PEG 400, but the immersion time should be increased to 3 months.

(ii) Ash and Fruit Woods:

Nearly all of the objects made from these woods, take the form of bowls or small utensils. They are usually very thin with a large surface area, so present few problems when freeze-drying from 10% PEG 400. However, as they have multiseriate rays, there is a tendancy for them to crack along the rays.

(iii) Maple, Alder, Birch and Hazel:

These are mainly woods with uniseriate rays, and when they fail to respond well to a treatment, cellular collapse rather than tangential cracking, takes place. Almost invariably they have freeze-dried successfully from 10% PEG 400 solution. However attempts to consolidate these woods after freeze-drying with a resin solution, may result in objects warping badly; even if freeze-drying produced a sound object with no apparent dimensional change. For badly degraded items the consolidating solution mentioned in (i) above would probably give better results.

(iv) Willow and Poplar:

These woods could also warp considerably if solvents are used in a consolidation treatment. However the freeze-drying method has always been very successful; and this is probably the most reliable method for treating objects made from these woods.

Fungal Degradation

Fungal decay is not always easily recognisable on the surface of an object, but can usually be seen on microscope slides mounted for the identification of the wood species. This decay has usually taken place in the object before it became waterlogged. Although much of the fungal hyphae may have disappeared while under burial, spore sacs often survive (**Figure 1**). Hopefully after treatment these are unable to become active.

Under high magnification large troughs can be seen in the walls of the wood fibres, which have been produced by fungal hyphae (4). In a piece of oak with severe fungal decay, it was observed that the cells were breaking away from one another (**Figure 2**).

Most woods which have suffered from fungal attack, do not respond well to the standard freeze-drying treatment. They tend to splinter and fragment considerably.

Iron Salts

Waterlogged wood is often contaminated with iron salts formed in-situ from ground water containing mineral ions. Sometimes this is obvious by the presence of orange red crystals on the surface of an object. In a neolithic bark object which had been sewn, all the stitching and stitch holes were located along the reinforced edge by using x-radiography , as iron salts had accumulated in the stitch holes.

In SEM examination the iron salts appear as spherical deposits (**Figure 3**), and these don't seem to be affected by the PEG 400/freeze-drying treatment. (However, these iron salts have not been observed in the wood when the consolidating solution was used. Either the higher molecular weight PEG 4000 complexes out the iron, or just obscures it. It is difficult to tell which of these two is happening.) Electron probe microanalysis was carried out at the British Museum Research Laboratory on these iron salt deposits, and revealed that the spherical deposits contained iron, sulphur and potassium. It is possible that their mineral form is one of several rare potassium, iron sulphides that have been identified recently. X-ray diffraction studies and quantitative microprobe analysis are required to characterise the mineral fully.

The presence of this mineral could well account for the large amount of sulphur often observed in waterlogged wood. If a cationic biocide is used on wood containing sulphides, sulphur will be precipitated. There is also concern over the long-term stability of PEG, since it is in such intimate contact with these iron salts.

In some cases iron salts are so plentiful, that mineral replacement of the wood has started to take place (5); (**Figure 4**). Usually this only occurs in the surface layers, but it means that fine surface detail is retained in a part organic, part mineral layer. The use of acids and chelating agents to remove iron salts could destroy the wood surface, with the loss of decoration and tool marks.

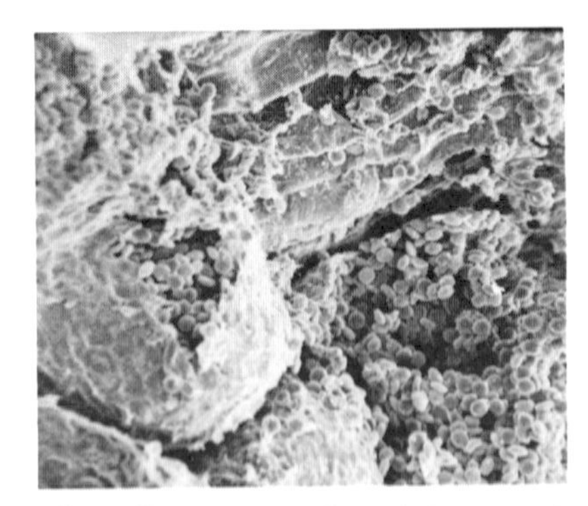

Figure 1: Spore sac found in wood. Mag. x300.

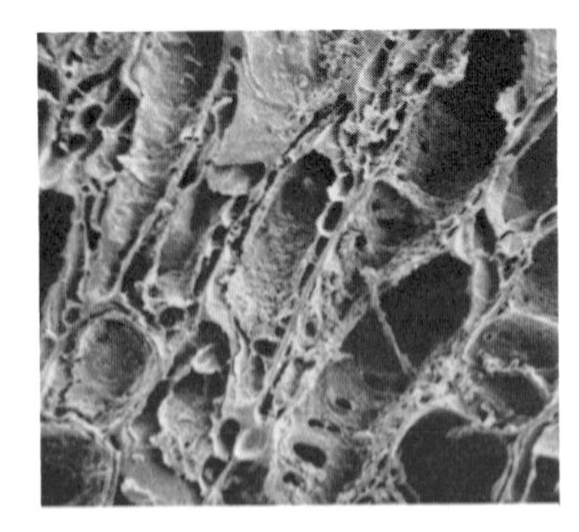

Figure 2: Section of oak freeze-dried from 10% PEG 400, with severe fungal attack, Mag. x1000.

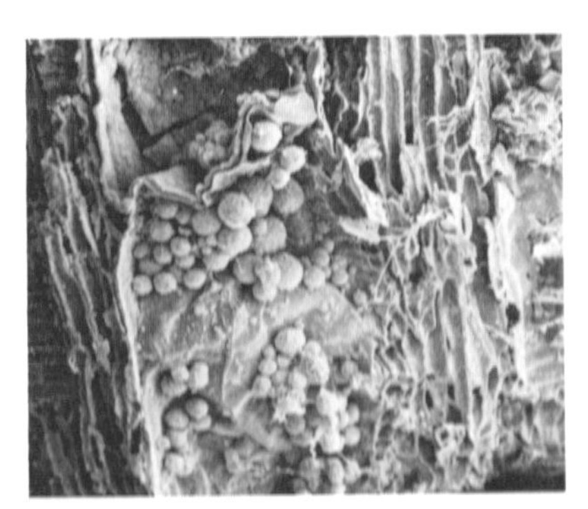

Figure 3: Vessel blocked with spherical shaped deposits of iron salts. Mag. x100.

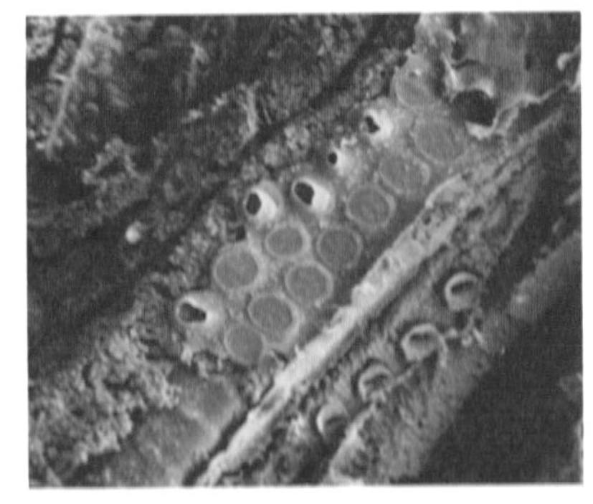

Figure 4: Oak fibres coated with iron salts. Mag. x1000.

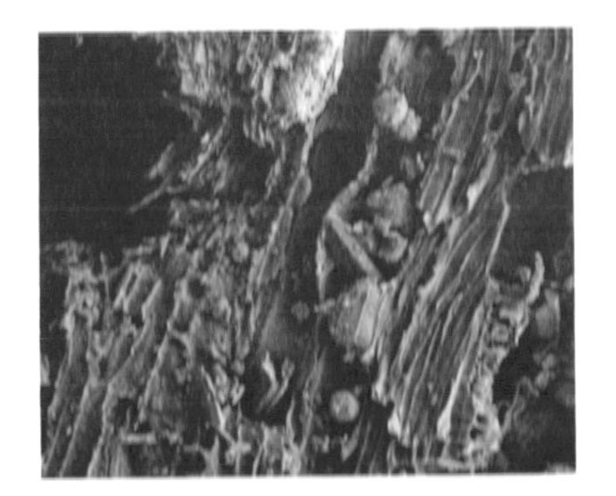

Figure 5: Oak after freeze-drying from hard water, displaying a crystal rupturing a vessel wall. May x100.

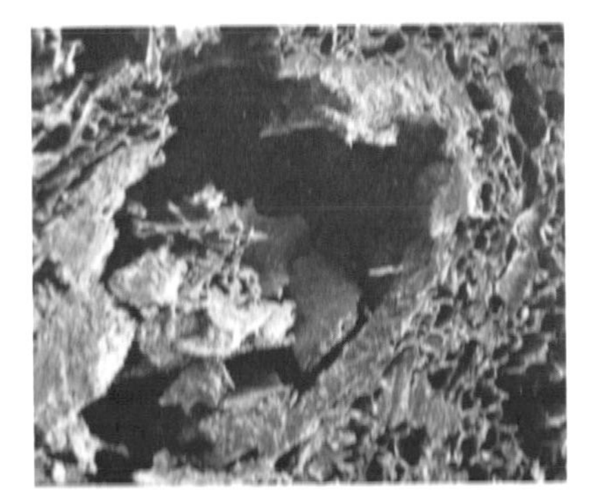

Figure 6: Transverse section of oak freeze-dried from the consolidating solution. PEG 4000 has coated the vessel wall and consolidated adjacent fibres. Mag. x300.

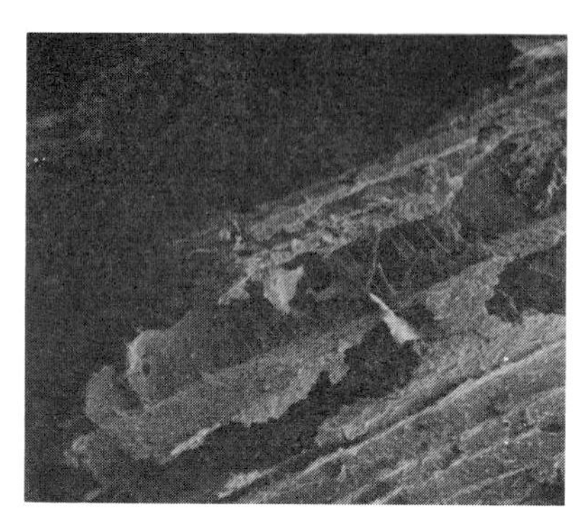

Figure 7: Alder freeze-dried from the consolidating solution, all surfaces are covered in a "furry" layer. Mag. x300.

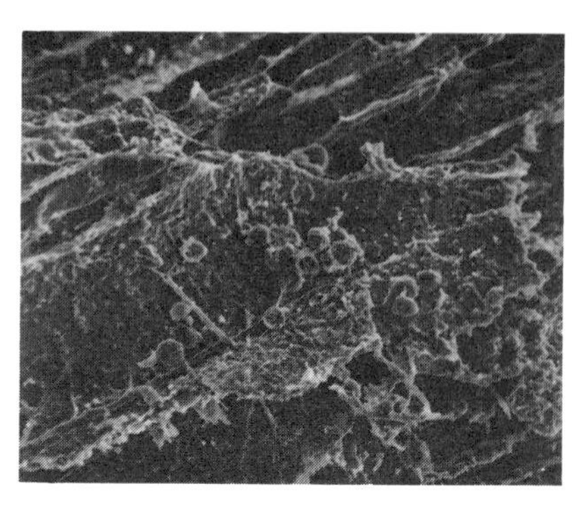

Figure 8: Section through alder showing the higher grade PEG covering everything including fungal spores and hyphae. Mag. x300.

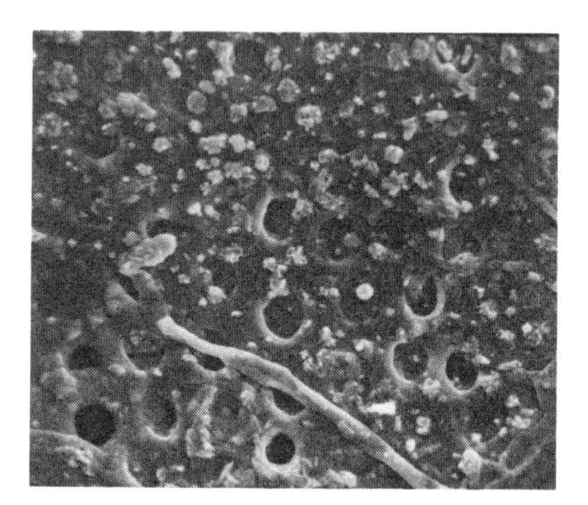

Figure 9: Ash consolidated with Acryloid B72 after freeze-drying. The vessel wall is coated in isolated patches of the resin. Mag. x1000.

Effects of Freezing Wood without Pre-Treatment

In order to see the effects of freezing wood without pretreatment, a piece of waterlogged oak was frozen in a domestic deep freeze and then freeze-dried. When removed from the freeze-dryer the sample appeared to be dimensionally the same as before freeze-drying. However, after leaving the sample on the bench over a damp weekend, it collapsed and split open. Later examination by SEM showed the presence of crystals which must have been precipitated out of solution during freezing, and as can be seen in **Figure 5** had ruptured vessel walls. Electron probe microanalysis, as before, showed that the crystals contained calcium and sulphur. It was concluded that the crystals were probably composed of gypsum deposited in the wood from hard water. The formation of these crystals has never been observed in wood treated with PEG 400, so it may be that the PEG complexes the inorganic salts, and thereby prevents such crystallisation.

Condition of the Wood after Treatment with the Consolidating Solution

The consolidating solution was used in two phases:

(i) The object was immersed in 10% PEG 400 in water for 1 month.

(ii) 5% PEG 4000 was added 3 times at intervals of two weeks to the previous solution, and the object allowed to soak in the final concentration of 10% PEG 400 and 15% PEG 4000 for a further month.

SEM examination of oak treated with this consolidating solution revealed the PEG had consolidated fibres adjacent to the large vessels (**Figure 6**), for which there was obviously easy access. The vessel walls were also coated. Similar examination of alder, revealed a "furry" layer or deposit on the inside of the cells, which covers everything (**Figures 7 and 8**).

This provides a small amount of consolidation, which is just enough to give structural support to the wood.

The advantage of using these dilute solutions is that the wood doesn't have to be cleaned of excess PEG on the surface after freeze-drying. The consolidated wood has a more compact surface, yet shows no obvious darkening compared to that treated with just the PEG 400 solution.

With this treatment it is essential that samples are taken for identificaion before soaking in PEG 4000, because pitting details in the vessels are invisible after treating in this way.

Consolidation with Resins after Freeze-drying

This was effected by capillary impregnation of objects with 5% Acryloid B72 (a copolymer of ethyl methacrylate and methyl acrylate) in toluene, and allowing the wood to dry suspended over the solvent, to avoid leaving patches of resin on the surface of the wood. Although it has been mentioned several times, this method of consolidation has now been discontinued because of the unpredictable results mentioned above.

SEM examination of ash treated in this way revealed the resin in isolated patches just resting on the vessel surface (**Figure 9**). It is unlikely that the resin gives much structural support in this condition. For the small amount of support this form of consolidation might produce, its use does not seem justifiable because of the negative effects of (i) darkening of the wood and (ii) possibility of cellular collapse.

Future Work

The use of mixed PEG solutions seems to offer the best possibilities for consolidating and chemcally stabilising wood when using a freeze-drying system. More work needs to be done using different grades of PEG at various concentrations, to try and find the optimum mixture for different species of wood in various states of degradation.

More consideration must be paid to iron salts remaining in the freeze-dried wood, to try and determine their long-term stability and any possible reaction they may have with PEG.

References

1. Ambrose, W.R., "Sublimation Drying of Degraded Wet Wood," in: Pacific Northwest Site Wood Conservation Conference Vol. 1, proc. of the conference, 19-22 September 1976, ed. Gerald H. Grosso (Neah Bay, Washington: 1976), pp. 7-15.

2. Plenderleith, H.J. and A.E.A. Werner, the Conservation of Antiquities and Works of Art, (London: Oxford University Press,

1971), p. 341.

3. Watson, J., "Some Botanical Considerations When Freeze-Drying Waterlogged Wood," proc. UKIC meeting on Freeze-Drying, London, May 7 1981, (to appear as a UKIC Occasional Paper).

4. Bravery, A.F., "The Application of Scanning Electron Microscopy in the Study of Timber Decay," Journal of the Institute of Wood Science, no. 30, vol. 5, part 6, 1971, pp. 13-19.

5. Keepax, C., "Scanning Electron Microscopy of Wood Replaced by Iron Corrosion Products," Journal of Archaeological Science, vol. 2, 1975, pp. 145-150.

Acknowledgements

I would like to thank Mr. J. Musty and Mr. J. Price of the Ancient Monuments Laboratory, for allowing this work to be undertaken and offering every encouragement. Also I would like to acknowledge the help given by Dr. M. Tite of the British Museum Research Laboratory for allowing the mineral samples to be analysed, and Mr. N. Meeks who kindly analysed them and helped me with their interpretation. I am very grateful to Miss E. Lawler for providing me with slides for my lecture and the photographs for this paper.

Questions

Howard Murray: Concerning the mineralisation of the wood. What you really have is a shale-like structure, and if you consider it as such, does it give you a way round the PEG treatment?

Jacqui Watson: I don't really know, because when you examine these pieces, the outside may be completely mineralised and yet the interior will be very much like wood. What you have is a material which is neither organic nor inorganic, but both at the same time.

James Argo: I am quite interested in the iron substance you found in the wood. What was the source of the iron in the wood?

Jacqui Watson: I presume it originated from ground water, since material from most British excavations has extraordinarily high iron content. Another strange phenomenon, is that even when there seems to be a low iron content in the soil, much may be found in the wood.

James Argo: How far into the wood do the iron salts penetrate?

Jacqui Watson: Very deeply, sometimes into the heart wood.

James Argo: Chelates such as those with acetyl acetone are very effective at removing ferric iron, although they producing a scarlet complex, readily soluble in alcohol. This may not be suitable for your conservation needs, but the red colour will tell you if you are removing iron.

Suzanne Keene: Is it possible that the iron is forming complexes with the tannins in the wood?

James Argo: If so, the iron would be much more stable, and there would also be large blue-black regions in the wood. The presence of free iron salts, indicates that iron tannates have not formed.

Colin Pearson: This again is an example of how the archaeological requirements dictate the conservation. Here the surface detail (e.g. lathe marks, tool marks, makers marks) is more important than the actual object.

A PRACTICAL COMPARATIVE STUDY OF TREATMENTS FOR WATERLOGGED WOOD PART II THE EFFECT OF HUMIDITY ON TREATED WOOD

David Grattan
Conservation Processes Research Division,
Canadian Conservation Institute,
1030 Innes Road,
Ottawa, Ontario,
K1A 0M8.

Introduction

Part I of this study reported the effect of various treatments on five groups of wood (1). Some of the information on these methods is listed in **Table 1**. This paper describes the effect of relative humidity change on samples of elm (group E) treated by various methods in the initial study. By so doing I hope to address the following questions:

(i) How does a particular treatment affect the hygroscopicity of the wood?
(ii) Is wood treated with low molecular weight grades of PEG such as 400 or 540 Blend excessively hygroscopic?
(iii) How should treated wood be stored or displayed from the point of view of humidity?

In the course of this work, other information about the interaction between PEG, wood and water was revealed which complemented some of the data reported in the first part of this work.

There have been earlier studies on the effects of humidity, however this is the first attempt to compare the differing effects of several treatments on one type of wood (2,3,4,5,6).

Experimental

For the first experiment (described here) samples of elm (Group E in (1)) treated in various ways were used. Elm was chosen because it is a fairly dense hardwood and would tend to have more dimensional response to moisture change than the other species employed in the comparative studies. Each sample was cut up as shown in **Figure 1**, to produce inner and outer semi-circular slices of about 2 mm thickness. Semi-circular slices were used to allow the wood maximum freedom of movement (without encouraging check formation) in the tangential dimension. The aim of taking two slices from each sample was to see if any obvious differences in behaviour were apparent comparing the inside to the outside for each treated piece of wood. Such differences might be expected near the surface if there were failures in penetration of impregnant or if the impregnant was concentrated at the surface by reverse flow brought about by the drying; indeed they were observed.

Pins were placed to follow movement in the tangential dimension i.e. that of greatest movement (the amount of movement in the other dimensions can be estimated from this) and vernier calipers used to measure the distance between them. The wood was then placed in a sealed glove box in which RH could be controlled precisely at any level from 0 to 100% within about 3%. This apparatus was devised by Lafontaine (7). The box was equipped with a balance to measure weight changes; spot measurements of RH were made with a Bendix psychrometer, and RH stability was followed with a hygrothermograph. The wood was brought down in several steps from ambient RH ca. 60%, to 0%. It was found that the wood took less than a week for the weight and dimension to stabilize at each new setting of humidity.

Measurements of change in the tangential dimension were made two weeks after each RH change. This allowed sufficient time for equilibration to take place. Changes in RH were never greater than 15% and normally about 10%. The results reported here are for water absorption only; the desorption data were obtained in a subsequent experiment, however, the results were not always useful because at high RH levels PEG leaches out of the wood. The experiments were performed between 25° and 30°C.

Results and Discussion

PEG 4000 (The polyethylene glycol used in this experiment was supplied by Union Carbide. Since it was carried out PEG 4000 has been renamed PEG 3350.

Regardless of the solvent used to transport the impregnant in the treatment; whether by water, t-butanol or methanol; the dimensional and weight change of wood treated by PEG

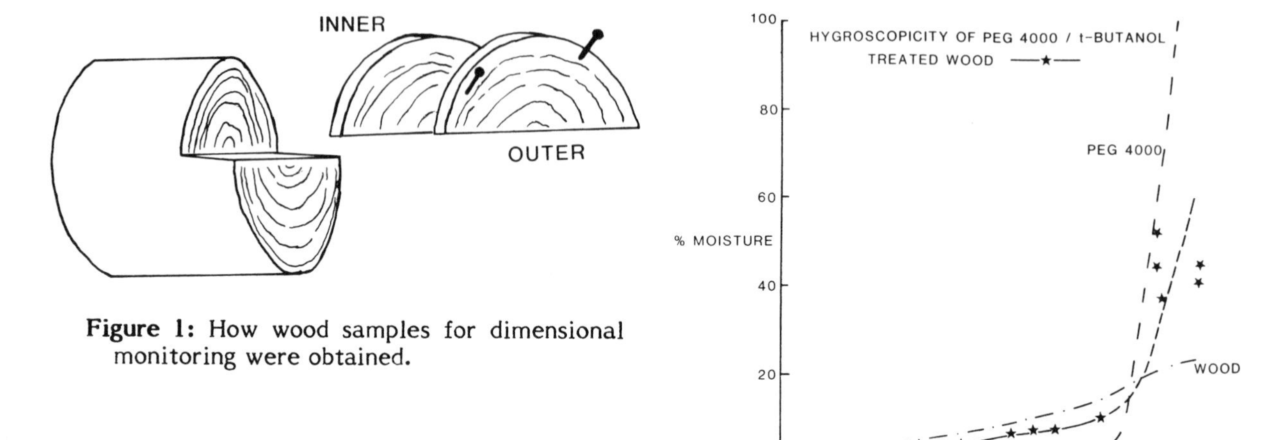

Figure 1: How wood samples for dimensional monitoring were obtained.

Figure 2: The hygroscopicity (water absorption) of wood impregnated with PEG 4000 solution. This diagram also shows the hyroscopicities of an untreated sample of the same piece of wood and pure PEG 4000.

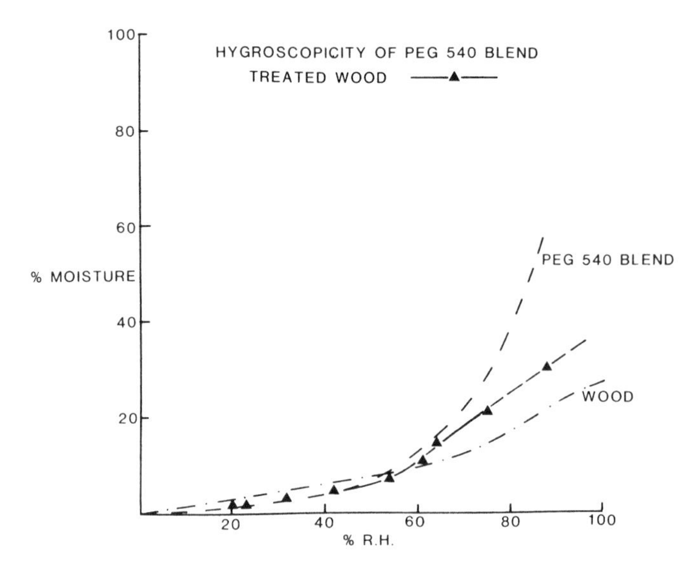

Figure 3: The dimensional changes accompanying the water absorption curves shown in **Figure 2.** The dimensional changes are the average of two determinations for each of the three different impregnation techniques which employed water, t-butanol and methanol as solvents. The "control" curve, shows the dimensional change of an untreated sample of the same piece of wood.

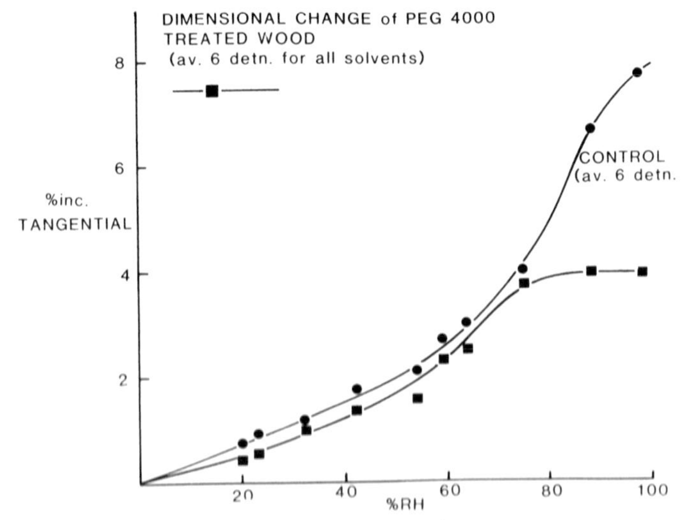

Figure 4: The hygroscopicity (water absorption) of wood impregnated with PEG 540 Blend (see legend to **Figure 2** for further explanation).

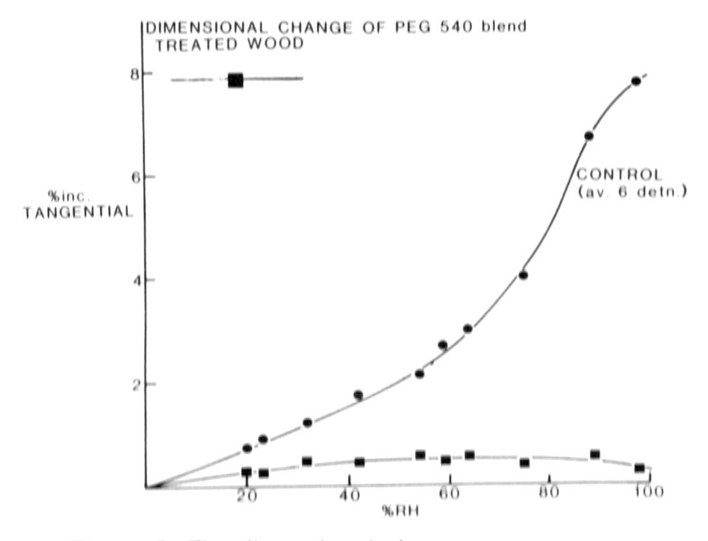

Figure 5: The dimensional change accompanying the water absorption curves shown in **Figure 4.** The "control" curve shows the change occurring with an untreated sample of the same piece of wood.

4000 were very similar to one another. The results can therefore be considered together. In **Figures 2,3** the results are plotted.

(here hygroscopicity is defined as the percentage moisture absorbed compared to the weight of the treated wood at 0% RH).

Below an RH of about 80%, hygroscopicity is reduced compared to untreated wood. Above this value, when PEG 4000 itself begins to absorb excessively, the wood "sweats" and PEG solution oozes slowly out of the wood. AT 100% RH the wood becomes a sticky mess. Hygroscopicity can be quantitatively accounted for by adding the separate known values for PEG 4000 and wood in the appropriate ratio.

The dimensional behaviour is surprising. Below 80% RH it differs little from untreated wood, but above this value, when water is absorbed in large quantities, movement ceases. Below RH 80%: The results suggest that water absorption by the wood is unaffected by the presence of PEG and agrees with Young's observation on the same wood that PEG 4000 has not entered the cell wall (8). At RH 80% and above, the hygroscopicity of PEG probably enables the wood to achieve the fibre saturation point, and the difference in the total dimensional movement from 0% to 100% RH between the treated and untreated wood reflects the extent to which PEG has associated with or penetrated the cell wall. i.e. not much! It will be interesting to test the effect of PEG 4000 impregnation with more degraded wood, in which better penetration of the cell wall might take place; movement may consequently be reduced compared to the control.

The consequences for storage and display are that the wood has to be treated in essentially the same way as untreated wood. Good humidity control is therefore important. The consequences for wood treatment are as follows: After the impregnation it is important to "condition" the wood carefully; it should be returned to equilibrium with ambient RH in a slow and carefully controlled fashion. Conditioning could well begin at 80% rather than 100% RH without giving the wood any dimensional shock, and this would avoid leaching out PEG solution.

PEG 540 Blend

The hygroscopicity of this impregnant is much higher than PEG 4000. This can be seen in **Figure 4**, where the hygroscopicities for PEG 540 Blend, control wood, 540 Blend treated wood are shown: **Figure 5** shows the accompanying dimensional movement. Up to RH 60% the hygroscopicity calculated from the known values of PEG 540 Blend, and the control wood in their appropriate ratio is between 20 and 40% greater than that of the measured hygroscopicity of the composite. This phenomenon is also observed for PEG 400 treated wood and is discussed more fully in the next section.

Dimensionally the wood is almost completely invariant with RH change. Therefore RH control is not very critical for wood treated in this way, and the advice for storage or display would be simply to maintain it below about 60% RH. Above this level mould growth could become a problem. Consequently, after the impregnation treatment it is not as important to have such careful conditioning as for wood treated with PEG 4000.

"Sweating" of the wood at humidities greater than 75% RH was a problem, and leaching out of impregnant occurred.

PEG 400

The behaviour of wood treated by this impregnant is shown in **Figures 6,7.** As for PEG 540 Blend the wood/PEG composites have reduced hygroscopicity compared to the values calculated from the hygroscopicities of the separate components. Here, the extent is more marked than for wood treated with PEG 540 Blend and is clearly illustrated in **Figure 6** in which it can be seen that the percent moisture absorbed lies below that of pure PEG 400, and control wood, in the relative humidity range below 60% RH.

For wood treated by PEG 400 with 25% and 35% v/v impregnation solutions, there is little or no dimensional movement with RH change, but when less PEG 400 is incorporated, as with the 15% impregnation solution, some dimensional movement is observed. (See **Table 1** for the amount of PEG incorporated in the wood). It appears that low molecular PEG, either molecular weight 400 or 300 (which is present in PEG 540 Blend) is absorbed into the cell wall, thereby severly limiting the absorption of water and hence preventing dimensional change. It is significant that the amount of PEG necessary to prevent dimensional movement is approximately volumetrically equivalent to the amount of water at the fibre saturation point.

To summarize, PEG 400 treated wood is

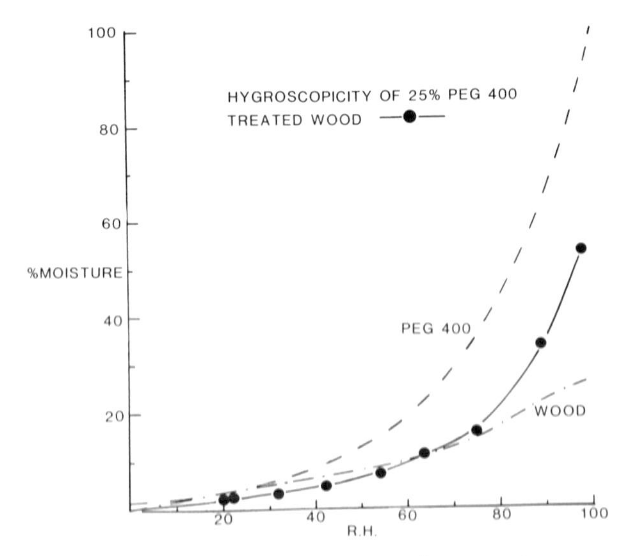

Figure 6: The hygroscopicity (water absorption) of wood impregnated with a 25% solution of PEG 400 before freeze-drying. Compared to the hygroscopicities of pure PEG 400 and wood, that of the PEG 400/wood mixture (or composite) is less than that either of these pure components, when the RH is below 60%.

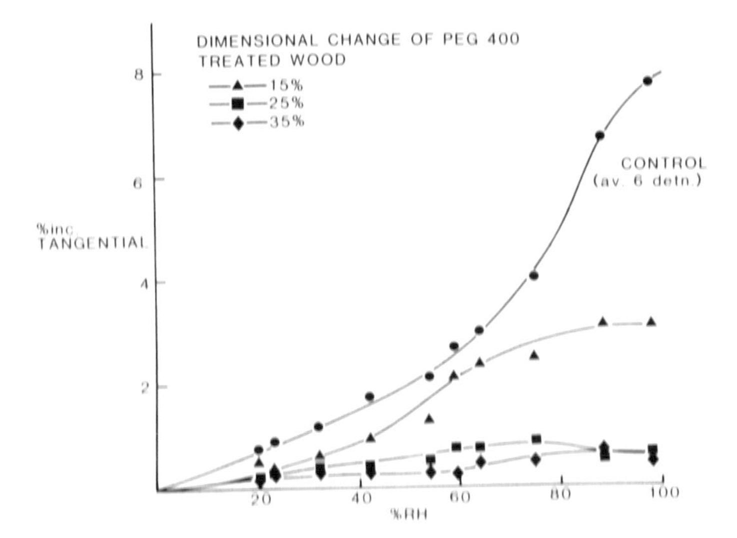

Figure 7: The dimensional changes of wood treated by impregnation at three different concentrations of PEG 400 (15%, 25% and 35%) before freeze-drying. As the concentration of PEG 400 is increased, dimensional change becomes less. The dimensional change for untreated wood is also shown.

TABLE I

Impregnant Content of Treated Wood

Impregnation Method & Solvent	Impregnant & Concentration (%v/v)	Content of Impregnant in Treated Wood (Based on Oven dry Weight of Wood)
Methanol	72% PEG 4000	114%
Butanol	48% PEG 4000	69%
Water	49% PEG 4000	74%
Water	54% PEG 540 Blend	72%
Water & Freeze-Drying	15% PEG 400	15.0%
Water & Freeze-drying	25% PEG 400	27%
Water & Freeze-drying	35% PEG 400	40%
Acetone	73% Rosin	102%

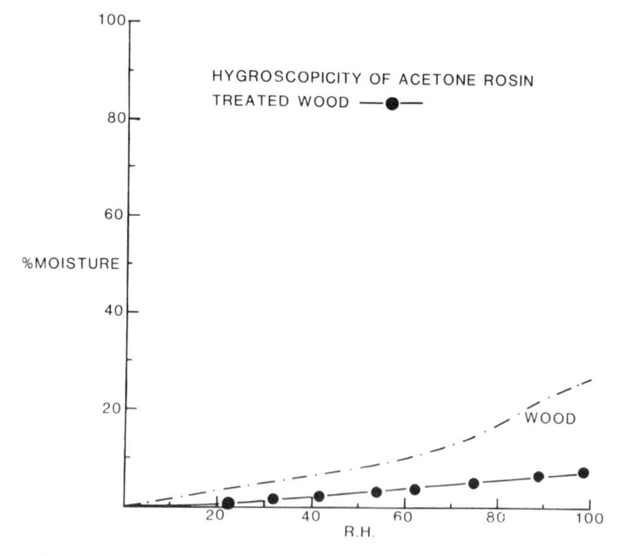

Figure 8: The hyproscopicity (water absorption) of Acetone Rosin treated wood, shown together with that of untreated wood.

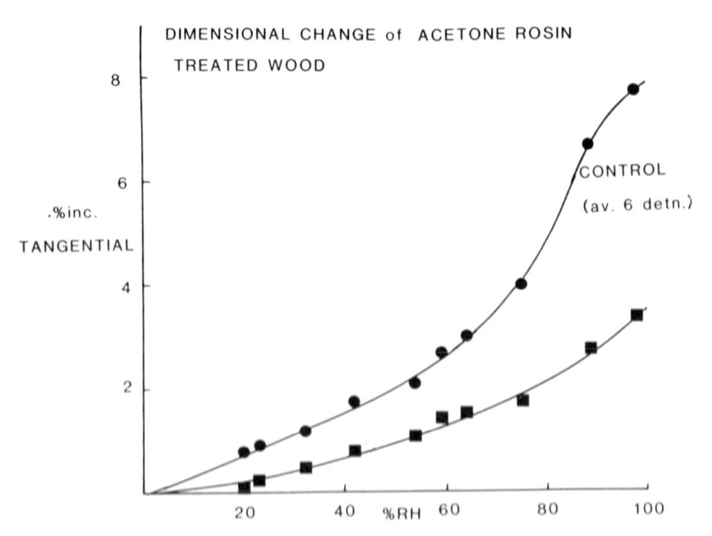

Figure 9: The dimensional change accompanying water absorption curves shown in **Figure 8** for Acetone Rosin treated wood.

much less hygroscopic than might be expected and is dimensionally very stable provided that enough PEG 400 is present, (See discussion in (1)).

Above an RH of 60% mould growth could become a problem, but this is also true for untreated wood. In this high humidity region the calculated moisture contents of the PEG/-wood composites (based on the hygroscopicities of the pure components) exceeds those of the observed values: For wood impregnated with 15, 25 and 35% solutions of PEG 400, the calculated values were 44%, 84% and 54% higher respectively. Since the absorption of large quantities of water has been prevented, this is evidence for an interaction between the PEG and wood at least as strong as that between water and wood. Association between the hydroxyl groups of cellulose and the PEG probably prevents the full absorption of water by each component.

Why is this behaviour not observed for wood treated with PEG 4000? Probably similar associative processes are present but are much less noticeable because of the lower hygroscopicity of PEG 4000 and the much lower concentration of hydroxyl groups. The possibilities for association are thus diminished. The more bulky molecule of PEG 4000 does not appear to penetrate the cell wall, and thus association may also be physically prevented (8). This association is also demonstrated by the dimensional response, shown in **Figure 7**; particularly by the different control exercised by PEG 400 at different concentrations: As the amount of PEG is increased, the dimensional movement becomes increasingly small.

Thus, the recommendations made above for PEG 540 Blend treated wood in regard to storage/display can be repeated here, i.e. the dimensions will be unaffected by RH change and therefore control of RH is not critical. Since the moisture is normally removed by freeze-drying for PEG 400 treated wood the remarks on conditioning do not apply. However the results also indicate, that if all of the water is removed by freeze-drying so as to produce a desiccated specimen, the wood will be dimensionally unaltered; thus cracking on desiccation or subsequent re-absorption of moisture is unlikely.

Other Methods

Other methods of treatment studied were, the Acetone/Rosin Method, the in-situ polymerisation of glycol methacrylate or of a melamine formaldehyde "Casco" resin, and the deposition of SiO_2 by means of tetraethoxy silane (TEOS) (1). All reduced the hygroscopicity and the dimensional response. In **Figures 8-9** typical results are shown for wood treated by the acetone/rosin method. Very similar results were obtained for the TEOS and Casco resin treated wood.

Conclusion

It is clear that the most dimensionally stable wood is produced with PEG 400 or 540 Blend. Acetone/Rosin treated wood is the next best.

It is also important to note that wood treated by low molecular weight PEG (400 or 540 Blend) should not be exposed to RH above 60%. Below this value, it matters little what the RH level is, or what RH variations take place. The wood should be perfectly stable.

These results also relate to the post-treatment conditioning of wood at high RH. Since wood treated with 540 Blend and PEG 400 is almost unaffected by RH change, it is pointless to expose it to high RH after treatment if impregnation has been effective. For wood treated with any molecular weight of PEG (unless it is continually recoated with PEG solution) exposure to high RH (60%) will leach out the impregnant. Unless it is hoped to use conditioning at high RH to impregnate the wood further with PEG 540 Blend it is hard to see the benefit. Wood impregnated with PEG 540 Blend or PEG 400 will not receive any dimensional shock if exposed to completely dry conditions. For PEG 4000 treated wood, slow conditioning below an RH of 80% may be quite important in prevention of checking.

Proviso: This study was carried out with sound elm. It remains to be seen whether the results are also valid to other types of wood and different levels of degradation.

Future Work

This study is only partly completed. It will be extended to an examination of more degraded wood, and of other methods of treating wood. It will be reported in full at a later date.

References

1. Grattan, D.W., "A Practical Comparative Study of Treatments for Water-

logged Wood; Part I," To appear in, Studies in Conservation (1982).
2. Rosenqvist, A.M., "Experiments on the Conservation of Waterlogged Wood and Leather by Freeze-Drying," Maritime Monographs and Reports, vol. 6, 1975. National Maritime Museum, Greenwich, London, p. 9.
3. Barkman, L., "The preservation of the warship Wasa," ibid., p. 65.
4. Sawada, M., "The Conservation of Waterlogged Wooden materials from the Nara palace site," proc. of the 1st ISCRP, Wood, 1977, pp. 49-58.
5. Kenaga, D.L., "Effect of Treating Conditions on Dimensional Behaviour of Wood during PEG soak treatments," Forest Products J., 1963, pp. 345-349.
6. Schneider, A., "Basic Investigations on the Dimensional Stabilization of Wood with PEG." Holz Als Roh Und Werkstoff., vol. 9, 1969, pp. 209-224.
7. Lafontaine, R.H., Personal communication.
8. Young, G., This publication.

Questions

Mary-Lou Florian: How did you determine the percent bulking?

David Grattan: The percent bulking is simply taken as the ratio of the volume of the impregnant, compared to the original volume occupied by the waterlogging water. In other words, it tells you how much of the void space in the wood is filled with impregnant. It can be calculated if you know the wet weight and final weight of the object, the water content of the wood in question, and some idea of the hygroscopicity of the treated object and the density of the impregnant.

For example:

A specimen S has an initial wet weight of Xg, a water content of Y% (based on the oven dry weight).

Thus S is composed of:

$$\frac{X\,Y}{100 + Y}\ \text{g water};\ \frac{100\,X}{100 + Y}\ \text{g wood}$$

The weight of S at ambient RH, after treatment is Z g, and it contains W% water.

Thus the weight of S, treated and under dry (i.e. RH = 0%) conditions is:

$$\frac{Z\,(100 - W)}{100}\ \text{g}$$

The weight of impregnant incorporated is given by:

$$\frac{Z\,(100 - W)}{100} - \frac{100\,X}{100 + Y}\ \text{g}$$

If the density of the impregnant is; D, then the volume of incorporated impregnant is:

$$\frac{Z}{D}\,\frac{(100 - W)}{100} - \frac{100\,X}{D\,(100 + Y)}\ \text{cc},$$

Since the volume of waterlogging water is numerically approximately the same as its weight, % bulking is given by:

$$\frac{Y + 100}{D\,Y\,X}\left\{\frac{Z\,(100 - W)}{100} - \frac{100\,X}{100 + Y}\right\}(100)$$

In other words by weighing, which is a little rough and ready.

Per Hoffman: The observation that wood treated with PEG 400 is less hygroscopic than might be expected, has been found for oak wood some years ago.

David Grattan: Yes, there are a few previous determinations of the hygroscopicity of treated wood. My aim here is principally to get a complete set of comparative measurements, on one set of wood treated in various ways. I want to collect a coherent and complete data base, to get real comparative information.

Victoria Jenssen: Have you measured directly how much PEG is in the wood?

David Grattan: No, but it is something that I must do. I intend to extract it out of the wood and determine it either gravimetrically or colourimetrically as described by Allen Brownstein earlier this week.

Victoria Jenssen: Why did you begin the isotherms at low RH?

David Grattan: Because if I had begun the wood at the wet end of the scale, much of the PEG would have leached out

Jamie Barbour: I want to comment on your observations with acetone rosin treated wood. During the impregnation the transient capillaries in the wood are swollen, and rosin is deposited in them. The treatment inserts rosin into the voids in the wood that the acetone makes available. It makes sense that you would not have very much dimensional change, because the rosin is non-hygroscopic and would not shrink or swell and thus would not allow the wood to move.

David Grattan: Yes, I simply don't know, and have not measured the hygroscopicity of rosin.

Jamie Barbour: It is a resin acid which is non-polar and non-hygroscopic. PEG, which is hygroscopic, will shrink and swell, and also do so presumably in the cell wall.

SOME EXPERIMENTS IN FREEZE-DRYING: DESIGN AND TESTING OF A NON-VACUUM FREEZE DRYER

J. Cliff McCawley
David W. Grattan
Clifford Cook
Conservation Processes Research Division,
Canadian Conservation Institute,
1030 Innes Road,
Ottawa, Ontario,
K1A 0M8

Abstract

The design and testing of a non-vacuum freeze dryer, using two domestic chest freezers, is described. Criteria for design included: cost, availability of materials, size of chambers, simple means for trapping water vapour, easy accessibility for addition or removal of samples. Preliminary trials on the suitability of the apparatus for drying degraded waterlogged wood were carried out.

Introduction

Over the past several years the Conservation Processes Research Division, of the Canadian Conservation Institute, has carried out research into three aspects of freeze-drying for waterlogged wood. Two of these, the use of the Canadian winter climate for freeze-drying (1,2) and an investigation of Ambrose's method of pretreatment (3), have been reported elsewhere, the third non-vacuum freeze-drying arose naturally from our successful work on exterior freeze-drying and is the subject of this paper.

The aim of the project was to see if a feasible and inexpensive alternative to vacuum freeze-drying could be devised. Essentially, we attempted to carry out atmospheric freeze-drying in the laboratory with an artificially created winter climate. The concept is hardly new, since in conservation both Ambrose (4) and Drocourt (5) have investigated techniques of non-vacuum freeze-drying, and in the freeze-dried food industry, there has been a great deal of interest in cheaper alternatives to the expensive vacuum methods.

The theory behind the system is very simply that, if you have two chambers, one very cold to act as a condenser, and one just below freezing to hold the frozen samples, then if it is a closed system, ice will sublime and migrate from the warmer chamber to the colder. That is, freeze-drying will take place. In this operation, a blower was used to transfer the moist air from the sample chamber to the condensing chamber. In normal operation it was found unnecessary to run the pump continuously since this tended to upset the temperature equilibrium established in each chamber.

Method

There were two basic criteria involved in the design and construction of the freeze-drying unit. The first was cost; it had to be as inexpensive as possible to put its possible use within the budgets of as many people as possible. The second consideration was simplicity of construction and ready availability of materials. This latter consideration was adhered to with only one exception; it was decided to incorporate a cooling coil with more limited availability. All other materials and equipment used are conveniently found in hardware stores and appliance dealers everywhere.

The basic components were:

(a) A domestic chest freezer (0.74 m^3), or as large a capacity as necessary, to act as the sample chamber.

(b) A domestic freezer (0.25 m^3), which may be considerably smaller than the sample chamber, to act as a condenser or ice trap. (Both of the freezers used were standard domestic appliances.)

(c) A Virtis Unicool Coldfinger (cooling coil) 2.8 $ca.^{-1}$ kg^{-1}.

(d) A small air blower (7.2 m^3 min^{-1}).

(e) Variable voltage power supply.

(f) Various materials such as plywood, polystyrene foam insulation, ABS (poly acrylonitrile butadiene styrene) plastic drainpipe (10 cm).

Design

The basic design is shown in **Figure 1** and **Photo 1**. To construct the dryer, both freezer doors were removed and replaced by specially constructed two piece plywood cover. One piece of each cover was permanently fixed to the end of the freezer opposite the compressor. The freezers were then lined up end to end, with the compressors at opposite ends. Initially, rubber weatherstripping was used as a gasket for the new lids, however, this proved ineffective, and was replaced by the original

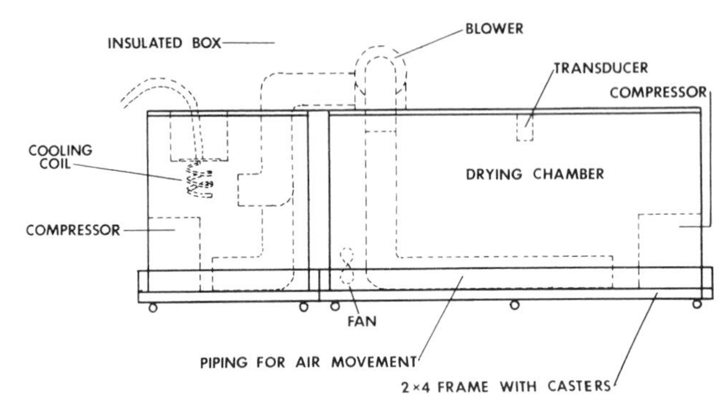

Figure 1; The basic arrangement of the freeze-drying unit. All external connecting pipes, and the blower were insulated with 2.5 cm thick neoprene foam rubber sheet.

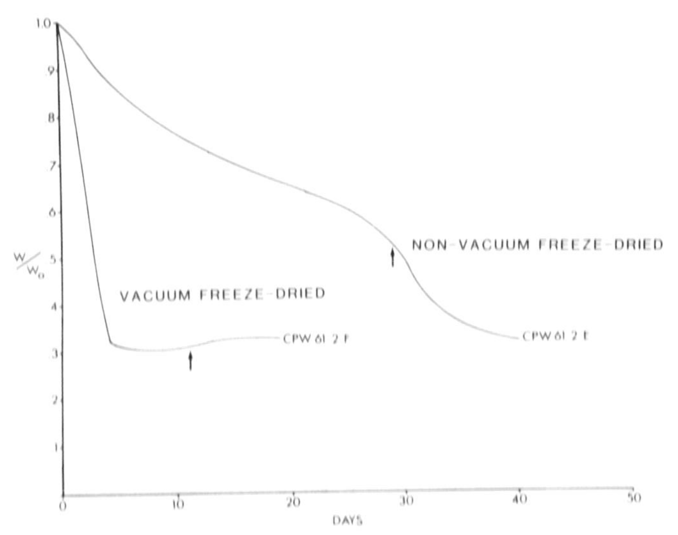

Figure 2: The loss in weight during freeze-drying of similarly sized pieces of water-logged wood in a vacuum-freeze dryer, and in the non-vacuum unit. The arrows mark the times at which the wood samples were removed. W is the weight at time t and Wo is the initial weight of the sample. Thus the ratio W/Wo represents the weight remaining as a fraction of the original, and the decrease in its value is caused by the sublimation of ice.

gaskets from the freezer doors. To have an effective vapour barrier was very important, especially in the summer months, when high ambient humidity can reduce the efficiency of the system through build-up of large quantities of ice. The fibreglass insulation, originally used to insulate the plywood lids, became saturated with water which froze, thereby reducing the R value of the insulation. Expanded polystyrene foam (styrofoam) did not suffer this defect. The lids were clamped down to ensure as tight a seal as possible. Holes were cut in the fixed part of each lid to accommodate the ABS pipe which was to conduct the moist air from the drying chamber to the small freezer/condenser. A small blower forced the air through the piping which was insulated with a neoprene foam. A variable voltage source was connected to the blower so that its speed could be reduced. A timer controlled the blower to cycle 25 minutes on, then 105 minutes off. To reduce the build-up of a moist air layer around the frozen samples, a small fan was placed in the sample chamber and run continuously. When the blower is activated this moisture laden air is passed through the condenser chamber where the moisture is deposited as ice.

The coil condenser (see **Photo 2**) was positioned in the centre of the small freezer by mounting it in a cylindrical well (diameter 28 cm) recessed into the lid of the condensing freezer. This allowed easy removal of the ice, and facilitated insulation of the area where the coil entered the freezer. Its central positioning also meant that the coil was in the main stream of the air current.

The cooling coil was used to efficiently trap water vapour as ice. Other methods could have been adopted for this purpose, such as: silica gel or a dry ice (solid carbon dioxide at -78°C) or liquid nitrogen trap. The problem with silica gel is that it requires regular reconditioning and would be no more effective than the vapour trap. Dry ice and liquid nitrogen traps both need replenishing and are, therefore, very expensive over the long term.

It was felt that even though the initial cost of the cooling coil was probably higher than that of other methods, the ease with which the system could be handled and maintained made it the most suitable. Ice is very easily removed from the coil by immersing it in a container of warm water for a short period (ca. 5 minutes). The coil is then quickly replaced in the small freezer.

To monitor weight loss for the samples a load transducer, suspended inside the sample chamber and connected to a digital readout, was used. This gave a continuous weight reading and eliminated the need to open the sample chamber. To measure the temperature at various points in the system and inside the wood, copper constantan thermocouples connected to a digital multimeter and chart recorder was used. Calibration of the recorder was carried out using a potentiometer to supply small accurately known voltages. In addition, continuous records of temperature and humidity for periods of 120 hours, were obtained by placing a thermohygrograph in each freezer.

To make the system moveable the two freezers were mounted on wooden frames equipped with castors.

Operation and Results

The cooling coil was capable of operation to -60°C. To provide maximum cooling in the condensing chamber, the thermostat for the freezer was removed from the circuit, forcing the compressor to run continuously. In the sample chamber, the thermostat was set at the warmest setting. This resulted in a frequent cycling of the compressor and a temperature was maintained which was just low enough to keep the samples frozen. During operation, the termperature of the sample chamber was kept at -6 ± 5°C and that of the condensing chamber at -40 ± 5°C.

Obviously, the first question to ask was; "Does the dryer work? To ascertain this, two samples of Populus sp., (258% water) were taken from the same piece of wood. Both pieces were sealed in polyethylene bags and frozen in a deep freezer. One was dried in the non-vacuum-freeze-drier (NVFD); the second was freeze-dried under vacuum at -20°C. The drying rates are compared in **Figure 2.** The arrow indicates the time at which the sample was removed from the sample chambers and allowed to equilibrate in the atmosphere. Note that the sample dried using NVFD continued to lose weight after removal. The NVFD took much longer to bring the wood to a state of dryness similar to that attained in a much shorter time using vacuum freeze-drying. However, there was no visible difference between the two pieces of wood.

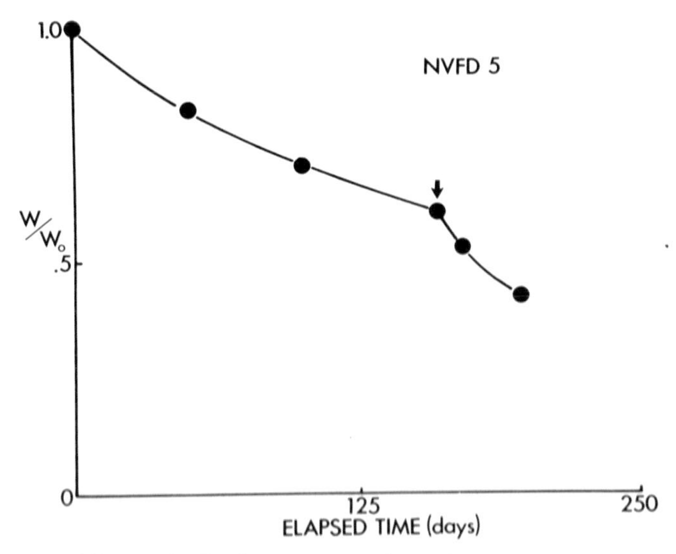

Figure 3: Drying curves when the non-vacuum freeze-drying unit was entirely filled with wood. (W/Wo is described in the legend to **Figure 2**).

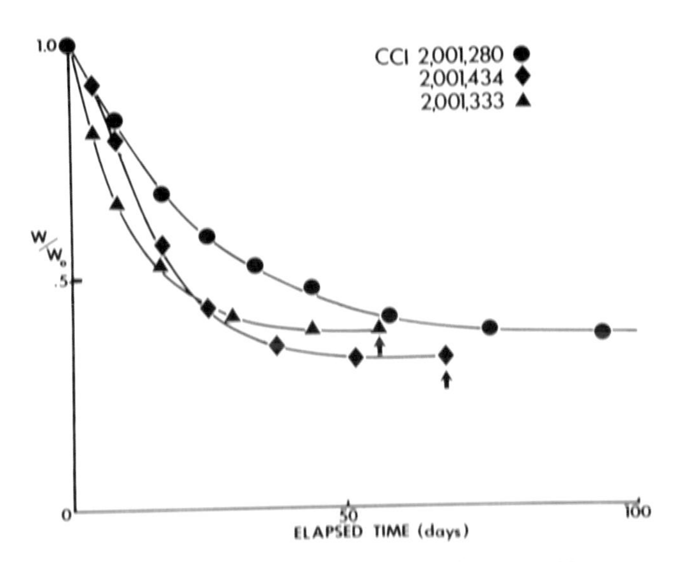

Figure 4: Drying curves for three artifacts treated in the non-vacuum freeze-dryer, showing complete sublimation of ice. The arrows denote the times at which the artifacts were removed from the drying chamber: note that W/Wo remained constant after this, no further loss of water was observed. (See legend to **Figure 2** for explanation of W/Wo).

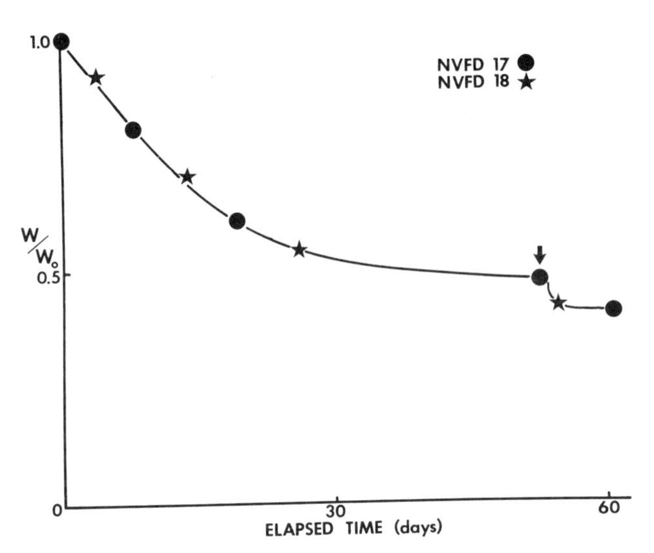

Figure 5: Drying curves for PEG treated wood. (see legend to **Figure 2** for explanation of W/Wo).

Because one of the attractions of this system is the large sample chamber, it was necessary to see how drying a large volume of waterlogged wood affected the drying rate. Sixteen pieces of frozen waterlogged wood weighing 14 kilograms, and occupying a volume of approximately 1 m^3, were placed in the sample chamber. The weight loss of one, NVFD 5, was followed by suspending it from the load transducer. The result is shown in **Figure 3** (Note that the elapsed time is measured in days). At the point of removal from the NVFD, the sample had lost 62% of its water. This fell to 93% water loss after 1.5 months of laboratory equilibration. The rate of weight loss for this sample was much slower than that in the previous experiment. During the first month, 12% of the initial frozen weight was lost, and at the same rate, it would take approximately 125 days to dry the piece. However, the reason for this was not the large volume of wood in the sample chamber, but rather the larger size of the sample (initial frozen weight of 4 kg). Two artifacts (weights 140 g and 170 g sample numbers 2,001,280 & 2,001,434 respectively) were dried at the same time as NVFD 5 and they dried much more quickly, see **Figure 4**. A third artifact (2,001,333 weight 119 g) was placed in the drying chamber 70 days after the other samples and also dried quickly. This difference in drying rate can be primarily attributed to the difference in the surface to volume ratios of the different pieces. As the object gets bigger, the surface to volume ratio gets smaller. Drying rate depends upon surface area and, of course, the larger the volume the more water there is to be removed. One would, therefore, expect larger pieces to dry more slowly than smaller pieces (2).

Two similar sized pieces of maple (water content 290%) were pre-treated in 15% PEG 400 for six months and then dried in the NVFD. **Figure 5** shows that these PEG treated woods dried effectively, and at the same rate. The shrinkage measurements for these two pieces were less than 1% and the anti-shrink efficiency was 91%, compared with 77% for vacuum freeze-drying and 74% for exterior freeze-drying for the same piece of maple.

This sample of wood is typical of many problem types of waterlogged wood. It has a relatively sound core with an extensively degraded surface zone extending 1 to 2 cm into the wood. NVFD like exterior freeze-drying and vacuum freeze-drying was effective in preventing shrinkage, however, vacuum freeze-drying was the most effective in preventing extensive collapse of the surface zone.

Conclusion

The preliminary trials have shown that the Non-Vacuum-Freeze-Dryer works well and can be used to dry waterlogged wood. Whilst the rate of drying is less than half that of a vacuum-freeze-dryer system, the cost is less than ten percent. This slower drying rate can be an advantage in conservation, because, any physical changes taking place in the samples takes place more slowly and can more easily be followed by the conservator. The system has also been shown to be effective at drying paper and books.

If the system is used to freeze-dry wood pre-treated with PEG, it is desirable to run the sample chamber as cool as possible. At -5°C for instance a 20% solution of PEG 400 will consist of about 23% ice in the form of small particles and 77% of a liquid PEG-400/water solution, 35% in concentration; whereas at -25°C the amount of ice increases to about 44% of the total volume and the remaining portion is a 56% solution of PEG 400 in water in the liquid phase (6). It is almost certainly desirable to maximise the proportion of ice in the early phases of drying so it is probably best to commence the freeze-drying at low temperature (-20°C) and gradually increase this as drying progresses. This will inevitably make the treatment slower, but the results of treatment will probably be superior. We hope to test this in future.

References

1. Grattan, D.W. and McCawley, J.C., "The Potential of the Canadian Winter Climate for the Freeze-Drying of Degraded Waterlogged Wood," Studies in Conservation, vol. 23, 1978, pp. 157-167.
2. Grattan, D.W., McCawley, J.C. and Cook C., "The Potential of the Canadian Winter Climate for the Freeze-Drying of Degraded Waterlogged Wood: Part II," Studies in Conservation, vol. 25, 1980, pp. 118-136.
3. Grattan, D.W., "A Practical Comparative Study of Treatments for Waterlogged Wood: Part I'" Studies in Conser-

vation accepted for publication (1982).

4. Ambrose, W.R., "Sublimation Drying of Degraded Wet-Wood," in: Pacific Northwest Wet-Site Wood Conservation Conference, proc. of the symp., ed. G.H. Grosso, 19-22 September 1976, Neah Bay, Washington, vol. 1, pp. 7-16.
5. Rossion, P., "La Mise en Bouteille D'un Vrai Bateau," Science et Vie, no. 750, vol. 131, March, 1980, pp. 73-74.
6. Carbowax Polyethylene Glycols, (Union Carbide Corporation, Ethylene Oxide Derivatives, Old Ridgebury Road, Dunbury, CT 06817, U.S.A., 1981).

Suppliers

Virtis Cooling Coil, Unicool Model 10-141.

Available from:

Virtis Co. Inc., Route 208, Gardiner, New York, 12525, U.S.A.;

Academy Instruments Inc., 705 Progress Ave., Unit 55, Scarborough, Ontairo, M1H 2X1, Canada.

Similar Units are available from:

Polyscience Corp., 6366 Gross Point Rd., P.O. Box 312, Niles, Illinois, 60648, U.S.A.;

Neslab Instruments Inc., 871 Islington St., Portsmouth, New Hampshire, 03802, U.S.A.

Neslab is represented in Canada by:
Acadian Insturments Ltd., P.O. Box 303, Etobicoke, Ontario, M9C 4V3.

Questions

(Tape change - beginning of question lost.)

Herman Heikkenen: If you took new wood, in which you could define the moisture content, number of rings per cm., extent of heartwood, resin content of heartwood, sapwood. Use uniform blocks with standard lengths, widths and heights, set that aside and go and work on the old wood.

Cliff McCawley: I think the problem we all have, is that it really is difficult if not impossible to have or to obtain standard pieces of wood, especially when you are talking about degraded wood. Your suggested approach would give us very accurate answers with sound wood, but what would you do then? I really don't see that we would benefit that much from that kind of methodology.

Mary-Lou Florian: Water doesn't have to freeze at -40°C in small capillaries. Are we sure that we are dealing with freeze-drying, or simply cooled water?

David Grattan: I don't know.

Per Hoffmann: Water will be frozen in the lumina.

Mary-Lou Florian: A capillary of diameter 1 mm is enough to prevent freezing.

Per Hoffmann: But only under very carefully controlled conditions.

Mary-Lou Florian: This is something we should be concerned with.

David Grattan: In the results that we presented, there is an empirical answer to your question; the effect of partial freeze-drying, on shrinkage and on collapse of wood whch had been pretreated in 15% PEG 400 solution was shown. We observed that freeze-drying under the conditions at -20°C, had an enormous effect in terms of the results actually observed. Does it then matter whether actual freeze-drying was occurring or not?

Jamie Barbour: Mary-Lou has a good point, Harvey Erickson (see Barbour & Leney, this publication) at the University of Washington carried out many investigations on freeze-drying because they wanted to use the process in the lumber industry. He discovered that the bound water does not freeze, and that when you freeze-dry a piece of sound wood, that the shrinkage is actually greater, than if you air dry it and subsequently oven dry that same piece of wood. The reason he gave was that as you lower the temperature the FSP increases, and at the lower temperature here bound water was present in the walls, and more bound water results in more shrinkage. The increase in moisture content at the FSP at say -50°C is of the order of 5-6%, and so is a marked increase. Another factor is that the bound water does not freeze, as Goring of McGill University showed in a recent paper. To me the freeze-drying allows shrinkage but prevents collapse.

David Grattan: I think that with freeze-drying

from a PEG solution, that the most important aspect is that the water is removed and, at the same time, migration of the impregnant to the wood surface has been prevented. Freeze-drying, enables the PEG to be evenly distributed throughout the wood.

Jim Spriggs: Another aspect of freeze-drying is that water, if it has anything dissolved in it, such as natural sugars and PEG, I think forms a eutectic . When it is cooled down to a low temperature, not all the water actually freezes out - to leave solid PEG behind. Depending on the type of PEG, the concentration and the temperature, different effects will take place due to the various eutectics present in the fibres in the cell walls. This is something which I think should be studied, to see how they can be manipulated to give the best results for our needs.

(ed. Comment: I would like to clear up the points raised above which are important for the understanding of the freezing behaviour of PEG solutions in water and the process of freeze-drying. Shown in the **Figure** is data taken from the Union Carbide Technical Bulletin "Carbowax polyethylene glycols" (New York: Union Carbide, 1976), p. 10.

This is essentially an example of the most simple type of freezing-point phase diagram, which is discussed in any textbook of physical chemistry. In this phase diagram there are two components which are completely miscible in the liquid phase, but have negligible solid-solid solubility.

Let us image that we have a solution which contains 25% by weight PEG, i.e. at point P on the curve. As the solution is cooled, pure water will start to freeze, and hence separate out. Thus, the liquid remaining will become more concentrated in PEG. As the temperature falls, the composition of the remaining liquid will be given by the curves P-E.

In a simple system such as this, a eutectic point E would be reached, at which the eutectic mixture of the two components would freeze out. However, in the PEG/water system a eutectic is not formed; at temperatures below minus 23°C a super cooled liquid mixture remains.

For a piece of wood which has been impregnated with a solution of 25% PEG 400 in water; at -20°C it will consist of finely divided crystals of pure ice, in a matrix of a super cooled water/PEG solution containing about 50% of each component. As freeze-drying takes place, the crystals of ice will gradually recede, until the overall concentration of PEG is at about 50%, at which point only the super cooled PEG/water mixture will be present and drying will <u>not</u> take place by freeze-drying. As more water is removed, solid PEG 400 separates out, until it alone remains.

One further point, is that only a small volume of the wood will be drying at any given time. On one side of the drying front there will be pure solid PEG and on the other a mixture of ice crystals in a 50% PEG 400 solution. All the processes described above will take place in the drying front. (How deep is this front is an interesting question!)

As for the bound water, this will certainly not be frozen, because it is in some respects in a condition which resembles the gaseous state. Presumably, as freeze-drying proceeds, bound water is lost and is replaced by the mobile super cooled water/PEG solution. Thus one reason why the PEG freeze-drying process works to avoid shrinkage is that the movement of the super cooled solution of PEG, enables PEG to replace the bound water, continuously as drying proceeds, leading to complete replacement when the wood is dry. Thus, throughout the freeze-drying process, the cell walls are kept in an expanded condition.

Is this really freeze-drying? This interpretation of the process suggests that pure freeze-drying may not even be a desirable aim!

One other item which may also be of great importance in the process was brought out in the next comment.)

Cliff McCawley: One other factor is crystal habit modification. How is the growth and ultimate size affected by the presence of PEG? Almost certainly PEG, modifies the crystals so that the ice formed is quite different from that formed from pure water. (ed. The reader is recommended to read several sections in the book <u>Advances in Freeze-Drying</u>, ed. Louis Rey (Paris: Hermann, 1966) where ice crystal formation, crystal habit modification and freezing damage by ice crystals are discussed. The following sections are recommended: "Thermal Analysis in Freezing and Freeze-Drying," J.D. Davies, pp. 9-20, "Nucleation and growth of ice crystals," J. Hallett, pp. 21-38 and "The importance of the prefreezing stage on the viability of freeze-dried organisms," R.I.N. Greaves, pp. 95-101.)

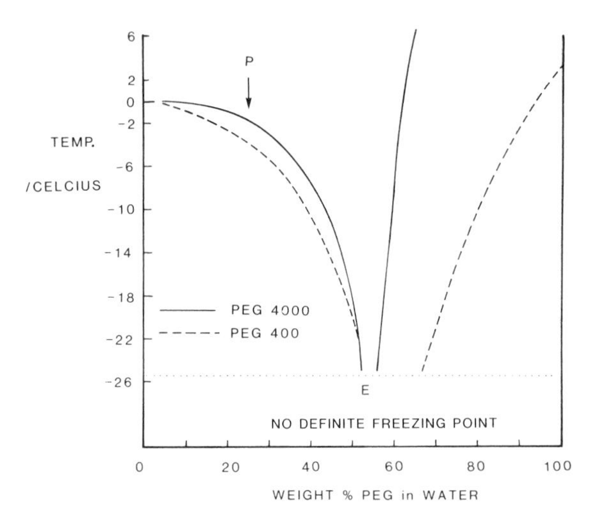

Figure: Freezing point phase diagram of PEG in water.

AGEING OF REINFORCED ANCIENT WATERLOGGED WOOD BY GAMMA RAY METHODS

E.G. Mavroyannakis
Nuclear Research Center,
Democritos,
Athens, Greece

Abstract

The acceptability of various methods used for conservation of ancient materials, depends mainly on their satisfactory performance for long periods of time after the treatment has been completed. This important point is discussed in this paper, taking into consideration waterlogged wood treated some years ago by gamma ray methods, and aged under various conditions. Ageing tests were performed by exposing treated wood to (i) internal ambient conditions (ii) outdoor exposure (iii) accelerated bio-deterioration by soil burial.

Introduction

The technique of reinforcement of ancient materials with in-situ polymerisation by the use of gamma ray methods, has already been established (1-7). However, some points require clarification because of their importance in conservation. Among these is the question of the ageing of treated materials under more or less well defined environmental conditions. Consolidation in general imposes two different classes of problems:
(i) the problems of treatment, and (ii) the problems of the durability after treatment.

The problems of durability are related to the physical, chemical and biological behaviour of the wood after consolidation. The life of the objects after treatment must be as long as possible with minimum depreciation of their value (8). This means that their ageing must be slower than that of untreated objects. Although it is not my purpose to describe the mechanisms of degradation, it is useful to have a comprehensive picture of them. The three main groups of degradation processes can be represented schematically as in **Figure 1**. This representation illustrates the mutual interrelationship between physical, chemical and biological processes. The important point to appreciate is that any process can produce changes to the other two classes of processes. For instance, biological processes produce physical and chemical effects and so on.

In the following results of ageing for wood treated by gamma ray induced in-situ polymerisation, or "radiation consolidated" wood under various conditions is reported.

General Considerations

The ancient waterlogged wood described in this report has been treated by gamma ray methods, as described elsewhere (1-3). The monomers used and the ageing conditions are described in **Table I.**

Materials consolidated with gamma ray methods show a remarkable improvement of their properties after treatment; they are more compact and less water absorbing than before. It is however difficult to perform quantitative measurements on these materials, so our results are mainly qualitative.

Dry and Waterlogged Wood

Wood consolidation with synthetic monomers gives very satisfctory results, even in cases of advanced deterioration, as happens for wood of ancient shipwrecks found under the sea. We have treated wood objects as described in **Table II,** and the results are aesthetically very good in all cases.

Ageing the consolidated material for more than eight years under ambient interior conditions has not produced any sign of alteration. Modern wood, radiation consolidated, exposed to the exterior conditions shows small cracks on the surface, due probably to internal stresses, after 6 months exposure. However, under ambient interior conditions, cracks have not been observed up to now for this material. For ancient wood no cracks have been observed after five years of ambient interior ageing.

The appearance of treated material is remarkably natural and there is no obvious sign of treatment, although the weight is increased by more than 50% compared to the dry material. Appreciable deformations have not been found after treatment and dimensional changes seems to be of the same order of magnitude as that produced by water absorption in dry wood (5%).

Up to the present no biological degradation of the treated wood has been observed. However in order to test the durability of consolidated wood under accelerated conditions of biological attack, we have buried a number of

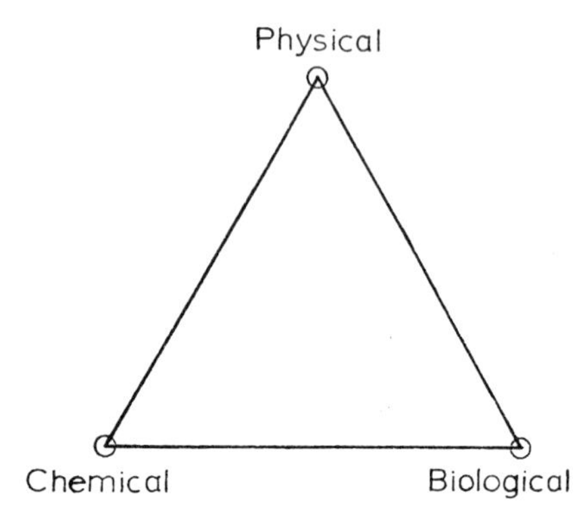

Figure 1

TABLE I

Matrix Material	**Monomer**	**Ageing Conditions**
Ancient Waterlogged Wood	MMA, VA, S	Ambient interior
Modern Wood	MMA, VA, S	Ambient interior, exterior expoure, Soil burial

S - Styrene Polyester
VA - Vinyl acetate
MMA - Methyl Methacrylate

TABLE II

Wood Object	Age	Origin	Water
Shipwreck	100 BC	Antikethyra*	Sea
Wood stick	Roman	Megalopolis*	Well
Plow piece	- BC	-	Soil
Shipwrecks	19th Cent.	Navarinon*	Sea

* Locations are shown on the map in **Figure 2.**

radiation treated and untreated modern wood samples for two years under the same conditions. After this period of time untreated wood was completely destroyed, while treated wood was nearly intact.

From the above results we can conclude that gamma ray consolidation of ancient wood, has valuable properties for conservation purposes. In comparison to other methods of treatment, there are few problems.

Concluding Remarks

Closing this paper it must be pointed out that the ageing studies of consolidated modern and ancient wood have only been carried out for a few years. The processes of ageing are still in the incipient stage. It is however evident that durability and other properties of dry and waterlogged wood when treated, are considerably improved. This increase of durability is therefore a very important effect, requiring much more attention. So, efforts are still necesary and more time is needed in order to obtain more reliable results.

Acknowledgements

The author is indebted to Mr. A. Antonopoulos and A. Kokkinos for their assistance.

References

1. Mavroyannakis, E.G., "Conservation Program of Ancient Terra Cotta Objects by Gamma-Ray Methods," in: ICOM Committee for Conservation, 4th Triennial Meeting, preprints of the conference, (Venice: 1975) working group 18, paper 3.
2. Mavroyannakis, E.G., "Experimental Results on Irradiated Terra Cottas," ibid., (Venice: 1975) working group 18, paper 3.
3. Mavroyannakis, E.G., "Radio Chemical Consolidation of Prehistoric Terra Cotta Potsherds," in: ICOM Committee for Conservation, 5th Triennial Meeting, preprints of the conference, (Zagreb: 1978) working group 17, paper 1.
4. Mavroyannakis, E.G., "The Conservation of Ancient Bones by Radiochemical Consolidation," ibid. (Zagreb: 1978) working group 17, paper 1.
5. Detanger, B., R. Ramiére, C. de Tassigny, R. Eymery and L. de Nadaillac, "Application de techniques de polymerisation au traitement des bois gorgés d'eau," in: Applications of Nuclear Methods in the Field of Works of Art, proc. of the conference Rome-Venice, 24-29 May 1973, (Rome: Accademice Nazionale Dei Lincei, 1976) pp. 637-643.
6. de Nadaillac, L. and R. Cornuet, "Considérations sur les developpement de l'utilisation des techniques nucléaires dan la conservation des biens culturels," ibid., pp. 645-652.
7. de Tassigny, C., "The suitability of gamma radiation polymerisation for conservation treatment of large size waterlogged wood," in: Conservation of Waterlogged Wood, International Symposium on the Conservation of Large Objects of Waterlogged Wood, in: proc. of the symposium Amsterdam, 24-28 September, 1979, ed. Louis H. de Vries-Zuiderboan, (The Hague: Government Printing and Publishing Office, 1979), pp. 77-83.
8. Toracca, G., "Nuclear Conservation Processes Examined in the Light of General Requirements of Conservation," in: Applications of Nuclear Methods in the Field of Works of Art, proc. of the conference Rome-Venice, 24-29 May 1973, (Rome: Accademia Mazionale Dei Lincei, 1976), pp. 543-551.

Questions

Kirsten Jespersen: What alcohol did you use for dehydration?

E. Mavroyannakis: Methanol.

Jamie Barbour: What radiation dosage do you use?

E. Marroyannakis: About 2 megarad for vinyl acetate, 2 1/2 to 3 for methyl methacrylate.

Mary-Lou Florian: I understand that there is a problem in keeping a monomer in very degraded wood, and that it just flows out, as quickly as it goes in. Is this a problem?

E. Mavroyannakis: Yes, this is a problem and there are two others. The monomers are toxic, and the operation must be carried out in a controlled atmosphere. Secondly, the monomer evaporates during polymerisation due to the heat of polymerisation. Temperature must therefore be controlled, but this is difficult with the high dose rate. Lower dose rates can be employed, but the polymer produced has lower molecular weight and thus poorer mechanical properties. However to avoid monomer flowing out during polymerisation, a mixture of pre-polymer and monomer is used.

David Grattan: Can you tell me about the practicalities, and possible costs of the radiation polymerisation method? Do you foresee a time when this process will be cheaply available to the average archaeologist?

E. Mavroyannakis: There are problems in the sizes of pieces that can be treated, because of the limit imposed by the irradiation chamber. The maximum size treatable at present is 4-5 metres in length. The large pieces are also very difficult to impregnate because they need a very large immersion tank capable of holding prepolymer and monomer mixture under pressure. The main advantage is its speed in comparison to other methods of treatment. Substitution of water with the continuous method of Cornuet and de Tassigny takes two weeks, this is followed by impregnation and irradiation.
There are many facilities in many countries, which easily handle smaller objects.
We have a new facility under construction, about 7 metres long by 2 metres across.
Regarding cost, it is a rather cheap method of treatment.

Tom Stone: What temperatures do you achieve during polymerisation?

E. Mavroyannakis: For vinyl acetate, about 80°C. Terra Cottas crack under these conditions, because of the pressure developing in the interior, and the increasing viscosity of the monomer. These effects are not so harmful in wood.
For methyl methacrylate, higher temperatures can be used, since the boiling point is 145°C, whereas for vinyl acetate it is only 73°C.

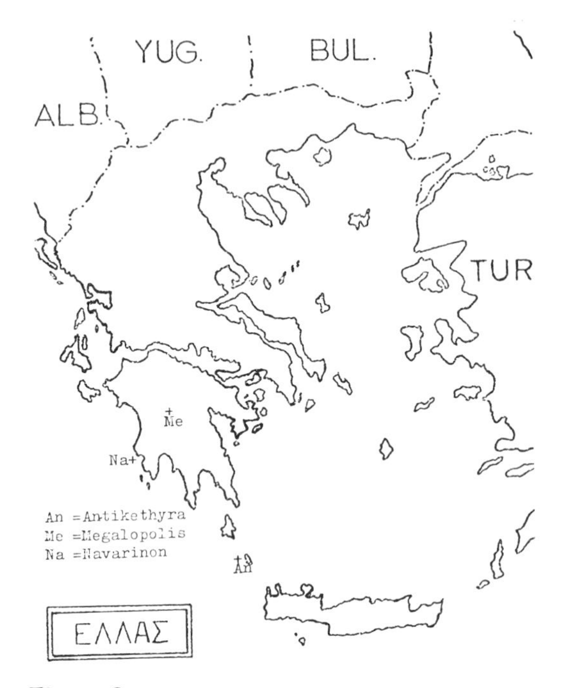

Figure 2

GENERAL DISCUSSION: PROGRESS IN THE TREATMENT OF WATERLOGGED WOOD: BIOCIDES

Introduction (ed.)

It was intended that there should be a general discussion on biocides, however what emerged from it was that we know very little about them. Some seem to work in some situations and not in others. Other workers found that no biocide at all was necessary, and it was generally agreed that this was the best solution. Of major concern was the toxicity of biocides. All those which employ some form of chlorinated phenol, will inevitably be contaminated by small amounts of dioxin, (an exceedingly toxic material which is also highly carcinogenic) irrespective of whether they are water soluble, i.e. the sodium salt form of the phenol, or water insoluble. Manufacturers do not always make this toxicity danger as clear or as prominent as it should be, furthermore health regulations vary from country to country. (When spraying with this type of material workers should have all skin surfaces covered and wear some kind of respiratory device, and it should be done outside in the open air.)

Another major problem was that we use manufacturers names in a very sloppy fashion. We are inclined to use the name "Dowicide" or "Cytox" without specifying which type. For instance there are at least eleven forms of Dowicide, most of which contain different chemical compounds. In reporting a biocide, a conservator should give (i) the full commercial name and (ii) the principal active components. (He or she should know this in any case for his or her own safety!).

Our knowledge of biocides has proven to be so inadequate that we tend to use the "next bottle off the shelf" approach, if something doesn't work we simply try the next available material without regard to the purpose for which it was intended for use by the manufacturer. An example was given of the mistaken use of a quaternary ammonium salt, which is a sterilant with little long term activity.

It became clear that we have a really serious problem in our use of biocides. Insufficient attention has been paid to them and clearly we must all put in some effort to rectify the situation.

To partially address this problem an extra paper was prepared by John Dawson, which follows on from the discussion.

Discussion

John Dawson: There was a question about Panacide, which is 1-4 dichloro phenol; according to Sax (Sax. N. Irving, Dangerous Properties of Industrial Materials 4th ed., (New York: Van Nostrand, 1975) p. 632), it has only moderate toxicity but this is based on limited animal experiments and the reader is then referred to the section on chlorinated phenols (p. 538), wherein he or she finds that they are rated as highly hazardous in every respect, and by every means of exposure.

Jim Spriggs: In the literature for Panacide, it is recommended for internal use in the treatment of worms.

Colin Pearson: It is dihydroxy, dichloro diphenyl methane, DDDM which has an LD_{50} = 1200 mg/kg.

John Dawson: That is not a very high LD_{50}. It would help if people using various biocides could specify them more exactly and describe their success. The presently available information is very sketchy. For instance, some people have good success within the standard 7/3 borax/boric acid mixture, others find it useless. Similar remarks have been made about quaternary ammonium compounds; which, according to the literature are supposed to be very good.

Barbara Purdy: We have been using Cytox, which seems to have been effective. If growth occurs we skim it off and add more Cytox. This seems to be effective, although to me it is the skimming off process which is most useful.

Suzanne Keene: The 7/3 borax/boric acid mixture was used on the water pipe I mentioned yesterday. It was totally ineffective in controlling fungus. We have recently begun to use Mystox, which is pentachlorophenyl laurate, a solvent soluble fungicide. A light spraying with it stops all of the growth.

John Dawson: Because of the toxicity of that material, the Canadian Department of Agriculture has banned it for interior use, because of the problem of dioxin micro-contaminants. It is sold for archival purposes. Even when using ortho phenyl phenol in a spray form, which is far safer than pentachlorophenol, gloves should be worn, i.e. skin protection is needed and a respirator should be used.

Richard Clarke: Storage of large quantities of pulp, as is done by the paper indusry, is a related problem. What does the paper industry use?

Jamie Barbour: They used to use mercury compounds, but they have now stopped. I know that they are having trouble now with slime in pulp and hence in finding a suitable fungicide.

Cliff McCawley: (to John Dawson) Can you say something about the activity of biocides, what affects it, and how long may the activity be expected to last? To give an example, in one lab that I worked in, wood was placed in a tank, a few handfuls of ortho phenyl phenol were thrown in and it was then left for four years.

John Dawson: I can't answer the question of longevity, some last for a long time and others don't. Other organic material in the water has been reported to affect the activity. The only way to find out is to test under specific circumstances.

Greg Young: I have gathered together a list of about 300 biocides. I think it would be useful to have a survey of these. By specifically defining the conditions of intended use, this can be reduced by a factor of ten, and then these 20 or 30 would have to be tested to see which works best for the application in question.

Victoria Jenssen: Many of us are using biocidal products without paying sufficient attention to what they are specifically intended for. During a consultation on our use of biocides to date, we were asked by Agriculture Canada why we had ever used them since they had a sterilant effect, and we were totally misguided if we thought that they would have a long lasting effect. This is an important point to consider in regard to the properties, the immediate effects versus the persistent activity.

Colin Pearson: I would like to propose the circulation of a questionnaire on the uses of biocides, which could perhaps be compiled by Grey Young?

Greg Young: In compiling this list of biocides; factors such as water temperature, oxygen content of the water, mineral content of the water, pH, volume would be important. It would also be useful to attempt to identify bacterial slimes and/or other infections.

Kirsten Jespersen: In sixteen years of working in this area, I have never used a biocide. Perhaps I should have my water tested? Does the stability come from the water, fish or the stainless steel tanks that I use? To me it seems that slimes tend to grow more in plastic tanks rather than stainless steel.

John Dawson: It is obviously much better to avoid biocides if you can, and many people such as Kirsten Jespersen, Barbara Purdy, and Robson Senior have successfully avoided using them. Another problem which should be considered are the possible problems arising from the interactions of biocides and materials used in treatment such as PEG, or acids, acetone rosin etc. There are other things which can be done to prevent growth. Gases such as nitrogen can be bubbled through to prevent the dissolution of oxygen. This would control the aerobic but not the anaerobic bacteria.

Jamie Barbour: But of course the aerobic are what are mainly present and the main problem; the anaerobic bacteria are not going to cause much of a problem. We have stored material in sealed jars containing distilled water, successfully without anything growing, although it caused the colour of the wood to be affected.

We also found when we were extracting wood with hot water at 50-60°C that the wood developed no fungus or bacteria. Only a little bit of algae. Although this would be a very expensive kind of approach.

John Dawson: There are several mentions in the literature to the heating of PEG impregnating solutions periodically, to about 40°C to get rid of any bacteria.

Roar Saeterhaug: In Trondheim, we have used the sodium salt of ortho phenyl phenol for wood storage. Now we have a circulating tank, which gives better results with no biocide at all.

John Dawson: It is important to be very precise and use the correct chemical name for the biocide.

What we have learned is that there is no single biocide which can be used in all cases, and it is better to avoid biocides.

SOME CONSIDERATIONS IN CHOOSING A BIOCIDE

John Dawson
Canadian Conservation Institute,
1030 Innes Road,
Ottawa, Ontario
K1A 0M8

Introduction

The discussion on biocides, revealed the need for a better understanding of biocides which are suitable for use in waterlogged wood storage. The following is an introduction to those which have found use, and explains some of the problems with them, and why it is difficult to make recommendations about biocides. This branch of the subject requires much greater attention than it has received hitherto; hopefully this review will be a start in the right direction.

General Considerations of Biocide Use

When terms such as fungicide, bactericide etc. are used it suggests some knowledge of the target organisms. These agents are collectively referred to as biocides. When dealing with the eradication of micro-organisms, the general term microbiocide has also been used (Baynes-Cope, 1975); Block however (1977) notes that although the term biocide encompasses all kinds of organisms, high and low from homo sapiens to bacteria, it is commonly used with reference to micro-organisms.

When referring to a biocide, particularly in the literature, the chemical name(s) should be used; if it was purchased under a trade/proprietary name, this should also be reported in full. Examination of **Table I**, reveals that four different compounds are referred to as a Dowicide (In fact Dow chemicals, the manufacturers, manufacture at least eleven under that general name). This can lead to dangerous confusion, particularly if they vary in toxicity. Even the tradename "Mystox" can refer to different formulations of different compounds. A trade name must be given in full for it to convey the vital information. A biocide should never be employed unless its chemical composition is known.

When considering the use of biocides, several factors need to be discussed:
(a) effectiveness, (b) resistance, (c) toxicity, and (d) compatibility.

(a) Effectiveness:

Often biocides tend to be more effective on some organisms than others. Before choosing a biocide, the target organism(s) should be identified as accurately as possible. The type of compound or concentrations suitable to control a bacterium may not be suitable to control an algae.

Furthermore, the effectiveness of a biocide depends on the conditions in which they are used. Factors determining effectiveness are normally given in the manufacturers data sheets and include temperature, pH and contaminants.

(b) Resistance:

Resistance refers to the "natural or genetic ability of an organism to tolerate the poisonous effects of a toxicant" (Ware, 1978). Since a biocide is a toxicant, this resistance is an extremely important consideration. Populations resistant to a biocide appear either as a result of selection of resistant individuals in organisms which pass their resistance on to future generations (this is normally the main course of resistance) or from a single gene mutation. This problem of developing resistant strains of micro-organisms, may occur, often necessitating a change in the type of biocide. The effectiveness of a biocide should therefore be assessed frequently. Identification of the micro-organism is important in making the choice of the subsequent biocide.

(c) Toxicity

Before using a biocide toxicity literature should be investigated. One should not only be aware of how poisonous a compound is, but also of the way(s) in which it acts on the body. Such information should be requested from the manufacturer, the appropriate government regulatory agencies or from toxicology sourcebooks such as Sax's (1975) "Dangerous Properties of Industrial Materials." Biocides should not pose health hazards to those applying them, or working on materials treated with them: Keep in mind that the sensitivity of individuals to them can be quite variable; even to biocides of low toxicity.

(d) Compatibility:

A biocide should be compatible with (i) the materials composing an artifact needing treatment; (ii) the materials employed for artifact storage; (iii) subsequent analyses or treatments e.g. the radiocarbon dating.

Sheppard (1976) reported that synthetic and

natural contaminants (e.g. from preservatives to organic materials in the soil) can affect radiocarbon dating unless physically or chemically separated from the artifact. Noting that the dates would be incorrect although still useful, he indicated that experiments would be needed to determine exactly the effect of carbon containing contaminants.

Another aspect of compatibility concerns that between biocides themselves. If it becomes necessary to change a biocide, as when for example a resistant micro-organism develops, the second biocide must be compatible with the first. This can only really be assessed by experimentation, but is a point to consider.

A Brief Review of Some of the Biocides used in Waterlogged Wood Conservation

A list of the major biocides that have received use in the storage of waterlogged wood is presented in **Table I**; for some of these trade names have been given.

What follows is a brief account of each biocide. For each type there is an enormous amount of literature, and in the space available, it is not possible to review this.

(a) Phenol:

Phenol or carbolic acid was one of the first materials used as a biocide. Its use is diminishing since it has been replaced gradually by phenolic compounds derived from it (Spaulding et al., 1977). It is still employed as biocide in the storage of waterlogged wood (Lucas, 1981) and this causes some concern. The acute oral LD_{50} (lethal dose 50) is 530 mg/kg in comparison to pentachlorophenol; LD_{50} = 54 - 140 mg/kg and orthophenyl phenol; LD_{50} = 21700 mg/kg (test animal - rats) and it is a serious local hazard as an irritant and a serious systemic hazard via inhalation, ingestion or skin absorption. Phenol has also been classed as a cocarcinogen; a compound which, although not carcinogenic itself, can increase the potency of carcinogens (Sax, 1975). Even if very low concentrations were employed, extreme caution would have to be exercised and any exposure avoided. In addition to the toxicity problems, phenolic compounds are considered to be corrosive to metals (Spaulding et al., 1977).

(b) Chlorophenols:

Numerous chlorophenols have been used in the storage of waterlogged wood (Stamm, 1970). However many are very toxic and their use should therefore be avoided. More recently, byproducts present in the technical grades of chlorophenols as microcontaminants (dibenzodioxins, dibenzofurans etc.), have been found to pose serious health hazards. In Canada this has led to a change in the regulatory status of chlorophenols (Agriculture Canada, 1980) which has meant:

(i) Suspension of chlorophenol products for use as wood preservatives for interior home use.
(ii) Suspension of products containing the chlorophenols and their sodium salts for use as slimecides in pulp and paper mill applications.
(iii) Suspension of all domestic class products for application by spraying.
(iv) No incorporation into material of which the end use will result in prolonged skin contact.

As noted for phenol, phenolic compounds are corrosive to metals and their use with composite artifacts must be avoided. Suther's (1975) reported that zinc plating on tank immersion heaters were attached when a 0.17 solution of sodium pentachlorophenate was present.

Clearly, the use of these compounds should be avoided.

(c) Orthophenyl Phenol:

Orthophenyl phenol and the sodium salt have been used for the storage of waterlogged wood and have received wide usage in conservation generally. A good discussion of their use has been given by Baynes-Cope (1975). Baynes-Cope found that orthohenyl phenol was compatible with PEG.

Compared to the chlorophenols and to phenol, these componds are much safer and are relatively non-irritating at the manufacturers specified concentrations for normal use. Even though orthophenyl phenol is only slightly hazardous as a skin irritant, precautions should be taken to avoid unnecessary exposure; particularly during the handling of the pure materials, or concentrated solutions.

Orthophenyl phenol is effective against a broad spectrum of fungi and bacteria (Manowitz and Sharpell, 1977), but is relatively ineffective against algae (Baynes-Cope, 1975). There have however been failures with it reported.

In the Somerset Levels Project, it was reported that sodium orthophenyl phenate was used in the waterlogged wood storage tanks but on occasion, it was not capable of preventing slime growth on the wood (Coles, 1979). In instances such as this, identification of the micro-organisms present causing the growth would be helpful. For this reason, and the fact that the biocide did not prevent the development of unpleasant odours, and the rotting of support bandages, its use was discontinued.

The unpleasant odours were probably caused by hydrogen sulphide resulting from the action of anaerobic bacteria. The sulphides produced could have reduced the effectiveness as described earlier herein.

(d) Thymol:

A few grams of thymol in a volume of 10-20 litres of water (the solubility of thymol in water is 1 gm/1000 ml Merck, 1968) was suggested by Muhlethaler (1973) to be a suitable 'fungicide' for the storage of waterlogged wood and that it would not interfere with the subsequent consolidants.

Thymol has been used for antiseptic and disinfectant purposes (Gump, 1979) and in the cosmetic industry, also as a preservative because of its strong anti-bacterial properties (Manowitz and Sharpell, 1977). In conservation, it has been frequently used as a biocide for library and archival materials; but it is limited in this respect because it is a good solvent for organic materials such as varnishes, paint, N-methoxymethyl nylon (Baynes-Copes, 1972). Reports that it can soften and discolour plexiglass have been confirmed by myself. Although this sort of behaviour is unlikely to cause problems in the storage of waterlogged wood, the question of thymol's compatibility with materials should always be considered.

The range of micro-organisms capable of being controlled by thymol is uncertain, especially for the little studied organisms which might be encountered in waterlogged wood. There are unpublished reports of species of fungi resistant to thymol. Cultures should be made with the specific organism(s) causing problems, to determine the effectiveness of the biocide.

Thymol poses a slight hazard as an irritant and an allergen, and as a moderate hazard by inhalation or ingestion. When heated it emits toxic fumes (Sax, 1975). Although generally considered to be a safe compound, exposure to thymol should be avoided.

(e) Tego:

A group of amphoteric surfactants have been produced commercially under the trade name Tego. Chemically they are made up of an amino acid, usually glycine, substituted with a long chain alkylamine group. The Tego compounds are reported to be bactericidal, fungicidal, viracidal, non-toxic, safe, effective in the presence of dirt, fat and protein, non-corrosive to metals and other materials. In Europe they have found extensive use in dairies, breweries, food plants etc.

Muhlethaler (1973) suggested a 1-2% solution of Tego 51B ($RND(CH_2)_2NH(CH_2)_2NHCH_2CO_2H$, R = C_8H_{17} to $C_{16}H_{33}$; 22.5% aqueous solution to pH 8) as a suitable fungicide for waterlogged wood. He reported that it did not affect the subsequent use of consolidants.

However as noted by Block (1977) and Gump (1979), the effectiveness of the Tego products are disputed because of contradictory reports on their activity. Furthermore, Tego 51 and Tego 103S have been reported to be corrosive to iron, copper and brass; in addition to these metals Tego 103S caused corrosion in zinc (Block, 1977). Specific information regarding the corrosiveness of Tego 51B is not available but its similarity to Tego 51 would indicate that its use on composite metal and wooden objects should be avoided until further data is available.

(f) Salicylanilides:

Ambrose (1971, 1976) reported adding a small amont of water soluble fungicide to waterlogged wood stored temporarily in a 10% aqueous solution of PEG 400. He used a 0.1% solution of sodium salicylanilide tetrahydrate (Shirlan Plus).

Salicylanilides are a group of fungicidal and antibacterial agents characterized by the presence of a grouping with the chemical formula:

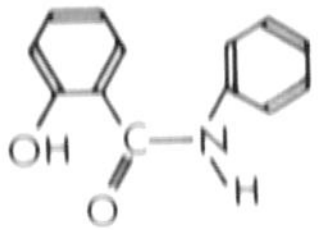

OH C—N H O

Salicylanilide itself is the condensation product of salicylic acid and aniline (Stecker, 1977). Although some of the substituted salicylanilides are comparatively safe to the user, salicylanilide itself, some of the halogenated derivatives and the sodium salts can cause skin irritation (Sax, 1975; Stecker, 1977). Caution should be exercised to prevent exposure of the skin to them.

(g) Quaternary Ammonium Componds:

The use of quaternary ammonium compounds has been discussed by Baynes-Cope (1975) in reference to their use in conjunction with orthophenyl phenol. As mentioned previously, the latter biocide was found to be of little effect in destroying algae. The compounds listed in **Table 1** are reported to be compatible with orthophenyl phenol. Baynes-Cope also noted the suggested use of cetyl pyridinium orthophenyl phenate as a microbiocide with wide applicability.

Quaternary ammonium compounds are in general, fungicidal, bactericidal, and algicidal at relatively low concentrations (several ppm to about 1000 ppm). They have received use in the sterilization of drinking water, the treatment of slime in paper mills, and of water in swimming pools etc. (Petrocci, 1977).

However, as a class, the quaternary ammonium compounds exhibit a realtively poor activity against gram negative bacteria, compared to that against gram positive species (Spaulding et al, 1977; Petrocci, 1977). Their activity can be reduced by the presence of soaps or other anionic surfactants as well as the hardness of the water.

The toxicity of quaternary ammonium compounds can vary from relatively harmless to very toxic. In general, concentrated solutions of 10% or more are toxic, causing severe irritation to the skin and to the conjunctional membranes (mucous membranes lining the eyelids and the front part of the eyes), and resulting in death if taken internally. Fortunately the concentrations at which they are used (effective-use-concentrations) are well below the concentration level causing serious toxic effects. However, even at low concentrations, some compounds may act as irritants and care should be taken to avoid unnecessary contact (Petrocci, 1977).

(h) Hypochlorites and Related Compounds:

Baynes-Cope (1975) discussed the possibility of using hypochlorites and related compounds (see **Table 1**) as microbiocides. From his preliminary studies, these compounds seemed promising. He felt that the small quantities required would be unlikely to harm the wood.

Chlorine and chlorine compounds are the most widely used drinking water disinfectants. Their action is associated with the hypochlorous acid formed and its subsequent dissociation. As with many biocides, these compounds are sensitive to temperature and pH. Their reactivity is increased by an increase in temperature. The dissociation of the hypochlorous acid is poor below pH 6, however above pH 8.5, chlorine has relatively little killing power. Furthermore most biocides suffer impairment in their activity in the presence of organic and inorganic material in the water and this can prevent an efficient biocidal action from occurring; this is very prevalent for chlorine compounds (Clarke and Hill, 1977; Dychdala, 1977). Gump (1979) goes as far as to suggest the removal of organic contaminants before the use of hypochlorites.

Blackshaw (1976) pointed out that lignin (an important component of wood) reacts with chlorine to form water soluble "lignin chloride" and therefore suggested that treatments with chlorine bleaches should be avoided. Moreover, all chlorine containing compounds are considered to be corrosive and to cause irritation to the skin (Clarke and Hill, 1977). Whether these problems would be apparent at the use-concentrations requires further investigation.

In conclusion, chlorine compounds generally exhibit control over a wide range of microorganisms. Various types of bacteria, fungi and algae however exhibit different resistances to hypochlorites under diverse practical conditions. Such selective resistance can often be compensated for by changes in the pH, temperature or concentration (Dychdala, 1977).

(i) Boric Acid: Borax

The 7:3 wt/wt mixture of boric acid and borax is a widely used biocide for the storage of waterlogged wood. It is frequently chosen because of its reported compatibility with PEG (Christensen, 1970; Barkman and Franzen, 1972; Muhlethaler, 1973). Muhlethaler describes the use of a 0.6-1.0% solution. Solutions of sodium borate alone, are strongly alkaline, and in spite of this alkalinity tend to make hardwoods slightly brittle. Neutral solutions containing sodium borate and boric acid together do not exhibit these detrimental effects (Barkman and Franzen, 1972).

Boron compounds are frequently used as wood preservatives and the borates combine fungicidal, insecticidal and flame retardant properties furthermore borates exhibit exceptional penetration (Richardson, 1978). Tests on the effectiveness of the 7:3 mixture and its penetrability have been carried out during its use on the Wasa (Barkman and Franzen, 1972). Boron concentrations of less than one percent in treated wood (calculated on the dry wt of the wood) have in general been found to be effective against insects and fungi; the exact concentration required is dependent on the species involved (Barkman and Franzen, 1972).

Borates are also reported to be extremely effective in the control of stain but relatively inefficient against superficial moulds such as Penicillium spp. and Trichoderma spp. (Richardson, 1978). However, Barkman (1976) stated that it did not control all fungi, and gave blue stain fungus as an example. Because most fungi are aerobic, they are less likely to be a major problem in submerged wooden objects (however since the Wasa is not submerged, this is not so and it is likely to be affected by aerobic bacteria).

Baynes-Cope (1975) has questioned whether penetration of the biocide is necessary in every case, particularly if the environment likely to be favourable to growth of the micro-organisms is confined to the surface of the wood. In most cases deep penetration is probably not necessary, however only identification of the type of organism causing the problem and its location in/on the wood will help to answer this. For the Wasa, where the biocide is used to prevent insect attack and attack by fungi, a penetration of the biocide below the surface might be advantageous.

Sax (1975) reports that boron compounds are not highly toxic and therefore not considered an industrial poison. However it is also noted that they represent a moderate hazard via ingestion and inhalation; although boric acid is not absorbed into intact skin, it can be absorbed via cuts, scrapes etc. and represents a health problem. Even though borax has been used as a skin cleanser, Sax (1975) recommends that the careless use of borax in this regard should be discouraged.

Grosso (1976) mentioned that the 7:3 mixture was effective "as far as its ability to kill things" but noted those working with it found it quite irritating to their respiratory systems and skin. Proper safety precautions (gloves, goggles, respirators etc.) should be taken during all stages of its use, particularly when it is concentrated.

(j) Copper Sulphate and Ammonium Carbonate:

Suthers (1975) reported the use of the fungicide, "Cheshunt Compound" in the storage of waterlogged wood; this product consists of copper sulphate and ammonium carbonate in a 2:11 wt/wt ratio.

Copper sulphate has had much usage as a biocide:

(i) As a water soluble wood preservative (Richardson, 1978).

(ii) In pulp mills in conjunction with chlorine at 1-2 ppm to control fungi not checked by chlorine alone (Block, 1977).

(iii) As the most common method of controlling algae (at a dosage of 0.5-2 mg/l) (Clarke and Hill, 1977).

There are difficulties however:

(i) Although possessing a high fungicidal activity, some fungi are resistant to copper preservatives (Richardson, 1978).

(ii) It is very corrosive to iron and steel (Richardson, 1978).

(iii) At concentrations of 2 mg/l or less, its algicidal effect appears to be selective rather than general (Clarke and Hill, 1977).

(iv) "Heavy metals other than mercury and silver exert only a feeble germicidal action on bacteria, " (Salle, 1977), copper is much more effective against algae and moulds than bacteria (Ware, 1978).

(v) Metallic salts can reduce the aging resistance of PEG (Muhlethaler, 1973).

(k) Fluralsil:

A 2% solution of Fluralsil BS has been reported to be a suitable 'fungicide' for storage of waterlogged wood (Muhlethaler, 1973). At the time of writing, the specific formulation of Fluralsil BS was not available, however Richardson (1978) reported that Fluralsil was a German product that consisted of a mixture of sodium fluorosilicate and zinc chloride, which was designed to precipitate zinc fluorosilicate in wood. The product was initially intended as a wood preservtive for the sterilization of Dry Rot (Serpula lacrymans).

The range of micro-organisms which this biocide can control is not known and would have to be investigated further as frequently the organisms causing most problems are not fungi. Furthermore Richardson (1978) commented, "The fluorosilicates have found only limited use in wood preservation as their relatively low activity (in reference to fungi) is unfortunately associated with low solubility and unacceptable corrosion."

The components of this biocide and the resultant precipitate are all considered to be very toxic, particularly via inhalation or as skin irritants (Sax, 1975). Perhaps at low concentrations, these problems would not be too serious but caution would still be required. Furthermore, in the presence of metallic salts, the aging resistance of PEG (if used) could be affected, as mentioned earlier.

(l) Formalin:

The potential health hazards posed by formalin (Sax, 1975; Spaulding et al., 1977) is such that its use in the storage of waterlogged wood should be avoided. Furthermore Muhlethaler (1973) reported that it accelerates the degradation of PEG.

(m) Cytox:

Cytox biocides are n-dodecyl guanidine salts (DGXs) and are potent bactericides, algicides, fungicides and virus inactivators (data on Cytox available from American Cyanamid Co.). The salts exhibit a low mammalian toxicity. Although the antimicrobial activity of the different DGXs is reported to be similar, they differ considerably in their physical properties. Since the DGXs have been successful for a wide range of purposes (domestic anti-mildew, textile rot proofing, algae control, etc.), they would appear to be ideal for waterlogged wood. However Grosso (1976) has mentioned an instance of failure of Cytox. The manufacturer's literature gives 30 ppm as the ceiling for the lethal dose for every strain tested. Accordingly, a concentration of 100 ppm was employed (Advice of U.S. Forest Products Laboratory, Madison, Wisconsin). This suddenly failed, and sulphate reducing bacteria developed. As a result the concentration was increased first to 300 and then to 500 ppm. This is an example of the development of a resistant or tolerant strain of bacteria.

It also must be taken into account that, the DGXs are cationic surface active agents which can form precipitates in conjunction with anionic substances; these might affect the activity adversely. High salt concentrations might result in salt exchange wth the DGX, itself an amine salt. This could also cause precipitation to take place (Could this have caused the above failure)? These compounds are also chlorine acceptors and although chlorine has little effect on the antimicrobial activity of DGXs, it will lose its own biocidal ability.

Disposal of DGXs can pose problems (as with most biocides), in large concentration they can interfere with oxygen exchange in the gills of fish, and should not be dumped at high levels in lakes or waterways.

Conclusions

Perhaps it is worth giving some positive advice on the approach to be taken before adopting a particular biocide. The conservator working on site or in a small laboratory may not have expert assistance, and those "experts" involved may not comprehend fully the unique problems of waterlogged wood conservation.

The first approach should avoid the use of a biocide altogether. Attempts should be made to stabilise the wood by techniques such as temperature control, agents such as snails, or perch etc. However if these methods fail or are unavailable, the use of biocides could be considered. The criteria for assessing a biocide have been outlined in the text, however it must be reiterated that the most important consideration in choosing one should be the toxicity towards those who might use it. Compounds such as othophenyl phenol or some of the quaternary ammonium compounds might be tried first since these are least toxic. However, no matter how safe a biocide is reported to be, exercise common sense and caution in its use.

References

Agriculture Canada, Changes in the Regulatory Status of the Chlorophenols, (Ottawa: Food Production and Inspection Branch, 1980).

Ambrose, W., "Freeze-Drying of Swamp Degraded Wood," in: Conservation of Wooden Objects: New York Conference on Conservation of Stone and Wooden Objects, preprints of the contributions, 7-13 June 1970, (New York: The International Institute for Conservation of Historic and Artistic Work, 2nd ed., 1971), pp. 53-58.

Ambrose, W.R., "Sublimination drying of degraded wet wood," in: Pacific Northwest Wet Site Wood Conservation Conference, ed. G.H. Grosso, 19-22 Sept. 1976, (Neah Bay: Washington), pp. 7-16.

Barkman, L. and Franzen, A., "The Wasa:

Preservation and Conservation," in: Underwater Archaeology, A Nascent Discipline, (Paris: UNESCO, 1972), pp. 231-242.

Baynes-Cope, A.D., "The choice of Biocides for Library and Archival Material," in: Biodeterioration of Materials, vol. 2, proc. of the Second International Biodeterioration symposium, Lunteren, the Netherlands, 13-18 Sept. 1971, ed. by A.H. Walters and E.H. Hueck-Van der Plas, (New York: John Wiley and Sons, 1972), pp. 381-387.

Baynes-Cope, A.d., "Fungicides and the Preservation of Waterlogged Wood," in: Problems of the Conservation of Waterlogged Wood, proc. of the symposium, National Maritime Museum, Greenwich, 5-6 October 1973, ed. W.A. Oddy, (London: National Maritime Museum, 1975), Maritime Monographs and Reports, no. 16, pp. 31-33.

Blackshaw, S.M., "Comments on the Examination and Treatment of Waterlogged Wood Based on Work Carried out During the Period 1972-1976 at the British Museum," in: Pacific Northwest Wet Site Wood Conservation Conference, 19-22 Sept. 1976, Neah Bay, Washington, ed. G.H. Grosso, vol. 1, pp. 17-34.

Block, S.S., "Amphoteric Surfactant Disinfectants," in: Disinfection and Sterilization, ed. S.S. Block, (Philadelphia: Lea and Febieger, 1977), ch. 18.

Christensen, B.B., The Conservation of Waterlogged Wood in the National Museum of Denmark, (Copenhagen: National Museum of Denmark, 1970).

Clarke, N.A. and Hill Jr., W.F., "Disinfection of Drinking Water, Swimming Pool Water and Treated Sewage Effluent," in: Disinfection and Sterilization, ed. S.S. Block, (Philadelphia: Lea and Febieger, 1977), ch. 35.

Coles, J.M., "Conservation of Wooden Artifacts from the Somerset Levels: 2," in: Somerset Levels Paper, no. 5, ed. J.M. Coles, (Cambridge: Dept. of Archaeology, Univ. of Cambridge, 1979).

Dychdala, G.R., "Chlorine and Chlorine Compounds," in: Disinfection and Sterilization, ed. S.S. Block, (Philadelphia: Lea and Febiger, 1977), ch. 10.

Grosso, G.H., Comments, Field Conservation Discussion Session, Pacific Northwest Wet Site Wood Conservation Conference, 19-22 Sept., 1976, Neah Bay, Washington, ed. G.H. Grosso, vol. 2, pp. 96-97.

Gump, W., "Disinfectants and Antiseptics," in: Kirk-Othmer Encyopedia of Chemical Technology, vol. 7, (New York: John Wiley and Sons, 1979), p. 793-832.

Lucas, Deidre, "On-Site Packing and Protection of Waterlogged Wood," this publication.

Manowitz, M., and Sharpell, F., "Preservation of Cosmetics," in: Disinfection and Sterilization, ed. by S.S. Block, (Philadelphia: Lea and Febiger, 1977), ch. 39.

The Merck Index, An Encyclopedia of Chemicals and Drugs, Eighth Edition, (Rahway (N.J.): Merck and Co. Inc., 1968), p. 1050.

Muhlethaler, B., Conservation of Waterlogged Wood and Wet Leather, (Paris: Editions Eyrolles, 1973).

Petrocci, M.S., "Quaternary Ammonium Compounds," in: Disinfection and Sterilization, ed. S.S. Block, (Philadelphia: Lea and Febiger, 1977), ch. 17.

Richardson, B.A., Wood Preservation, (Lancaster: The Construction Press, 1978), ch. 4.

Salle, A.J., "Heavy Metals Other Than Mercury and Silver," in: Disinfection, Sterilization and Preservation, ed. by S.S. Block, (Philadelphia: Lea and Febiger, 1977), ch. 408.

Sax, N.I., Dangerous Properties of Industrial Materials, (New York: Van Nostrand Reinhold Co., 1975).

Sheppard, J.C., "Influence of Contaminants and Preservatives on Radiocarbon Dating," in: Pacific Northwest Wet Site Wood Conservation Conference, 19-22 Sept. 1976, Neah Bay, Washington, ed. G.H. Grosso, vol. 1, pp. 91-96.

Spaulding, E.H., Cundy, K.R. and Turner, F.J., "Chemical Disinfection of Medical and Surgical Materials," in: Disinfection and Sterilization, ed. S.S. Block, (Philadelphia; Lea and Febiger, 1977), ch. 33.

Stamm, A.J., "Wood Deterioration and Its Prevention," in: Conservation of Wooden Objects: Preprints of the Contributions to the New York Conference on Conservation of Stone and Wooden Objects, proc. 7-13 June, 1970, (New York: The International Institute for Conservation of Historic and Artistic Work, 2nd edition, 1971), pp. 1-11.

Stecker, H.C., "The Salicylanilides and Carbanilides," in: Disinfection and Sterilization, ed. S.S. Block, (Philadelphia: Lea and Febiger, 1977), ch. 14.

Suthers, T., "The Treatment of Waterlogged

Oak Timbers from Ferriby Boat III Using PEG," in: Problems of the Conservation of Waterlogged Wood, proc. of the symposium, National Maritime Museum, Greenwich, 5-6 Oct., 1973, ed. W.A. Oddy, Maritime Monographs and Reports, no. 16, 1975, pp. 115-119.

Ware, G.W., The Pesticide Book, (San Francisco, W.H. Freeman and Co., 1978).

TABLE 1

Biocides

	Chemical Name	Trade Name or Common Name
(a)	Phenol (Carbolic Acid)	
(b)	Chlorophenols	
	(i) Pentachlorophenol	Penta Santobrite, Dowicide 7
	(ii) Sodium pentachlorophenate	Dowicide G
	(iii) Pentachlorophenol laurate	Mystox (Mystox LPL 1007)
(c)	Ortho phenyl phenol	
	(ortho-phenylphenol	Dowicide 1, Topane S
	Sodium orthophenyl phenate	Dowicide A, Topane WS
(d)	Thymol	
	Thymol, 3-p-cymenol	
(e)	Tego	
	R $NH(CH_2)_2$ $NH(CH_2)_2$ $NHCH_2$ CO_2 H	Tego 51B (22.5% aqueous solution)
	R = $C_8 H_{17}$ to $C_{16} H_{33}$	
(f)	Salicylanilides	
	Sodium salicylanilide tetrahydrate	Shirlan Plus
(g)	Quaternary Ammonium Compounds	
	(i) Cetyl pyridinium chloride	
	(ii) Cetyl trimethyl ammonium bromide	Cetab
	(iii) Specific formula not available	Saequartyl B
	(iv) Specific formula not available	Benzalkon B
	(v) Cetyl pyridinium orthophenyl phenate	
(h)	Hypochlorites and Related Compounds	
	(i) Calcium hypochlite	Bleaching Powder
	(ii) p-carboxy benene sulphone dichloramide	Sterotabs
	(iii) para-toluene sulphon chloramide mono sodium salt	Chloramine T
(i)	Boric Acid: Sodium borate (7:3)	
(j)	Copper sulphate, ammonium carbonate (2:11)	Cheshunt Compound
(k)	Sodium fluorosilicate, Zinc Chloride	Fluralsil
(l)	Formaldehyde Solution (A solution of about 37% by wt of formaldehyde gas in water.)	Formalin
(m)	n-Dodecyl guanidine salts	Cytox (2013,2160,2240,2340)

THE CHEMISTRY OF POLYETHYLENE GLYCOL

Dr. Allen Brownstein
Union Carbide Company
New Jersey, New York

An informal discussion chaired by Dr. Colin Pearson.

Colin Pearson: We tend to forget that polyethylene glycol (PEG) came into conservation quite a few years ago. For example, the Wasa began treatment some twenty years ago. However, like many other things that are not perfect, PEG went out of favour and people started to look for new alternatives. But it is interesting to see how it never disappeared, and we find that whatever processes we seem to use these days, PEG is still incorporated into waterlogged wood at some stage or other. Therefore, to understand its chemistry is obviously very important. So I will open the discussion and call on Dr. Brownstein.

Allen Brownstein: Thank you for the invitation to attend this conference.

I feel the best way to conduct this session is to answer any questions you might have about the chemical and physical properties of PEG, and of any concerns you have about its use.

Gregory Young: I would be interested to know more aobut the size of the PEG molecule and the size when hydrated.

Allen Brownstein: The size of a PEG molecule in aqueous solution is difficult to predict with complete certainty. PEGs are not rigid chain polymers for which one can easily calculate the end-to-end distance from the C-C and C-O bond lengths and bond angles, and the average number of each bond in the polymer chain. On the contrary, PEGs are flexible polymers capable of random coiling with numerous folds and twists throughout the chain. The extent to which PEGs fold and twist, of course, increases with increasing molecular weight as the chain becomes longer. For flexible polymers, such as PEG, the average end-to-end distance is usually proportional to the square root of the molecular weight, resulting in unusually compact molecules. An excellent example is the reported average end-to-end distance of 11.0-12.0 nm for a PEG 20,000 molecule in aqueous solution. If the polymer were rigid a distance of 100 nm would be more appropriate.

Another factor to consider when assessing the size and mobility of the polymer in aqueous solution is complexation with water molecules, which is, of course, why PEGs are water soluble. That is to say, water solubility results from hydrogen bonding of water molecules to the electron-rich oxygen atoms in the polymer chain. Structurally speaking, PEGs are polyether-diols possessing two terminal hydroxyl groups and many alternating ether linkages. PEG 200, for example, has considerable diol character that contributes significantly to the water solubility, whereas, high molecular weight PEGs possess largely polyether character and would not be water soluble if it were not for hydrogen bonding of water to the ether oxygen atoms. Since each bound water molecule is capable of hydrogen bonding to another, it is reasonable to suspect a rather large solvent shell as opposed to a monolayer of associated water molecules. In fact, an average of three water molecules per ether oxygen has been reported, based solely on interpretation of viscosity data. Further evidence of complexation between water and PEG is the inverse solubility characteristics of PEGs in aqueous solution That is, at temperatures slightly above 100°C, PEGs become insoluble in water as a consequence of rupturing the hydrogen bonds involved in complexation.

Gregory Young: Is there any logic in assuming that the size of the hydration complex will be different for different solutions? With Butanol for example.

Allen Brownstein: Yes, there is.

Gregory Young: So some could be larger than others?

Allen Brownstein: Yes.

David Grattan: What about methanol? What differences would you expect between methanol, butanol and water?

Allen Brownstein: Studies have shown that the binding of cations to PEG is considerably greater in methanol than in aqueous solution. This would appear to indicate that fewer methanol molecules are bound, leaving available binding sites for cations to occupy. From these studies it may be concluded that the PEG-water complex is larger due to the greater number of bound water molecules.

The PEG/t-butanol complex, in comparison, may be the smallest of the three due to the steric bulk of the t-butyl group which can severely limit the number of bound t-butanol molecules.

Mary-Lou Florian: In talking about complexing, could you say something about phenol, borax and boric acid.

Allen Brownstein: To the best of my knowledge, association reactions of boric acid or borax with PEG have not been observed. There has been, however, a number of reports confirming the formation of PEG-phenol association complexes. The mechanism of phenol complexation appears to be similar to that of water, that is, association through hydrogen bond formation. In several instances the complex has been isolated as an oily, or sometimes solid precipitate which is insoluble in water. The water-insoluble nature of the complex is undoubtedly due to the attachment of phenyl groups which impart sufficient hydrophobicity to render the complex water-insoluble. The pharmaceutical industry is particularly cognizant of PEG-phenol complexes since the effectiveness of drugs containing phenol groups may be greatly diminished in the presence of PEG. As many of you are aware, PEGs are used extensively as bases for ointments and suppositories. Another well known PEG association complex is that with polyacrylic acid. In this case, however, complex formation is pH dependent. Above pH 3.5, where a substantial number of carboxyl groups are ionized, no complex is formed. This observation is again consistent with the view that complexation occurs via hydrogen bond formation, except in this instance, the hydrogen atom of a carboxyl group is involved, hence the extreme pH dependence.

It would not be appropriate to conclude a discussion on PEG assocation reactions without mention of a very different type of complex that has been of commercial value for years and may hold some promise in the conservation of waterlogged wood. That is, namely, the inclusion or clathrate complex formed between urea and PEG. The urea crystal has parallel channels or capillaries into which PEG molecules will lodge. The product formed is a hard solid, much harder than PEG, that melts at 133°C. Perhaps, addition of urea to a PEG bath may serve to further strengthen the degraded wood through complex formation within the interior of the wood.

Gregory Young: Does the molecular weight of PEG make a difference?

Allen Brownstein: There is no lower molecular weight limit that can be used. For example, PEGs of molecular weight 200 to 8,000 have been used successfully.

Colin Pearson: Is the degradation or breakdown of PEG related to pH?

Allen Brownstein: To a small extent.

Colin Pearson: We have talked alrady this week of pH levels of around 3.5 in peat bogs. Can we use PEG at these pH levels?

Allen Brownstein: PEGs exhibit excellent stability at low pH under moderate conditions. In one study, 5% hydrochloric acid was added to PEG 1000 and the mixture was heated to 80°C under nitrogen for several hours. Reduced viscosity measurements were recorded and showed little or no degradation had occurred.

Colin Pearson: Is heat more important than pH?

Allen Brownstein: PEGs are susceptible to thermal degradation which is to some extent pH dependent. That is to say, thermal degradation can be accelerated or the temperature at which degradation occurs can be lowered at low or high pH levels.

Colin Pearson: What about the long term degradation of PEG?

Mary-Lou Florian, has just mentioned to me that nobody seems to have studied the long term effects of treating a soft, degraded piece of wood with PEG. What is going to happen in 50 years time? Because of the hygroscopic nature of PEG is it going to attract moisture into the wood which could, over the long term, result in the breakdown of the wood components.

Allen Brownstein: No chemical is stable indefinitely. Long term degradation of PEG will occur. Degradation may be accelerated by such factors as heat, air, moisture content, and even perhaps, the fluorescent lights used for display purposes. Certain decomposition products will be corrosive toward badly degraded wood. formic acid, for example, a known PEG decomposition product, will have an adverse effect. Yes, I believe this is a legitimate concern; PEG treatments may not be the perfect solution to a difficult problem.

Per Hoffmann: How is the oxidative degradation process dependent upon temperature?

Allen Brownstein: The process through which PEGs undergo thermal-oxidative degradation is quite complex and not fully understood. If oxygen is completely excluded PEGs can withstand temperatures of 250°C before thermal degradation occurs. An excellent example is the use of PEGs in gas chromatography columns. These columns are usually heated to temperatures approaching 250°C. In order to obtain good analytical results, it is crucial that PEG decomposition products are not formed, since these products will interfere with the analysis of the sample. A carrier gas, usually helium or nitrogen, is used to sweep the sample through the column. Since oxygen is completely absent from the column, PEG degradation does not occur. The situation changes dramatically, however, in the presence of oxygen. A free radical site on the PEG chain, caused by loss of a hydrogen atom from carbon, can react with molecular oxygen to generate a peroxy radical. The resulting peroxy radical abstracts another hydrogen atom from carbon to give a hydroperoxide which, in turn, is very unstable and at relatively low temperatures causes chain scission via several possible decomposition pathways. The decomposition products are typically smaller molecular weight fragments and one and two carbon acids, aldehydes and esters. Decomposition can be detected visually by a darkening in colour, or by a drop in solution viscosity and pH. The use of antioxidants will help to prevent oxidative degradation.

Mary-Lou Lorian: Is there any chance of ethylene glycol forming?

Allen Brownstein: I have not seen any reports indicating ethylene glycol to be a decomposition product of PEG. It may be present in trace amounts, however.

Per Hoffmann: I would just like to clarify that the degradation of PEG does not mean the cleavage of long chains into two short ones. Is it rather a "peeling-off" reaction from one end of the chain producing all these species that you mentioned?

Allen Brownstein: Actually, degradation of PEG occurs by random chain scission. This explains why a measurable drop in viscosity results from a very small amount of degradation.

Per Hoffmann: I would like you to comment on what, from your earlier comments, would seem to be a problematic situation. In a technical installation for large scale PEG impregnation the circulation from a PEG tank is fed through an evaporator unit to evaporate water continuously and achieve the required PEG concentration. I don't know at what temperature the evaporator will be run, but air will be passed through the PEG solution. What is the best approach for this type of situation to avoid degradation of the PEG?

Allen Brownstein: They should use nitrogen rather than air and evaporate at reduced pressures, if possible.

Jamie Barbour: Would it help if they used some of these anti-oxidants which are free-radical scavengers? Kodak makes one that is water soluble.

Allen Brownstein: Yes.

Per Hoffmann: Could you recommend any?

Allen Brownstein: Is this a solution, and at what concentration?

Per Hoffmann: A solution in dilute alcohol starting at 10% and getting up to 60% PEG.

Allen Brownstein: I would suggest additives such as p-methoxyphenol or phenothiazine, or, perhaps, a safer food grade antioxidant such as BHA, BHT, or propyl gallate. ((BHA) Butylated hydroxy anisole; (BHT) Butylated hydroxy toluene).

Jamie Barbour: Isn't isopropyl alcohol also effective?

Allen Brownstein: Yes, in some applications, it has been effective, however, it is much too volatile and flammable.

Suzanne Keene: In what concentration would you need to use the anti-oxidants?

Allen Brownstein: The amount varies depending on the severity of the conditions, but usually concentrations of 0.05% to 0.10% are adequate. Under severe conditions, as much as 1-2% can be used.

David Grattan: Can I ask a question about BHT. My understanding is that it is generally considered to be slightly poor as an anti-oxidant because it is a rather small molecule and tends to be slightly volatile, or sometimes even because it is slightly water soluble, when you don't want the latter property in many applications. Is this actually test data or is it just based on suppositon? Because of the solution work I am aware of, it has been shown to be a very effective anti-oxidant, in fact one of the most effective.

Allen Brownstein: The effectiveness of various antioxidants is dependent, of course, on the operating conditions used. I believe BHT is ineffective at temperatures above 150°C due to its volatility. Test data can be obtained from the major producers such as Shell and Eastman-Kodak.

David Grattan: I have another question about the hydroperoxide mechanism. You suggested earlier that it might be a good idea to carry out hydroperoxide or peroxide determinations. It occurred to me that, if for instance, you had PEG at 60°C, it would rapidly destroy the hydroperoxide and tend to make the concentration rather small, because you would accelerate the rate of decomposition. Even though the hydroperoxide was playing an important part in the degradation mechanism.

Allen Brownstein: Studies in our laboratories and others have shown that peroxide levels, initially determined at room temperature, drop to an undetectable level after heating.

David Grattan: So, if you have a tank of hot PEG solution, hydroperoxide is not going to be a good monitor of what is going on. Could you suggest other species?

Allen Brownstein: PEG degradation can be monitored by determining carbonyl concentration. As I mentioned previously, the decomposition products are carbonyl containing compounds such as aldehydes. Alternatively, total acidity can be monitored since organic acids are formed upon PEG degradation. The latter method, however, is often unreliable since the acids formed may be esterified by the hydroxy-containing species present in the mixture causing the initial downward shift in pH to drift upward.

Solution viscosity measurements is perhaps the most convenient and reliable method to use since degradation produces an appreciable concentration of smaller molecular weight fragments that will cause a lowering of solution viscosity.

Colin Pearson: As oxygen and ultraviolet radiation are so important, is there any point in trying to exclude them after treatment by lacquering or lowering U.V. illumination, that is, to exclude oxygen and ultraviolet. After all, some people have said that they use surface finishes after treatment to consolidate the surface, could they not also incorporate a U.V. absorber?

Allen Brownstein: That is certainly possible.

Colin Pearson: How would they behave over the long term? Would it be worth it?

Allen Brownstein: My concern is that U.V. absorbers may photosensitize a degradation process, particularly in the presence of oxygen.

Jamie Barbour: As far as wood is concerned, you wouldn't want to allow much U.V. to get to it; since it would degrade lignin, and that's why wood turns so white when it's outside.

Cliff McCawley: Could you say something about the effect of metal salts on the degradation of PEG. I think that at the Amsterdam conference they began to ask questions about the effect of metal salts, and recently de Jong wrote to me about the same thing. I do know that Robert Organ, some years ago, told me that he had observed that when he had used metal tanks, particularly copper, for PEG solutions, a dark brown residue was left in the tank. Unfortunately, he couldn't say whether this was from PEG degradation or from the phenol that had been used in the storage tanks.

Lorne Murdock: The Union Carbide literature does not recommend the use of copper plumbing or copper built tanks. Christensen tin plated his copper parts just for that reason.

Allen Brownstein: Trace metal ions may accelerate the decomposition of PEG hydroperoxides. Ferrous, ferric and cupric salts have been cited.

Richard Jagels: Is the formation of the various breakdown products, such as aldehydes, the reason why PEG gives biocidal protection. In a stable form do you have any data as to its biocidal properties?

Allen Brownstein: PEGs are not effective biocides although they do not support bacterial growth. Any biocidal activity exhibited by PEGs is believed to result from competition with bacteria for available water sites. In high concentration, PEGs may dehydrate or even rupture bacteria cells by causing a drastic change in osmotic pressure.

Per Hoffmann: Does the degree of polymerisation have any influence on the ease with which the chains are degraded?

Allen Brownstein: No, not until extremely high molecular weights of 100,000 to several million are reached. Aqueous solutions of these polymers are subject to shear stability problems. Thermal and oxidative stability, however, is generally independent of the degree of polymerization.

Richard Clarke: When we use large amounts of PEG for treating waterlogged wood, for reasons of economy, we tend to re-use it. If, as you say, it is breaking down, is the breakdown complete or only partial? Or are we using say PEG 900 or lower, when we started with 1500? To what extent can it be safely re-used?

Allen Brownstein: Again, solution viscosity measuements may be reasonably informative. However, if more quantitative results are needed, gel permeation chromatography (GPC) may be considered. GPC gives a graphic picture of the molecular weight distribution of polymers. Appropriate standards are used to quantitate the distribution curves. Chromatograms taken before and after PEG treatments will reflect changes in the molecular weight distribution and determine if the material is suitable for reuse.

Many of you, most likely, do not have access to GPC, in which case, published procedures are available for determining average molecular weight by wet chemical methods.

I think an interesting experiment would be to see if GPC will aid in determining absorption rates of various molecular weight PEGs into different woods, or, when one particular molecular weight range PEG is used, if selective or preferential absorption occurs. The latter would be seen as a narrowing of the molecular weight distribution curve from one end. It would not be surprising if the curve narrows from the low molecular weight side, indicting a faster absorption rate for low molecular weight PEGs. A GPC experiment on CARBOWAX® PEG 540 Blend (an equal mixture of molecular weights 300 and 1450) may show preferential or even sequential absorption starting with the band centered at 300 M.W.

Richard Clarke: We do our impregnation treatments at 60°C and when you put the wood into the PEG solution, it quickly turns black. However we know that there is much sulphide present from bacterial action. What is the reaction between PEG and sulphide? I am wondering if the discoloration could be caused by the deposition of PEG/sulphide in the wood. Would that be possible?

Allen Brownstein: The hot PEG solution is probably leaching the sulfides from the wood. Sodium sulfide is known to discolor upon exposure to air and light to first a yellow and then to a black color.

Lorne Murdock: There are also tannates and gallates emanating from the wood as well.

Allen Brownstein: Tannins and gallates are probably washed out as well. Both are known to darken considerably upon exposure to air and light.

Barbara Purdy: If the artifact had been shaped in a charcoal fire, like much of our material, would charcoal have any effect?

David Grattan: The mention of gallates brings to mind something that I have often wondered about, that is, does the lignin in the wood act as an anti-oxidant. After all it is a polyphenol. Is it possible that you have a self stabilizing system?

Allen Brownstein: That may be possible although the tannins and gallates present should also contribute antioxidant properties. Propyl gallate, for example, is a commonly used food grade antioxidant. I would think hot aqueous PEG solutions will be particularly effective at leaching out these products since they are phenolic species and, as such, should be capable of forming association complexes with PEG.

David Grattan: This possibly answers the question raised earlier by Mary-Lou Florian, as to whether lignin associates with PEG.

Mary-Lou Florian: My question is that the main purpose of using PEG was to stabilize normal wood, but the chemistry depends on its association with cellulose in the wood. Now we are dealing with wood which has lost some or all of the cellulose.

David Grattan: You are saying that PEG associates strongly with phenol. Well lignin is a polyphenol, and on that basis could PEG not associate strongly with lignin?

Allen Brownstein: Lignin sulfonates and alkali lignin form association complexes with polyethylene oxides (PEO), which are simply high molecular weight PEGs. These lignin derivatives are by-products of the papermaking process and are present in wood pulp liquors. The paper industry uses PEO to remove lignin derivatives from wood pulp. Usually, the lignins are sulfonated to form lignosulfonic acids which are removed by flocculation or precipitation upon addition of PEO. Although the sulfonic acid groups are believed to take part in complexation with PEO, I would be very surprised if naturally occurring lignins did not form association complexes with PEG. But I'm curious; would association hinder or help the conservation process? After all, the mobility of PEG within the wood may be slowed down.

David Grattan: I think that association between the substrate and PEG is exactly what we are seeking.

Richard Clarke: We have been talking a lot this week about the penetration of PEG into wood. If the PEG complexes with cations and is deposited, can this deposited material act as a barrier to the penetration of fresh PEG, and could this be the reason you quite often don't get penetration, especially with the problems of size, cell condition, etc.?

Allen Brownstein: I think many factors are involved. I've heard mention this week of prewash treatments using EDTA. Do they seem to help?

Richard Clarke: I don't think it really matters how much you wash, the ions are really deep into the wood and the wood will be saturated with them. I wonder could you be getting penetration into the cell spaces and deposition of PEG itself causing a barrier?

Colin Pearson: Nobody seems to be doing much pretreatment of wood now. At one stage people used to use the acid pretreatment for iron contaminated samples, but they don't appear to be bothering now. Is this because the iron acts to consolidate the wood anyway.

David Grattan: There are problems with the acid pretreatment. It is impossible to restore the wood to neutrality by washing alone.

Howard Murray: The acid treatment bleaches the wood, but I would definitely never do a PEG treatment without EDTA or acid pretreatment. I don't think you get proper impregnation unless you do. In addition to getting back to the original colour of the wood, it also cleans out much of the insoluble material. We have examples which have an iron stain and fresh wood, and if you don't remove the iron stain the wood just cracks right open.

Colin Pearson: Is it the disodium salt of EDTA that you use?

Howard Murray: Yes.

Suzanne Keene: Have you observed how far the EDTA penetrates?

Howard Murray: No. I am not particularly interested in the interior of the wood; as long as the surface is good from a "viewing" point of view, that's all I really want. I have no inclination to cut open a priceless Tudor artifact, to see if the interior is damaged or not. As long as the surface is alright I am satisfied.

Jacqui Watson: I have found with the surfaces of many of my artifacts that they are actually becoming minerally replaced with some of these iron salts, and if I start taking them out, after treatment a lot of surface detail is lost.

David Grattan: I am planning some experiments to cure up some PEGs with isocyanates. A little of this has been done before for treated waterlogged wood, although nothing extensive, and there are no reports in our literature. I wonder if you could give me your thoughts on the subject?

Allen Brownstein: That's an interesting idea. PEGs are used as the polyol component for the preparation of polyurethanes. Perhaps you might consider experimenting with trifunctional polyols to introduce crosslinks that will provide a very hard and durable finish.

David Grattan: Thats what I mean. As we heard the other day, the problem with freeze-drying is that, yes, you can prevent shrinkage and collapse, but what you end up with, if you have a very degraded piece of wood is something very fragile. And as Jim Tuck said the other day, what he needs is an artifact as hard as a cribbage board. PEG cured by isocyanates would be one way of producing such a hard surface.

Suzanne Keene: We try and avoid that kind of

treatment because of the lack of any sort of reversibility.

Allen Brownstein: I want to take a moment to respond to earlier discussions on problems concerning the penetration of PEG into wood and on methods to accelerate the process. I would like to suggest the use of nonionic surfactants to improve the wetting properties of the PEG solution. Wetting refers to the ability of an aqueous system to spread over and subsequently penetrate into a solid substrate. Poor wetting results when a high interfacial tension exists between the substrate and the liquid phase. Surfactants are substances designed to lower these interfacial tension forces and in turn accelerate penetration. Of the many different types of surfactants available, I believe nonionic surfactants will be particularly well suited for your application. They provide excellent wetting properties at concentrations of less than 1%, and have a PEG-like structure, the only difference being that one end of the PEG chain bears a hydrocarbon group.

David Grattan: In the recent literature, there was an article in Nature which suggested that there was an interaction between sugars and PEG. Do you know anything about this?

Allen Brownstein: I've not seen that paper.

David Grattan: It was suggested there could be some kind of liquid crystal formation between sucrose and PEG and diffusion rates/-interdiffusion rates betwen PEG and sugars were very much higher than predicted. And it seems to me that this could quite possibly have an application for the interaction of PEG and cellulosic materials.

Allen Brownstein: Sometimes polymer/polymer or polymer/solvent association reactions can create an unusually large concentration gradient that will increase the diffusion rate of one or more of the components in the system. Once equilibrium is reached, however, diffusion rates should return to normal.

End of discussion and vote of thanks to Dr. Allen Brownstein.

*1. Preston, B.N., Laurent, T.C, Comper, W.D. and Checkley, G.T., "Rapid Polymer Transport in concentrated solutions through the formation of ordered structures," Nature, vol. 287, 9th October 1980, pp. 499-503.

APPENDIX I

This was the address given at the banquet, and is included here because of an overwhelming number of requests!

A SERMON FROM THE REVEREND BARCLAY

I was approached by two of my parishioners while I was standing at the church last Sunday and asked if I would come here today and address this waterlogged banquet.

It seemed to me appropriate to offer you a sermon today on the subject of consolation, standing as you do at the crossroads of your understanding of a complex problem. For it is true to say that into every life a little rain must fall - sooner or later in all our lives there must come upon us all a Mary Rose, or a Hamilton, or a Scourge. In effect, any theoretical square peg which does not fit into a practical round hole. A seemingly insurmountable problem then, which, while it blocks the light of the future, sorely taxes us.

How do we find consolation in time of woe? There are those who find it in their daily intercourse with their colleagues. There are indeed those who find solace in introspection. And, verily I say to you, there is even now one amongst us who finds consolation in the bottle.

But believe me when I say to you that much consolation may be found here in the scriptures. Both in the Old Testament and The New may be found copious references to the preservation of waterlogged wood - have we but the wit to find them. (and the scholarship to re-translate faulty modern texts.) Why, in the very opening phrases of the Gospel According Unto Saint John (John 1, Verse 1) are to be found the following words:

> "In the beginning was the wood, and the wood was with water, and the wood was wet. The same wood was in the beginning with water; all things were wetted with it and without it was wetted nothing that was wetted. In it was water and the water was the water of wet wood. And the wood was dug up and brought amongst us and we saw the excavating of it, and the excavating of it was in the beginning."

So you see we may indeed seek deeper and find comfort. Here are the words of Job (Chapter 25, Verse 4):

> "Then spake Zadock the Priest (I should point out a pronounciation error here - the "Z" is better pronounced as "M". Thus it should read Murdock the Priest.) So - "Then spake Murdock the Priest: though thy wood be all logged with water, though thy cellulose be all leached away, though thy lignin remain and thy wood be therefore like unto black cottage cheese; even so shall it all be made straight in thine own time, and thy moisture shall be dried up like as an well in the wilderness."

And it says further:

> "There shall be no checking, nor splitting, nor cracking, neither shall it collapse. It's colour shall be neither too light, nor too dark: there shall be no blackening nor bleaching: and the colour will be neither to the one nor to the other. Neither shall it snap, nor scrunch and neither shall it squish or be yuk. And it will be seen to be good."

And in York: thy Spriggs shall bring forth fruit; and they shall ride through thy place in chariots; and their eyes shall be amazed.

And the New Testament, echoing the Old, brings forth comfort where timbers must be recorded in situ: (Luke 8, Verse 13):

> "Let he who is without sin cast the first boat. (Or oven. The translators disagree on this point.) And though the pieces of his mould be legion, and the plans thereof cast away, verily I say unto you his mould shall be again made whole, and he diggest not his own gravery."

And Studies is witness to these things.

And here again from the New Testament - once more Luke, (Chapter 2, Verses 8-10):

> "And there were in that same country archaeologists grubbing in the fields, keeping watch over their finds by night. And lo, the angel of the Lord stood hard by them and saith unto them: "Hi, though your moisture content be low.......Lo, though your moisture content be high I bring good tidings of great joy which shall be unto all, for

there shall come unto you a Saviour which is called PEG."

Was it not our good Lord who said: "There was one came before me who baptised with alum, but I baptise with 540 blend."

So then, in the scriptures we find the words of wisdom in which we might find solace. So let me conclude today by offering a prayer in which, I am sure, we can all find much food for thought. I refer, of course, to the Psalms of David (No. 23) and I would like you all to imagine the sound of a softly murmuring Organ and repeat after me:

"Wet wood is my shepherd; I shall not want.
It maketh me to bring up wet timbers.
It leadeth me beside still waters.
It restoreth my budget; it leadeth me in the paths of scholarship for publication's sake.
Yea, though I walk through piled racks of ship's timbers, I will fear no shrinkage.
For thou art with me; thy tanks and thy pumps they comfort me.
Thou preparest a laboratory before me in the presence of mine colleagues.
Thou anointest my head with PEG; my tank runneth over.
Surely, funding and support shall follow me all the days of my career; and I shall dwell in the wet wood lab forever.
Amen."

APPENDIX II

List of Participants

Alex Barbour
Senior Engineer
Marine Restoration
R.S.D., Parks Canada
Les Terrasses de la Chaudière
Ottawa, Ontario
K1A 0H4

R.J. Barbour
College of Forestry Resource
Mail Station Ar. 10
University of Washington
Seattle, Washington 98195
U.S.A.
C/O Dr. Ramsey-Smith

Stephen W. Brooke
Chief Conservator
Maine State Museum
Station 83, State House
Augusta, ME. 04333
U.S.A.

Allen D. Brownstein
Research Chemist
Union Carbide Corp.
Tarrytown Technical Center
Tarrytown, New York 10591
U.S.A.

Richard W. Clarke
Senior Scientific Officer
Archaeological Research Centre
National Maritime Museum
Greenwich, London SE10 9NF
England

J.M. Coles
Professor
Dept. of Archaeology
University of Cambridge
Cambridge
England

Clifford Cook
Conservation Scientist
Canadian Conservation Institute
1030 Innes Road
Ottawa, Ontario
K1A 0M8

Charles Costain
Conservation Scientist
Parks Canada, Conservation Div.
1570 Liverpool Court
Ottawa, Ontario
K1A 1G2

Janey Cronyn
Lecturer
University of Durham
Dept. of Archaeology
Durham, England

Tom Daley
Senior Conservation Technician
Parks Canada, Conservation Div.
1570 Liverpool Court
Ottawa, Ontario
K1A 1G2

John E. Dawson
Conservation Scientist
Canadian Conservation Institute
1030 Innes Road
Ottawa, Ontario
K1A 0M8

Mary-Lou E. Florian
Conservation Analyst
British Columbia Provincial Museum
675 Belleville St.
Victoria, British Columbia

Louise Fox
Assistant Conservator
Fortress of Louisbourg
Conservation Lab.
Louisbourg, Nova Scotia
B0A 1M0

Jack Fry
Conservation Officer
National Museum of New Zealand
Private Bag
Wellington, New Zealand

Helen Ganiaris
Conservation Officer
Archaeology
Museum of London
London Wall
London EC2Y 5HN
England

D.W. Grattan
Senior Conservation Scientist
Canadian Conservation Institute
1030 Innes Road
Ottawa, Ontario
K1A 0M8

Herman J. Heikkenen
Consultant
American Institute of Dendrochronology
Box 293
Blackburg, Virginia 24060
U.S.A.

Charles E.S. Hett
Chief, Archaeology/Ethnology Division
Canadian Conservation Institute
1030 Innes Road
Ottawa, Ontario
K1A 0M8

Per Hoffmann
Conservator
Deutsches Schiffahrtsmuseum
D-2850 Bremerhaven
Germany

Richard Jagels
Assistant Professor
University of Maine
School of Forest Resources
Orono, Maine
U.S.A.

Victoria Jenssen
Head, Wet Organic Materials Section
Parks Canada
Conservation Division
1570 Liverpool Court
Ottawa, Ontario
K1A 1G2

Kirsten Jespersen
Nationalmuseet
Konserveringsafdelingen
For Jordfund
Traekonserveringslab.
Brede - 28 Lyngby
Denmark

B.A.H.G. Jütte
Central Research Laboratory for
Objects of Art and Science
G. Metsustraat 8, P.O. Box 5132
1007 AC Amsterdam
The Netherlands

Suzanne Keene
Senior Conservation Officer
Museum of London
London Wall
London ECZY 5HN
England

Diana Komejan
Conservation Technician
Fortress of Louisbourg
Box 45
Louisbourg, Nova Scotia
BOA 1M0

Ernest A. Lahey
Assistant Conservator
Conservation Laboratory
Atlantic Region
Parks Canada
Fortress of Louisbourg NHP
Box 160
Louisbourg, Nova Scotia

Eric Lawson
Conservator
Nautical Archaeology-Ethnography
Lawson Conservation Services
25 Cardena Road
Snug Cove, Bowen Island
British Columbia
V0N 1G0

Barry Lord
Curator
Wentworth Heritage Village
Rockton, Ontario
L0R 1X0

Dierdre A. Lucas
Senior Conservation Technician
Parks Canada
1570 Liverpool Court
Ottawa, Ontario
K1A 1G2

George F. MacDonald
Senior Archaeologist
National Museum of Man
Ottawa, Ontario
K1A 0M8

E.G. Mavroyannakis
N.R.C. Democritos
Aghia Paraskevi Attikis
Greece

Cliff McCawley
Chief
Conservation Processes Research Div.
Canadian Conservation Institute
1030 Innes Road
Ottawa, Ontario
K1A 0M8

Lorne Murdock
Archaeological Conservator
Parks Canada
Conservation Division
1570 Liverpool Court
Ottawa, Ontario
K1A 1G2

Howard J. Murray
Organics Conservator
Mary Rose Trust
Old Bond Store
48 Warblington St.
Old Portsmouth
Hants, England

Daniel Nelson
Director
Hamilton Scourge Foundation
City Hall
71 Main St. W.
Hamilton, Ontario

Charlotte Newton
Student
1-60 First Avenue
Ottawa, Ontario
K1S 2G2

Colin Pearson
Senior Lecturer
Materials Conservation Section
Canberra College of Advanced Education
P.O. Box 1, Belconnen, ACT 2616
Australia

Mary Peever
Assistant Conservator-Archaeology
Canadian Conservation Institute
1030 Innes Road
Ottawa, Ontario
K1A 0M8

Barbara A. Purdy
Associate Professor
Department of Anthropology
GPA 1350, University of Florida
Gainesville, Florida 32611
U.S.A.

R. Ravindra
Consultant
Canadian Conservation Institute
1030 Innes Road
Ottawa, Ontario
K1A 0M8

Roar Saeterhaug
The Royal Norwegian Society of
Sciences and Letters, Museum
Archaeological Department
Erling Skakkesgt. 47
7000 Trondheim, Norway

Marius St. Pierre
Technical Representative
Bach-Simpson Ltd
1255 Brydges Street
London, Ontario
N6A 4L6

Masaaki Sawada
Conservator in charge
Nara National Cultural
Properties Research Institute
2-9-1, Nijo-cho, Nara-shi 630
Japan

Fritz H. Schweingruber
Swiss Federal Forestry Research
Institute
CH-8903 Birmensdorf
Switzerland

Martha Segal
Senior Ass't Conservator
Archaeology
Canadian Conservation Institute
1030 Innes Road
Ottawa, Ontario
K1A 0M8

Bob Senior
Ass't Conservator
Archaeology

Canadian Conservation Institute
1030 Innes Road
Ottawa, Ontario
K1A 0M8

Katherine Singley
Conservator
Institute of Archaeology & Anthropology
University of South Carolina
Columbia, S.C. 29208
U.S.A.

James A. Spriggs
Conservator
York Archaeological Trust
47, Aldward
York, England

Willis Stevens
Marine Archaeologist
Parks Canada, Red Bay Project
1600 Liverpool Court
Ottawa, Ontario

Alfred Terfve
Technicien-chémiste
Institut royal du Patrimoine Artistique
(I.R.P.A.)
1 parc du Cinquantenaire
1040 Bruxelles, Belgique

Larry Titus
Graduate Student
Dept. of Archaeology
Simon Fraser University
Burnaby, British Columbia
V5A 1S6

James Tuck
Anthropology Department
Memorial University
St. John's, Newfoundland

Peter Van Geersdaele
Chief Conservation Officer
National Maritime Museum
Greenwich
London SE10 9NF
England

Ian N.M. Wainwright
Conservation Scientist
Analytical Research Services Division
Canadian Conservation Institute
1030 Innes Road
Ottawa, Ontario
K1A 0M8

Roderick Wallace
Doctoral Research Student
c/o Chemistry Department
Waikato University, Private Bag
Hamilton, New Zealand

Jacqui Watson
Conservation Officer
Ancient Monuments Laboratory
Fortress House; 23 Savile Row
London W.1.J.
England

Judy Wight
Conservator, Archaeology
Canadian Conservation Institute
1030 Innes Road
Ottawa, Ontario
K1A 0M8

Brian Yorga
4-468 Clarence St.
Ottawa, Ontario

Gregory S. Young
Research Consultant
Canadian Conservation Institute
1030 Innes Road
Ottawa, Ontario
K1A 0M8

APPENDIX III

PEG NOMENCLATURE

The following changes in CARBOWAX polyethylene glycols, from Union Carbide nomenclature have been made.

CARBOWAX 1540 becomes CARBOWAX 1450
CARBOWAX 4000 becomes CARBOWAX 3350
CARBOWAX 6000 becomes CARBOWAX 8000

This is purely a name change and the product will continue to be exactly the same.

The correct chemical nomenclature, as used in Chemical Abstracts, is polyoxy. 1-2 ethanediyl. People searching the chemical literature for information on polyethylene glycols, should use this name.